The Encyclopedia of
ROCKS
and
MINERALS

The Encyclopedia of

ROCKS

and

MINERALS

Nicola Cipriani

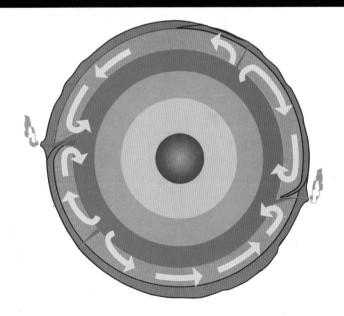

BARNES
&NOBLE
BOOKS
NEW YORK

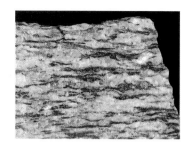

Copyright © 1996 Arnoldo Mondadori Editore S.p.A., Milan
English translation copyright © Arnoldo Mondadori Editore S.p.A., Milan

This edition published by Barnes & Noble, Inc.,
by arrangement with Arnoldo Mondadori Editore S.p.A., Milan

1996 Barnes & Noble Books

ISBN 0-7607-0291-8

Printed and bound in Spain by Artes Graficas Toledo, S.A.
D.L.TO: 884-1996

CONTENTS

PREFACE	7
INTRODUCTION	9
What are crystals?	10
The wonders of symmetry	11
The properties of minerals	12
More effects of light	15
Magnetism	16
The classification of minerals	17
Relationships among minerals	19
Origin of minerals	20
Rocks	22
TRANSPARENT MINERALS	**31**
OPAQUE MINERALS	**83**
IGNEOUS ROCKS	**99**
METAMORPHIC ROCKS	**123**
SEDIMENTARY ROCKS	**141**
Classification Table	156
Glossary	159
Bibliography	164
Index of entries	165

Legend

RARITY

(MINERALS)   (ROCKS)

     very common

common

rare

INDUSTRIAL AND COMMERCIAL USE

(ROCKS)

very frequently used

commonly used

not commonly used

COMMERCIAL VALUE

(MINERALS)   (ROCKS)

     none

low-medium

high

GEMOLOGICAL VALUE

(MINERALS)

no value

little value

intermediate value

high value

HARDNESS

(MINERALS)

soft

moderately hard

hard

PREFACE

The purpose of this book is to provide a reference guide with essential information about a considerable number of rocks and minerals present in the earth's crust. The fundamental physical and chemical properties of minerals are described in the introduction to supply the reader with general knowledge that will prove very useful when reading about the individual rocks and minerals. The main environments of formation for both minerals and rocks are also discussed in the introduction. In order to provide more extensive information on certain well-known minerals, often used as gemstones and in ornamental objects, and on some topics useful for the understanding of rocks, boxes exploring these subjects in greater detail have been included among the individual mineral and rock descriptions. While the subdivision of rocks into three main types (igneous, metamorphic, and sedimentary) follows convention, the author has chosen to divide the minerals into two informal groupings (transparent and opaque), in order to limit the number of sections in the book and to assist the reader in identifying specific minerals.

Within each of the two mineral sections, traditional mineralogical classification has been followed. The minerals are organized according to the classification scheme proposed by Hugo Strunz, but modified with the addition of a special class for the borates (which Strunz had included with the carbonates and nitrates). The description of each mineral includes its physical characteristics as well as the environment in which it formed; information which should greatly facilitate identification. Finally, to complete the picture, information on the mineral's use and a few notes and observations about its peculiarities have also been provided.

In general, the structure of the individual rock descriptions follows that for the minerals. The description and organization of the rocks in this section follows the most widely used classification schemes. For both intrusive and extrusive igneous rocks, the classifications proposed by Albert Streckeisen have been adopted. For metamorphic rocks, classification based on both texture and the presence of index minerals is now standard and the form presented in *Magmatismo e Metamorfismo* by Claudio D'Amico, Fabrizio Innocenti and Francesco Paolo Sassi has been adopted for this book. Because of the many variables involved in their mineralogical, chemical, and textural composition, the sedimentary rocks have been described on the basis of their depositional environment and mineralogical composition.

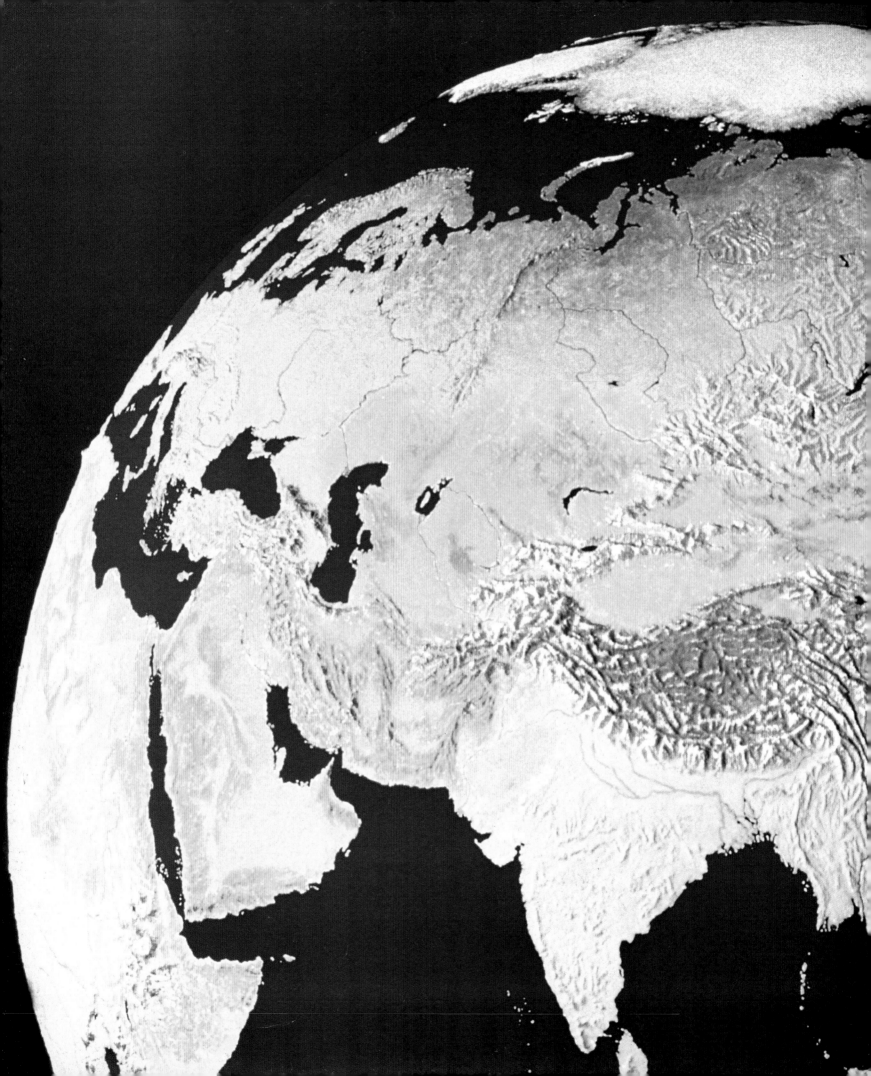

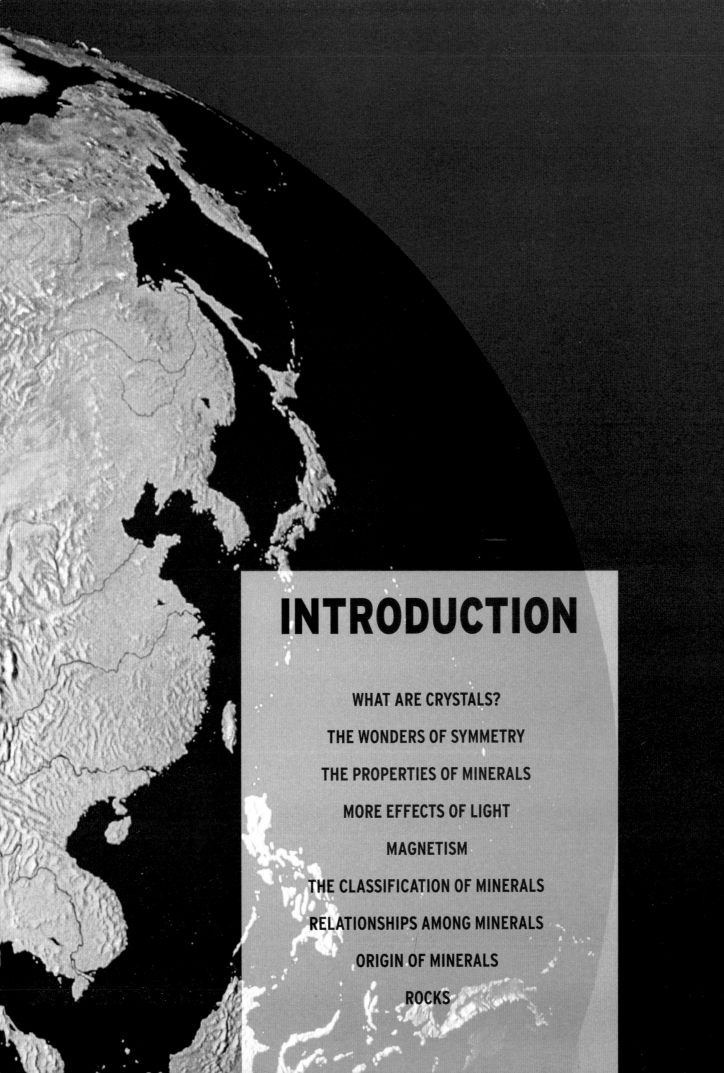

INTRODUCTION

WHAT ARE CRYSTALS?

THE WONDERS OF SYMMETRY

THE PROPERTIES OF MINERALS

MORE EFFECTS OF LIGHT

MAGNETISM

THE CLASSIFICATION OF MINERALS

RELATIONSHIPS AMONG MINERALS

ORIGIN OF MINERALS

ROCKS

WHAT ARE CRYSTALS?

Center: amethyst quartz crystal (Mexico).
Right: diagram of the arrangement of atoms
in a crystal (above) and in glass (below).

Bottom: portrait of Luca Pacioli (1445-1515)
by Jacopo de' Barbari.

In common terms, the word *crystal* means a valuable product of the glassmaking industry or craft, characterized by exceptional qualities of transparency, purity, and brilliance. The adjective *crystalline* is also frequently used to describe exceptionally clear water or to describe clarity of thought or expression. This meaning derives from our mental image of rock crystal, a variety of quartz. The ancient Greeks called this type of quartz "crystal" which literally means "formed from ice," since they were convinced that crystals were a form of extremely hard ice. This belief was strengthened by the fact that crystals were generally found high in the mountains. Pliny himself affirmed that "crystal is found nowhere else except where the winter snows accumulate in large quantities, and it is assuredly ice; it is for this reason that the Greeks gave it that name." He also added the adjective *sexangulum* in order to call attention to the hexagonal prismatic appearance which generally characterizes rock crystal.

The word *crystal* was subsequently used to denote any transparent mineral with a geometric form. Today this idea is not tied so much to the external form of a mineral as to the arrangement of atoms of which it is composed.

In order to understand the concept of the crystal, let's take a very simple example. We know that natural substances can be solid, liquid, or gaseous; if we take a container of water and bring it to the boiling point, it will evaporate, changing into steam; if, on the other hand, we expose it to temperatures lower than 32 °F (0° C), it will solidify, becoming ice. Therefore, there are ranges of temperature in which water appears in different states. Let's take another example – mercury: it is a solid at temperatures below -38 °F (-39° C) and gaseous above 673 °F (356° C). As everyone well knows, under normal conditions mercury is a liquid. Beginning with the solid form of mercury or water or some other substance and administering heat, we observe when the melting point is reached; continuing to add heat, we notice that the temperature of the substance does not rise during melting. Once melting is complete, however, adding more heat raises the temperature of the substance until the boiling point is reached. Administering more heat merely maintains the boiling point, while the temperature of the substance remains constant until the liquid is completely evaporated. Therefore, the change from one state to another is determined by a specific fixed temperature for each substance.

The atoms of ice and solid mercury have a very orderly geometrical arrangement; in the liquid states the atoms move apart into a less orderly arrangement, while in the gaseous states they are widely separated and their arrangement is chaotic. Thus, in the change from one state to another, the temperature of the substance is not increased because the heat that is administered is absorbed in order to overcome the atoms', ions', or molecules' resistance to separate from one another. The temperature of the substance remains constant until the point of complete *separation* of its components. All crystals behave in this fashion and, therefore, *solid* and *crystal* can be considered synonyms.

Now let's take a small glass rod and heat it: a candle flame will suffice. The rod will begin to bend, becoming flexible. With time, the temperature of the rod will increase as it becomes more and more fluid. In this case, there is no fixed temperature for melting. Although the cold glass rod seemed solid, its atoms had a disorderly arrangement similar to that of the liquid state. Since the atoms do not have to be moved from an orderly arrangement, the heat that is administered during melting produces a continuous increase in the temperature of the rod.

THE MYSTERIES OF CRYSTAL HABIT

When the science of mineralogy was still to be developed, scholars applied themselves to understanding the laws that governed the geometric forms (or habits) of crystals. Among the first scholars who considered natural forms in a rational manner, Luca Pacioli is worthy of mention. In 1494 he wrote the *Summa de arithmetica, geometria, proportioni et proportionalità*. Modern crystallography has, however, abandoned this path, since it is now recognized that the faces which define a crystal's external habit are a direct consequence of its internal structure. In other words, the atoms, ions, or molecules which make up a mineral occupy defined and symmetrically arranged positions in space within a three-dimensional array called a crystalline lattice. The *unit cell* is the smallest part of the lattice which, when repeated in all three dimensions, forms the entire crystal structure. The unit cell is described by three principal axes whose dimensions are characteristic for a given mineral. This means that two crystals of the same mineral have identical structures, while two minerals of different composition have different crystal structures.

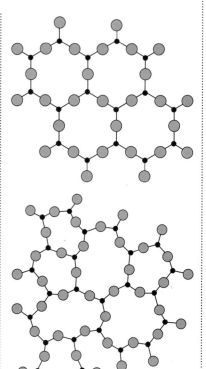

THE WONDERS OF SYMMETRY

Top: representation of the seven systems of crystal symmetry.

Bottom: elements of symmetry (a. plane of symmetry; b. axis of symmetry; c. center of symmetry.

Examples of the systems of symmetry:
1. cubic crystals of pyrite (cubic system);
2. prismatic crystals of apatite (hexagonal system); 3. prismatic crystal of phosgenite (tetragonal system); 4. smoky quartz crystals (trigonal system); 5. prismatic crystals of stibnite (orthorhombic system); 6. tabular crystal of gypsum (monoclinic system); 7. tabular crystal of kyanite (triclinic system).

On a first encounter with crystals, we immediately notice that we have before us solids of an often simple and regular geometric shape and that the position of the faces and the angles between them are repeated in the crystal giving it a symmetrical appearance. In order to understand this concept, we can give a few examples. The human body has a bilateral symmetry such that the elements which make up the right side are repeated on the left as if a mirror were passed vertically between the two sides. We have two legs and two arms each of which is not the same as the other; rather they are mirror images. This mirror is known as a plane of symmetry and has, in fact, the peculiarity of reproducing an object on the other side in a precisely mirror-like fashion. The corresponding limbs are similar to one another but not identical. Let us take another example: a starfish with five equal legs. We turn the animal with the center as the pivot and stop each time that the tip of an arm is in front of us. It is obvious that we will turn the star five times before we come back to the first arm and that each rotation has the same angular distance. Furthermore, the arm considered on each turn will assume the position of the previous one and we will not be able to perceive differences between the five positions. In this case, it is said that the star has a symmetry resulting from a vertical axis passing through its center and capable of repeating the same figure five times by a rotation of 360° / 5 or 72°. The axis which we have identified is a five-fold axis of symmetry.

We can repeat this operation with any object. For example, a normal pack of cigarettes has a twofold axis parallel to the position of the cigarettes inside it. It has two equal and parallel faces (front side and back side) and each assumes the position of the other after a rotation of 180°. The same pack may be cut in half, either horizontally or vertically, along three planes of symmetry all mutually perpendicular. In many objects, therefore, we can identify elements of symmetry.

In crystals, only the twofold (binary), three-fold (trigonal), fourfold (tetragonal) and sixfold (hexagonal) axes of symmetry can be found. Besides axes and planes of symmetry, another symmetry element some crystals possess is a center of symmetry. These three elements are defined as follows:

Plane of symmetry: plane passing through the crystal's center which cuts it into two mirror-imaged halves.
Axis of symmetry: a straight line passing through the crystal's center and around which occurs, during rotation of the crystal through 360°, the repetition of a morphological element (face, edge, corner) as many times as its degree of symmetry. This repetition occurs every 180° for a binary axis (360° / 2), every 120° for a trigonal axis, every 90° for a tetragonal axis, and every 60° for a hexagonal axis.
Center of symmetry: internal point in the crystal coincident with its center which correlates two diametrically opposite morphological elements.

Natural crystals are not always completely regular in form. Often some faces are more developed than others and the crystals take on a misshapen appearance which does not, however, alter their internal crystalline lattice. On the basis of which symmetry elements are present, crystals are subdivided into 32 classes, which are grouped into 7 systems that are further gathered into 3 groups:

The monometric group has the greatest symmetry and is called this because the edges of the unit cell of the crystalline lattice are the same size in the three principal mutually perpendicular directions. The dimetric group is so called because the vertical axis of the unit cell is different in length to the two horizontal directions. The crystals belonging to this group have a medium to high degree of symmetry. The trimetric group is so defined because the three principal axes are all of different lengths. Symmetry is always low in this group.

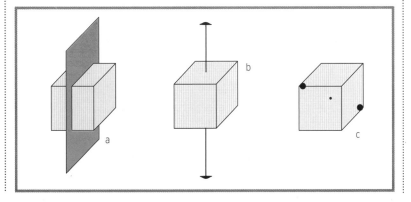

THE PROPERTIES OF MINERALS

Left: examples of rhombohedral cleavage in a calcite crystal (above) and octahedral cleavage in a fluorite crystal (below).

Bottom: examples of twinned crystals and their ideal geometric shapes for reference.

The characteristics or properties of crystals that readily attract our notice are varied. Some are more striking than others. If observed with a little attention, these characteristics can help us to distinguish between different crystals. Without a doubt the most striking properties are transparency, color, luster and hardness; the last determinable with the aid of simple instruments. Let us look at these and other properties.

CLEAVAGE

As we have seen, a crystal's structure is formed by a regular arrangement of atoms. The strengths of the bonds between the atoms change in relation to which atoms are present and to their spatial distribution. Since the atoms are

arranged in a symmetrical pattern, points of greatest weakness between atoms may also occur in a repetitive arrangement so as to define a plane within the lattice structure. These are called *cleavage planes* and crystals which have them break preferentially along these smooth surfaces. Cleavage planes may or may not be parallel to the faces of crystals. In the case of the rhombohedral crystals of calcite, the cleavage planes are parallel to the natural faces. Therefore, when one of these calcite crystals is broken, the smaller fragments are also rhombohedral in shape. In fluorite, on the other hand, which generally crystallizes as cubes, the cleavage planes are perpendicular to the axes which unite opposite corners of the cube. Thus an octahedron is obtained from cleaving the original cubic crystal. Recognition of the geometrical shapes or the arrangement of cleavage planes can help in identifying the mineralogical species.

AGGREGATES AND TWINNING

Crystals are not always found in isolation. They often appear as aggregates which may be either regular or irregular. This distinction is not always clear-cut since there are also *subregular* associations. Furthermore, the aggregates can occur either with only one mineral species or with several different minerals. One type of regular association is called *twinning*. Twinned crystals are aggregates of one mineral species in which the crystals' internal structures are related to each other by a symmetry element. Both simple aggregates and twins produce minerals of very pleasing, and sometimes downright odd, shapes.

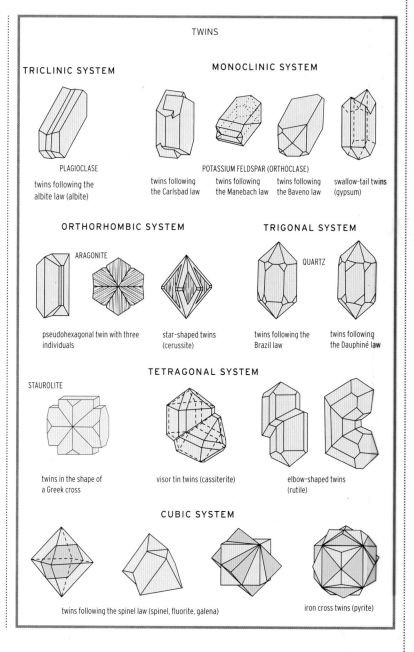

TWINS

TRICLINIC SYSTEM

PLAGIOCLASE

twins following the albite law (albite)

MONOCLINIC SYSTEM

POTASSIUM FELDSPAR (ORTHOCLASE)

twins following the Carlsbad law

twins following the Manebach law

twins following the Baveno law

swallow-tail twins (gypsum)

ORTHORHOMBIC SYSTEM

ARAGONITE

pseudohexagonal twin with three individuals

star-shaped twins (cerussite)

TRIGONAL SYSTEM

QUARTZ

twins following the Brazil law

twins following the Dauphiné law

TETRAGONAL SYSTEM

STAUROLITE

twins in the shape of a Greek cross

visor tin twins (cassiterite)

elbow-shaped twins (rutile)

CUBIC SYSTEM

twins following the spinel law (spinel, fluorite, galena)

iron cross twins (pyrite)

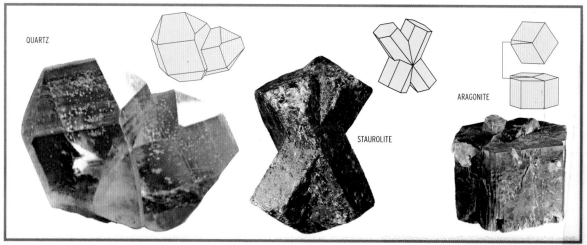

QUARTZ

STAUROLITE

ARAGONITE

Top: faceted gems and stones of various colors and shapes.

Bottom: phenomenon of light dispersion. When a ray of white light (sunlight) passes through a prism, it separates into the seven colors of the rainbow.

For example, staurolite twins are formed by two intersected crystals in the shape of either a Greek cross or a St. Andrew's cross.

TRANSPARENCY AND COLOR

Transparency is easily observable and is one of the characteristics that makes minerals extraordinarily fascinating. It is the degree to which light is able to pass through a mineral. As regards chromatic variations, we cannot help but recognize the extraordinary fantasy of nature. We have all noticed that minerals have an extraordinary variety of colors and tones. In many minerals, color is strictly dependent upon a major constituent in their chemical composition and consequently remains almost constant even when ground into powder. For example, malachite produces a green powder; azurite, a blue one; cinnabar, a red one, and so forth. These minerals are called *idiochromatic*.

Many others, however, are colorless when pure, but they can assume various colors depending on the impurities they contain. Fluorite, for example, can be colorless, yellow, brown, rosy, blue, green, violet, and blackish. Beryl, tourmaline, corundum, and zircon are also examples of minerals rich in chromatic variety. These minerals are called *allochromatic* and their powder is, in general, light and white.

In order to understand the interaction of light with minerals to produce color, let us first consider the nature of light. We know that sunlight is made up of all colors since we see these colors dispersed by rain drops into a rainbow. Thus, sunlight, also called white light, is produced by the mixture of all colors.

When minerals interact with white light, some colors may be absorbed so that only some of the light is transmitted. Let us better clarify this concept: when light penetrates a mineral and it appears colorless, the mineral has not absorbed the rays which have passed through it. If, on the other hand, it tends to absorb a portion of the colors which make up sunlight, then it will appear to be colored and the color we see corresponds to the sum of the unabsorbed colors. If instead, we see a black mineral, this means that it has absorbed all of the light. There are also minerals which do not permit the passage of light rays, some of which are absorbed, and some reflected outwards. These are minerals composed of metallic elements and whose powder is commonly dark.

LUSTER

Luster is the appearance of a mineral when it reflects light.

How many times have we noticed the pleasant twinkling of crystals, but we have not always paid attention to the type of luster? In order to understand the concept of luster, let us take a few examples. A polished steel plate reflects light in a typically metallic way: this is the luster that is, in fact, called metallic; glass on the other hand reflects light in a different way and in this case we speak of a vitreous luster. The natural mineral which reflects light most brilliantly is the diamond and this luster is called adamantine. There are other kinds of luster characteristic of other well-known materials and thus described as silky, pearly, or resinous.

HARDNESS

Hardness is defined as the resistance of a body to scratching or abrasion. It has been observed that minerals have very varied behavior with respect to this property. There are soft ones and hard ones, as well as unlimited intermediate possibilities. In 1822 Carl Friedrich Christian Mohs formed the idea of using a

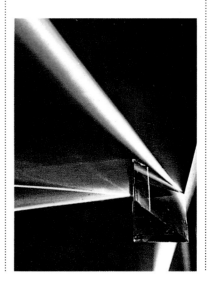

Left: examples of luster.

Right: mineral specimens in order of increasing hardness (Mohs scale).

Metallic luster: molybdenite.

Silky luster: gypsum.

Vitreous luster: quartz.

Adamantine luster: diamond.

succession of 10 common minerals (listed in the table) of increasing hardness as reference points in order to estimate the hardness of unknown minerals. This succession was called the Mohs scale of hardness or, simply, the Mohs scale.

It is very easy to use, as each mineral on the scale scratches the preceding one and is scratched by its successor. In this volume, for practical reasons, we have simplified the scale; defining as soft those minerals which can be scratched by apatite, as semi-hard those that can be scratched by quartz, and the rest as hard. In the mineral discussions the terms used are, respectively, low, medium, and high hardness.

SPECIFIC GRAVITY

Another property often used to identify minerals is specific gravity. This is defined as the ratio between the weight of a mineral and the weight of an equal volume of distilled water at a temperature of 39° F (4° C). It is numerically equivalent to the density when appropriate units are specified. Simple as it is conceptually, the determination of specific gravity requires a certain amount of equipment. It depends primarily upon composition.

MOHS SCALE	
1	TALC
2	GYPSUM
3	CALCITE
4	FLUORITE
5	APATITE
6	ORTHOCLASE
7	QUARTZ
8	TOPAZ
9	CORUNDUM
10	DIAMOND

1. Talc

6. Orthoclase

2. Gypsum

7. Quartz

3. Calcite

8. Topaz

4. Fluorite

9. Corundum

5. Apatite

10. Diamond

MORE EFFECTS OF LIGHT

Top: the phenomenon of double refraction in a calcite crystal.

Bottom left and right: specimens, respectively, of scheelite and willemite in natural light (above) and exposed to ultraviolet rays (below) under which they fluoresce.

Light rays allow us to observe other fascinating properties of minerals some of which may seem surprising or peculiar to us.

DOUBLE REFRACTION

When light rays pass through a mineral, they undergo important changes that are dependent upon the mineral's structure. There are some minerals in which light rays do not undergo any such change. In the majority of cases, however, a light ray which passes through a mineral is split into two rays with differing characteristics. The exact nature of these rays is different for each mineral species.

In order to understand the phenomenon of splitting a light ray, let us take a transparent calcite crystal of a thickness of at least two centimeters, put it on a written page, and read through the mineral. Before laying the crystal on it, the writing would appear to be made up of a single line; after placing the crystal on it, however, two slightly distorted lines appear.

If on the other hand, we place a transparent crystal of rock salt (kitchen salt) on the writing, we will continue to see a single line. It is easy to perceive that in calcite, in contrast with rock salt, light rays undergo the splitting which is defined as double refraction.

INDEX OF REFRACTION

One of the most fundamental optical properties of minerals is the index of refraction. This is defined as the ratio between the velocity of a light ray in a vacuum (in practice, in air) to that in a particular mineral. The velocity of light in a vacuum is always much greater than inside a solid, and, therefore, the index of refraction of minerals is always greater than 1.

This property is characteristic for each mineral species and so it is frequently used to identify them. As we know, one mineral species can exhibit multiple variations in color so that identification on the basis of color alone is difficult. A simple determination of the index of refraction eliminates any doubt. The determination is carried out by the use of a refractometer on which is placed a smooth face of a crystal or a gem. The value of the index is read directly from a graduated scale observed through an eyepiece.

LUMINESCENCE

Luminescence is the emission of light by some minerals when they are stressed by one of a variety of treatments. For example, if the emission occurs through crushing or rubbing, it is called triboluminescence. This quality is exhibited by fluorite, sphalerite, and lepidolite, among others. Some minerals emit light if heated at moderate temperatures. This phenomenon is called thermoluminescence, and it occurs in celestite, barite, apatite, spodumene, and others. Luminescence may also be produced by chemical means (chemiluminescence) or by excitation with X-rays or ultraviolet light (UV).

Among the types listed above, the one most often seen is certainly luminescence produced with ultraviolet light by means of a Wood lamp. This is a special lamp with very dark violet glass whose light can only be easily seen in the dark. The irradiated minerals emit characteristic colors completely different from those seen in normal light. The minerals scheelite, zircon, fluorite, and some calcites and aragonites exhibit this phenomenon in the most spectacular fashion, emitting almost unreal colors. Luminescence is defined as fluorescence when the emission of light ceases simultaneously with the excitation. It is called "phosphorescence" when, on the other hand, the emission continues for a certain period of time after irradiation.

MAGNETISM

Top: specimen of lodestone, a variety of magnetite; its magnetic character is shown by the iron filings which line up along the flux lines of the magnetic field.

Bottom: a nuclear power station in New Hampshire (United States).

How many times as children did we play with a magnet by attracting pieces of iron, and were astonished by this strange phenomenon? Well, if we bring certain minerals near a magnet, we see that they are also influenced by a magnetic field. These minerals contain iron. In many cases the phenomenon is not very obvious unless we use a magnet capable of producing a very strong magnetic field. With normal magnets, such as the kind found in the home, we can only attract magnetite and pyrrhotite; these are defined as *ferromagnetic* minerals. Hematite is weakly attracted to a magnet and only if it is in small fragments; minerals of this type are called *paramagnetic*. Those minerals that are slightly repelled in the presence of a magnetic field are called *diamagnetic*.

RADIOACTIVITY: AN IMMENSE SOURCE OF ENERGY

We all know that nuclear power is a source of energy that is in common use despite the risks involved. It is produced in nuclear power stations which use radioactive material extracted from natural minerals.

Radioactivity is a property linked to the so-called unstable elements; those elements transform themselves to more stable elements, and in the process emit radiaton. They are rare elements and only uranium is a major component of natural minerals. The most common of these minerals, although not particularly widespread, is pitchblende. The majority of the radioactive elements either form very rare minerals or, more commonly, enter the crystal lattice during the formation of other minerals to replace a few of the fundamental atoms. For example in zircon, zirconium can be replaced by uranium and thorium to give the mineral a certain degree of radioactivity. Radioactive disintegration occurs when the nucleus of an unstable atom spontaneously decomposes and emits an elementary particle. This particle is one of three types: alpha, beta, or gamma.

Alpha particles have a low speed (about 5% that of light). They are relatively large and only slightly penetrating.

They can travel only a few inches through the air and, therefore, a thin leaf of any material is sufficient to stop them completely. Beta particles are much smaller. They have a speed about equal to that of light and, having a rather high penetrating power, they can easily pass through a sheet of aluminum. Gamma rays are a form of electromagnetic radiation. They are not, therefore, nuclear particles like the others. They have a very high penetrating power and can pass through materials of great thickness. It is customary to protect environments in which radioactive substances are used with a shield of lead plates a few millimeters in thickness. Lead, in fact, is the most common of the metals of high atomic weight which succeed in blocking the passage of gamma radiation.

THE CLASSIFICATION OF MINERALS

The properties of minerals which have so far been described are very useful in describing and thus identifying them. These properties are not, however, useful for classifying them in a systematic fashion. For this purpose, it is better to use their chemical composition.

Therefore, to facilitate comprehension of the mineralogical formulas used in the mineral descriptions, we will quickly consider some characteristics of the chemical elements. They are listed in the table which is arranged in alphabetical order according to the symbols representing each element. These symbols appear in the mineral formulas. In the table the symbols are followed by the name, the atomic number, the atomic

In the table, the atomic weights are relative to carbon 12. The symbols and names are the most commonly accepted ones. For the radioactive (*) elements, the assigned atomic weight is that of the isotope either with the longest half-life or whose half-life is best known.

weight, and an estimate of the relative abundance of each; this calculation is based upon the average composition of the most exterior part of the earth (lithosphere, hydrosphere, and atmosphere).

The abundance of each element is given in relative terms by one of three categories: common, not very common, rare. The term *very common* means that an element is found in quantities over 1%, *common* in quantities between 1% and 0.001%, and *rare* in quantities less than 0.001%.

In the lithosphere common elements are those which constitute the fundamental minerals found in rocks. The other elements are distributed in a variety of ways. Of the less common elements, some are essential components of minerals which occur concentrated in a few specific localities called deposits. Otherwise, these elements either form rare minerals or partially replace the fundamental elements in more common minerals.

A chemical element is an atom composed of a positively-charged nucleus around which rotate the negatively-charged electrons. As a whole, the atom is neutral since the positive charges are equal in number to the negative. There are some atoms which lack one or more electrons and thus are positively charged. Others have an excess of electrons and, therefore, are negatively charged; these atoms are called "ions." In combining to form minerals, ions follow rules which depend on the type of bonds between the different ions and on two fundamental characteristics of each ion: radius and charge. Only a few minerals are solids composed simply of one element (native elements). As previously stated, the composition of a mineral is dependent upon the charge and the radius of the constituent ions. Varying the number of electrons in an ion involves a change in its radius and, as we have ascertained that the structure of a crystal is a well-ordered framework, the space available for the ions is, in consequence, very well defined.

The premise that a crystal structure is controlled by ion charge and size is necessary in order to comprehend the outline adopted for the systematic classification of minerals. The one followed at present, with a slight modification, was devised by a German, H. Strunz. It

TABLE OF CHEMICAL ELEMENTS

Symbol	Name	Atomic Number	Atomic Weight	Abundance	Symbol	Name	Atomic Number	Atomic Weight	Abundance
Ac	Actinium	89	(227)(*)	rare	Mn	Manganese	25	55	common
Ag	Silver	47	107	rare	Mo	Molybdenum	42	96	rare
Al	Aluminum	13	27	very common	N	Nitrogen	7	14	common
Am	Americium	95	(243)(*)	synthetic	Na	Sodium	11	23	very common
Ar	Argon	18	40	rare	Nb	Niobium	41	93	rare
As	Arsenic	33	75	rare	Nd	Neodymium	60	144	common
At	Astatine	85	(210)(*)	rare	Ne	Neon	10	20	rare
Au	Gold	79	197	rare	Ni	Nickel	28	59	common
B	Boron	5	11	common	No	Nobelium	102	(259)(*)	synthetic
Ba	Barium	56	137	common	Np	Neptunium	93	237	synthetic
Be	Beryllium	4	9	rare	O	Oxygen	8	16	very common
Bi	Bismuth	83	209	rare	Os	Osmium	76	190	rare
Bk	Berkelium	97	(247)(*)	synthetic	P	Phosphorus	15	31	common
Br	Bromine	35	80	rare	Pa	Protactinium	91	231	rare
C	Carbon	6	12	common	Pb	Lead	82	207	common
Ca	Calcium	20	40	very common	Pd	Palladium	46	106	rare
Cd	Cadmium	48	112	rare	Pm	Promethium	61	(145)(*)	synthetic
Ce	Cerium	58	140	common	Po	Polonium	84	(209)(*)	rare
Cf	Californium	98	(251)(*)	synthetic	Pr	Praseodymium	59	141	rare
Cl	Chlorine	17	35	very common	Pt	Platinum	78	195	rare
Cm	Curium	96	(247)(*)	synthetic	Pu	Plutonium	94	(244)(*)	synthetic
Co	Cobalt	27	59	common	Ra	Radium	88	226	rare
Cr	Chromium	24	52	common	Rb	Rubidium	37	85	common
Cs	Cesium	55	133	rare	Re	Rhenium	75	186	rare
Cu	Copper	29	64	common	Rh	Rhodium	45	103	rare
Dy	Dysprosium	66	163	rare	Rn	Radon	86	(222)(*)	rare
Er	Erbium	68	167	rare	Ru	Ruthenium	44	101	rare
Es	Einsteinium	99	(252)(*)	synthetic	S	Sulphur	16	32	common
Eu	Europium	63	152	rare	Sb	Antimony	51	122	rare
F	Fluorine	9	19	common	Sc	Scandium	21	45	rare
Fe	Iron	26	56	very common	Se	Selenium	34	79	rare
Fm	Fermium	100	(257)(*)	synthetic	Si	Silicon	14	28	very common
Fr	Francium	87	(223)(*)	rare	Sm	Samarium	62	150	rare
Ga	Gallium	31	70	rare	Sn	Tin	50	119	rare
Gd	Gadolinium	64	157	rare	Sr	Strontium	38	88	common
Ge	Germanium	32	73	rare	Ta	Tantalum	73	181	rare
H	Hydrogen	1	1	common	Tb	Terbium	65	159	rare
He	Helium	2	4	rare	Tc	Technetium	43	(98)(*)	synthetic
Hf	Hafnium	72	178	common	Te	Tellurium	52	128	rare
Hg	Mercury	80	201	rare	Th	Thorium	90	232	common
Ho	Holmium	67	165	rare	Ti	Titanium	22	48	common
In	Indium	49	115	rare	Tl	Thallium	81	204	rare
Ir	Iridium	77	192	rare	Tm	Thulium	69	169	rare
I	Iodine	53	127	rare	U	Uranium	92	238	rare
K	Potassium	19	39	very common	V	Vanadium	23	51	common
Kr	Krypton	36	84	rare	W	Tungsten	74	184	common
La	Lanthanum	57	139	rare	Xe	Xenon	54	131	rare
Li	Lithium	3	7	common	Y	Yttrium	39	89	common
Lu	Lutetium	71	175	rare	Yb	Ytterbium	70	173	rare
Lr	Lawrencium	103	(260)(*)	synthetic	Zn	Zinc	30	65	common
Md	Mendelevium	101	(256)(*)	synthetic	Zr	Zirconium	40	91	common
Mg	Magnesium	12	24	very common					

MINERAL CLASSIFICATION

1ST CLASS	NATIVE ELEMENTS, ALLOYS, CARBIDES, NITRIDES AND PHOSPHIDES.
2ND CLASS	SULFIDES, SELENIDES, ARSENIDES, TELLURIDES AND SULFOSALTS.
3RD CLASS	HALIDES.
4TH CLASS	OXIDES AND HYDROXIDES.
5TH CLASS	CARBONATES, NITRATES, SELENITES, TELLURITES AND IODATES.
6TH CLASS	BORATES.
7TH CLASS	SULFATES, TELLURATES, CHROMATES, MOLYBDATES AND TUNGSTATES.
8TH CLASS	PHOSPHATES, ARSENATES AND VANADATES.
9TH CLASS	SILICATES.
10TH CLASS	ORGANIC COMPOUNDS.

is based on crystallochemical criteria, which take into account both the chemical composition and the structure of the mineral. Our modification consists in adding a separate borate class which Strunz had included as part of the carbonate class. A more detailed discussion is necessary for the silicate class, as many of these are fundamental components of rocks and represent almost 80% of the minerals present in the earth's crust. Their classification is based not on chemistry but on structure since all of these minerals have the fundamental unit $[SiO_4]^{4-}$ in common. This unit is composed of one silicon ion with four positive charges and of four oxygen ions each with two negative charges. In this way, the unit has a total of four negative charges. The geometric form of the unit is also important: it is a tetrahedron in which silicon occupies the central part and the four oxygens occupy the vertices or corners.

The fundamental silica tetrahedral unit bonds with positively charged ions or with other tetrahedra. In bonding with the latter, groupings can be formed that define the subclasses into which silicates are divided (see illustration).

The *nesosilicates* are characterized by isolated tetrahedral units whose four negative charges are balanced by the positive charges of other ions. The minerals that crystallize in this subclass have a stocky appearance.

The *sorosilicates* have two tetrahedral units bonded together with one apex in common. Thus, three negative charges are in excess for each unit, for a total of six. These are balanced by an equal number of positive charges associated with other ions. Minerals belonging to this subclass show a variety of shapes.

The *cyclosilicates* have three, four, or six tetrahedral units joined to one another with one apex in common between every two units. There is an excess of two negative charges for each unit for a total negative charge of six, eight, or twelve, respectively. These are balanced by an equal number of positive charges from ions of other elements. The most common grouping is that with six units. Minerals belonging to this subclass have a prismatic appearance with triangular, tetragonal, or hexagonal cross sections.

The *inosilicates* are characterized by tetrahedral units joined together by apices so as to form an indefinitely long chain. The chains may be single or double (two parallel chains joined together). The first are typical of pyroxenes, the second of amphiboles. In both cases there are two excess negative charges for each tetrahedral unit, and these are balanced by an equal number of positive charges from other ions. Minerals belonging to this subclass are generally elongated or fibrous.

The *phyllosilicates* have an indefinite number of tetrahedral units joined to form a planar or sheet-like arrangement where each unit shares a total of three apices with contiguous units. There is a single excess negative charge for each unit that is positioned in a similar location for every tetrahedron so that the sheet exhibits the charges on only one side. These are balanced by an equal number of positive charges from other ions, also in a planar arrangement. The minerals belonging to this subclass have a flattened appearance and are easily flaked along planes parallel to the sheets.

The *tectosilicates* have an indefinite number of tetrahedral units, each of

which has all four apices joined to other tetrahedra. They form a three-dimensional structure or network inside of which are relatively more open spaces delineated by the arrangement of the tetrahedral units. With this type of structure, there are no excess charges and the resultant structure is charge-balanced. This can, therefore, generate only one mineral: quartz. However, numerous minerals belong to this subclass since it is relatively easy to substitute the silicon ion, Si^{4+}, with the aluminum ion, Al^{3+}. This exchange destroys the charge balance of the structure, producing one excess negative charge for each Al^{3+}. The excess charge is compensated for by the introduction of positive ions which occupy the spaces described above.

There are over 2,000 mineralogical species and every year new ones are discovered. These are usually very rare minerals, almost always in very small crystals which were easily missed by researchers in the past. Their discovery has often been facilitated by mineral collectors who, inspired by great passion and curiosity, bring their specimens to universities or other research centers. In the majority of cases, the samples are known minerals, but sometimes one may happen to run into an absolutely novel mineral, as did this author. The minerals treated in this volume, as may be guessed, are not a complete set of all the known ones. But a large number are described, and certainly a sufficient number for the novice who is just entering the fascinating world of minerals.

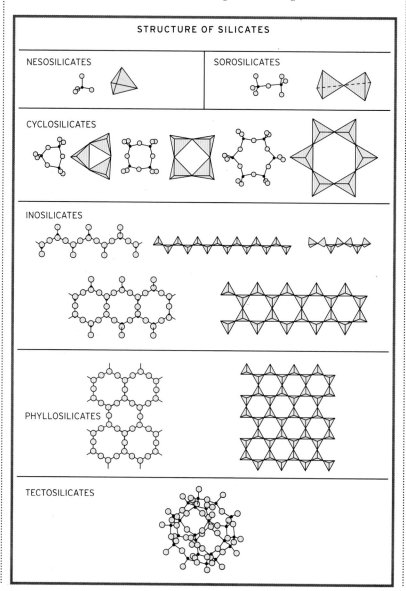

STRUCTURE OF SILICATES

NESOSILICATES

SOROSILICATES

CYCLOSILICATES

INOSILICATES

PHYLLOSILICATES

TECTOSILICATES

RELATIONSHIPS AMONG MINERALS

Top, in order: prismatic calcite crystals (Sardinia) and a specimen of witherite (Great Britain).

Bottom: specimens of mammillary smithsonite (above) and of cerussite (below), both from Sardinia.

We have seen how the systematic classification of minerals is based in part on the arrangement of atoms, ions, or molecules in crystals. Within a class or subclass there are different minerals that have very similar arrangements or internal structures. In some cases, the different ions that characterize each species have a similar charge and radius, and therefore can substitute for one another. These structural similarities cause some minerals to be, as it were, very close relatives. There exist groups of isostructural minerals (that is, having the same structure) that constitute a series, and in more complex groups that we call families there are more than one series. Among the most common and important groups, we can cite those of calcite, aragonite, plagioclase, garnet, pyroxene, amphibole, and olivine. To more easily understand this concept, let us look at the mineral formulas for some groups. The first two of the listed groups have minerals with the same composition, $CaCO_3$, which, depending on the conditions of crystallization, takes on one of two different types of structures: a trigonal (calcite) or an orthorhombic one (aragonite). In other words, the calcium ions, Ca^{2+}, and the carbonate groups, $(CO_3)^{2-}$, may assume two different arrangements and each of these is characterized by a different set of symmetry elements. In general, other conditions being equal, calcite crystallizes at lower temperatures and pressures than does aragonite. In other cases, for example with amphibole and pyroxene, the different structures are generated by variations in composition, as we will see later.

The calcite group is characterized by the carbonate ion, $(CO_3)^{2-}$, which can bond with a variety of ions having a 2+ charge and a radius similar to that of calcium. They are fairly common minerals with the exception of spherocobaltite and otavite.

$CaCO_3$	calcite
$MnCO_3$	rhodochrosite
$FeCO_3$	siderite
$ZnCO_3$	smithsonite
$MgCO_3$	magnesite
$CoCO_3$	spherocobaltite
$CdCO_3$	otavite

The aragonite group is characterized by a more compact structure than calcite which results in a greater specific gravity for aragonite (2.93) compared with calcite (2.71).

$CaCO_3$	aragonite
$SrCO_3$	strontianite
$BaCO_3$	witherite
$PbCO_3$	cerussite

The plagioclase series is very important as the minerals are fundamental constituents of many rocks. This series is made up of two endmembers:

albite $Na(AlO_2)(SiO_2)_3$
and anorthite $Ca(AlO_2)_2(SiO_2)_2$

A continuous series of other minerals exist with compositions intermediate between those of the two endmembers. Their compositions are expressed as a percentage of anorthite which describes the amount of calcium that substitutes for sodium.

albite $(Na,Ca)(AlO_2)x (SiO_2)y$
$1 < x < 1.1$ $3 > y > 2.9$
can contain up to 10% anorthite

oligoclase $(Na,Ca)(AlO_2)x (SiO_2)y$
$1.1 < x < 1.3$ $2.9 > y > 2.7$
contains 10% to 30% anorthite

andesine $(Na,Ca)(AlO_2)x (SiO_2)y$
$1.3 < x < 1.5$ $2.7 > y > 2.5$
contains 30% to 50% anorthite

labradorite $(Na, Ca)(AlO_2)x (SiO_2)y$
$1.5 < x < 1.7$ $2.5 > y > 2.3$
contains 50% to 70% anorthite

bytownite $(Na,Ca)(AlO_2)x (SiO_2)y$
$1.7 < x < 1.9$ $2.3 > y > 2.1$
contains 70% to 90% anorthite

anorthite $(Na,Ca)(AlO_2)x (SiO_2)y$
$1.9 < x < 2$ $2.1 > y > 2$
contains 90% to 100% anorthite

In the formulas, x and y indicate that the amounts of Al and Si change with the variation in the ratio between sodium and calcium. In particular, from albite to anorthite the quantities of AlO_2 increase from 1 to 2, while those of SiO_2 decrease from 3 to 2. The amount of SiO_2 is used as a measure for the degree of acidity; therefore, albite is an acid mineral since it contains a large amount of SiO_2 (68.74% by weight), while anorthite is a basic mineral (SiO_2 = 45.83%). The other plagioclases are of intermediate acidity. The concept of acidity is very important for describing magmatic rocks, as we will see later.

The olivine series is similar to that of plagioclase; in this case, the mineral endmembers are forsterite (Mg_2SiO_4) and fayalite (Fe_2SiO_4). The general formula for olivine is $(Mg, Fe)_2 (SiO_4)$ and describes the substitution of magnesium and iron for each other. The crystal structures of the two endmembers as well as those of the intermediate compositions are the same. Mineral series also occur in the garnet group although the substitutions are more complex. Once again the crystalline structure remains constant for all minerals in the series. The pyroxenes and amphiboles are even more complex. In these two large groups there are several series, due to the greater variety of substitutions that can occur among ions. In order to furnish a simple and easily understood picture, let us outline the principal series. In each of the two groups, there exist one series that crystallizes in the orthorhombic system and one in the monoclinic system. This variability in structural symmetry is produced by the presence of an extra major element (calcium) in the monoclinic series in comparison with the orthorhombic one. In order to understand this concept, let us take the pyroxenes as an example. For these minerals the orthorhombic series has the endmembers:

enstatite $Mg_2Si_2O_6$
and orthoferrosilite $Fe_2Si_2O_6$

The introduction of calcium into the structure lowers its degree of symmetry from the orthorhombic to the monoclinic system. We then have a different series with the following endmembers:

diopside $CaMgSi_2O_6$
and hedenbergite $CaFeSi_2O_6$

ORIGIN OF MINERALS

Nocturnal eruption of Mount Etna. There are three environments in which minerals form: igneous, metamorphic, sedimentary.

ond case occurs in minerals belonging to the same series when diffusion of the new substituting element towards the crystal interior is slow. In this situation, the crystals show a compositional change, often uniform from the inside outward like the layers of an onion, and it is said to be zoned.

It is interesting to note that different minerals under similar environmental conditions grow with substantial differences in speed. For example, gypsum, rock salt, and in general, all easily soluble minerals, have a speed of growth verifiable on a human scale. Many others (and these are in the majority) take geological ages to grow.

Finally, it is not possible to discuss the origins and formation of minerals without also considering that of rocks since they are assemblages of minerals. Minerals and rocks are formed in three different environments: igneous, metamorphic, and sedimentary. The majority of minerals exist under a wide range of conditions, and crystals of one mineralogical species can be found in more than one environment. Thus, the minerals listed below under each of the three types of genesis are not necessarily unique to that environment. Rather, it is the most common environment in which that species is found.

The igneous environment immediately calls to mind volcanoes, and we can, therefore, infer that high temperature plays an important role in the formation of these minerals. The sedimentary environment is associated with both fresh and salty water systems so temperature and pressure are more or less those in which we live. The metamorphic environment is a bit more difficult to understand. To simplify things, let us say that the temperature is intermediate to that of the two other environments and the pressure may be very high (up to about 10 kilobars which corresponds, in depth, to over 18 miles (30 km)).

IGNEOUS ORIGIN

As a magma cools beneath the earth it begins to solidify and minerals start to crystallize. Crystallization is subdivided into four stages that are characterized by different temperatures and different ratios of magma, gas, and water. In order of decreasing temperature the stages are: orthomagmatic, pegmatitic, pneumatolytic, and hydrothermal.

We have mentioned that plagioclase has a variable composition. In a rock, the particular composition of the plagioclase grains is in equilibrium with the compositions of the other mineral species and any melt present. This means that at the time of crystallization and during growth, each mineral is in physical and chemical balance with the other minerals or melt. Any variation in temperature, pressure, or composition in the surrounding environment produces a change in the minerals that are forming.

The most common phenomena are the interruption of growth, which can be followed by a partial or total dissolution of the initial crystals, or an internal restructuring of the crystalline lattice into which new elements or larger quantities of one already present may enter. The lattice change may affect the entire crystal or only areas near the surface. In the first case, sometimes a completely different mineral is generated. The sec-

In the drawing are shown some processes that characterize the continuing evolution of the earth: formation of new oceanic crust along a mid-ocean ridge, which offsets the consumption of crust in a subduction zone.

The first crystals to form are generally of minerals rich in iron and magnesium and poor in silica such as olivine, pyroxene, amphibole, biotite, calcium-rich plagioclase, and zircon, and nonsilicate minerals such as apatite, magnetite, chromite, titanite, and pyrrhotite.

In the pegmatitic stage: quartz, feldspar, beryl (emerald and aquamarine), topaz, tourmaline, apatite, spodumene, uraninite, gadolinite, etc.

In the pneumatolytic stage: quartz, molybdenite, cassiterite, scheelite, ilvaite, vesuvianite, fassaite, grossular, spessartine, etc.

In the hydrothermal stage: many sulfides including galena, sphalerite, pyrite, stibnite, and cinnabar, as well as fluorite, barite, siderite, quartz, etc.

SEDIMENTARY ORIGIN

The minerals formed in sedimentary environments are subdivided into two categories: those minerals that crystallize from the ions dissolved in water and those that are eroded from other preexisting rocks, transported by rivers and then deposited either along a river system, in a lake, or on the sea floor. Minerals which crystallize from a solution are said to *precipitate*, and some of these can give rise to very useful mineral deposits. The phenomenon occurs in environments of various kinds. The classic example is given by lagoon environments in hot climates in which the concentration of dissolved salts is increased due to high rates of evaporation; when saturation is reached, small crystals are formed that fall to the lagoon bottom. Among the most common and important minerals in this environment are calcite, halite (kitchen salt), gypsum, sylvite, and carnallite.

Included in the second category are all the rock-forming minerals, provided that they do not dissolve easily in water and are not particularly easy to abrade (excessive reduction of volume during transport by mechanical means such as chipping and grinding). During the process of transport some minerals are selectively deposited. This is due either to high specific gravity, a strong resistance to abrasion, or both characteristics, and results in concentrations of these minerals. These particular deposits are called *placers*. If the minerals in the deposits are useful to humans, they are mined.

METAMORPHIC ORIGIN

In metamorphic enviroments, changes occur in the solid state. This means that the conditions of temperature and pressure are such that the preexisting minerals in the rocks undergo compositional and structural changes. In this way, new mineral species are formed in the rock.

Many minerals are typical of metamorphic environments. Among the silicates we can list: chlorite, some garnets, muscovite, kyanite, andalusite, sillimanite, and vesuvianite. Typical non-silicate minerals include: magnesite, rutile, ilmenite, spinel, brucite, and graphite.

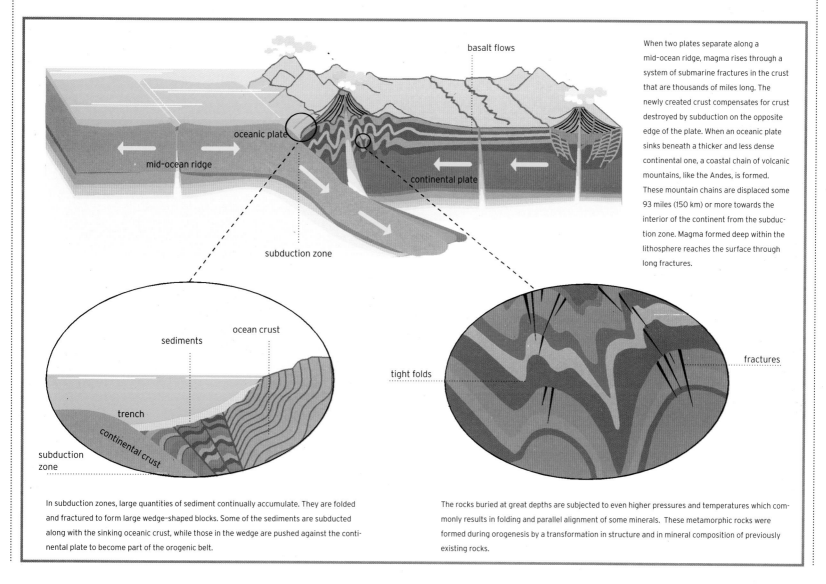

basalt flows

oceanic plate

mid-ocean ridge

continental plate

subduction zone

When two plates separate along a mid-ocean ridge, magma rises through a system of submarine fractures in the crust that are thousands of miles long. The newly created crust compensates for crust destroyed by subduction on the opposite edge of the plate. When an oceanic plate sinks beneath a thicker and less dense continental one, a coastal chain of volcanic mountains, like the Andes, is formed. These mountain chains are displaced some 93 miles (150 km) or more towards the interior of the continent from the subduction zone. Magma formed deep within the lithosphere reaches the surface through long fractures.

ocean crust

sediments

trench

continental crust

subduction zone

tight folds

fractures

In subduction zones, large quantities of sediment continually accumulate. They are folded and fractured to form large wedge-shaped blocks. Some of the sediments are subducted along with the sinking oceanic crust, while those in the wedge are pushed against the continental plate to become part of the orogenic belt.

The rocks buried at great depths are subjected to even higher pressures and temperatures which commonly results in folding and parallel alignment of some minerals. These metamorphic rocks were formed during orogenesis by a transformation in structure and in mineral composition of previously existing rocks.

ROCKS

Rocks are the solid material of which the most exterior part of our planet, called the earth's crust, is composed. As we have mentioned previously, rocks are aggregates of minerals and are formed in the same environments as minerals: igneous, metamorphic, or sedimentary. In order to understand how rocks are formed, let us first take a quick look at the structure of the earth displayed in an idealized cross-section of the planet. The earth has a structure made up of concentric layers like that of an onion. At the center is the core, subdivided into inner and outer, outside of which is the mantle, subdivided into lower and upper, and finally there is the crust. The

uppermost part of the upper mantle is rigid, as is the crust with which it forms the lithosphere. In turn, there are two types of crust: oceanic and continental (the one on which we live). The oceanic crust has a great deal of iron- and magnesium-rich minerals (mafic minerals) and is denser than the continental crust which consists predominantly of aluminum- and silicon-rich minerals (felsic minerals).

As has been made clear through seismological research, of these concentric layers the outer core is liquid whereas the others are solid. However, parts of the mantle behave as though they are in a very viscous liquid state. In the mantle,

the transmission of heat from the interior to the earth's surface happens through very slow movements, and the displacement upwards is balanced by overlying, less hot, masses that are pushed to the bottom. In this way, thermal convection cells are created. If two adjoining cells diverge near the earth's crust, they will produce a rift: this is an imposing system of fractures in the overlying crust that divides it into two plates which move away from each other. Where a cell cools and sinks back down into the mantle, a subduction zone is formed in which one lithospheric plate sinks beneath the other. In the areas of divergence or rifting, new oceanic crust is produced when

magma rises from the mantle, erupts through the rift, fractures, and solidifies. The continuous contribution of material causes a submarine mountainous chain, called a mid-ocean ridge, to rise along the edges of the main fracture.

In the areas of convergence, however, different situations are produced according to the characteristics of the colliding plates. If the edges of both plates consist of oceanic crust, a system of volcanoes parallel to the subduction zone and called an island arc (like Japan and the Aleutians) is generated. If instead the colliding edges are oceanic and continental crust, or both continental crust, then a continental mountain

Top: the drawing shows the distribution of heat inside the earth. The heat that instigates the formation of a magma comes in large part from the decay of radioactive elements in the crust and upper mantle. The arrows indicate the convection currents.

Center: drawing in which the possible origins of magmas are shown.

Bottom: principal mineralogical composition of the most important igneous rocks.

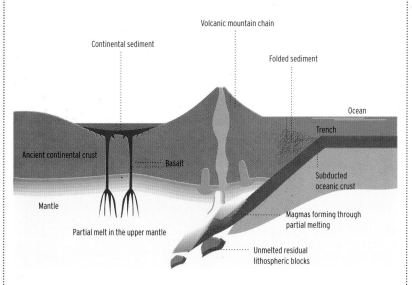

Crust

Upper mantle 2642°F (1450°C)

Rift Zone

molten outer core 7232°F (4000°C)

Solid iron core 7592°F (4200°C)

Subduction zone

Lower mantle 5432°F (3000°C)

Divergent convection currents

Volcanic mountain chain

Continental sediment

Folded sediment

Ocean

Trench

Ancient continental crust

Basalt

Subducted oceanic crust

Mantle

Partial melt in the upper mantle

Magmas forming through partial melting

Unmelted residual lithospheric blocks

rior of the crust cools more slowly and solidifies only after a long time. These slowly-cooled subsurface rocks are called magmatic, intrusive, or plutonic. The magma that remains within fractures in the crust solidifies at an intermediate speed relative to that of the extrusive and intrusive rocks, and produces dikes. Let us now move on to a more detailed discussion of the three fundamental kinds of rocks: igneous, metamorphic, and sedimentary.

IGNEOUS ROCKS

That igneous rocks are derived from the solidification of a molten rock is a well-known fact. Less well understood are the origin and the precise nature of magmas since, we cannot sample them directly unless they are erupted as lavas; even then sampling them is challenging. We can infer the nature of magmas that produced intrusive and dike rocks by studying those rocks that are now exposed on the earth's surface or that are found at slight depth.

These are, therefore, *a posteriori* studies concentrating on the rocks that resulted from magmatic processes in the geological past. Through studies of ancient and recent rocks an attempt is made to interpret the geological context and characteristics of magmas. It was thus discovered that some magmas are generated in the depths of orogenic chains; some have their origins directly from the mantle, while others result

from combined processes. In recent decades the increase in knowledge about the crust of our planet has dramatically clarified many geological processes.

Igneous rocks are subdivided into intrusive, dike, extrusive, and pyroclastic. As we have seen, intrusive and dike rocks are those that crystallize from a magma within the earth's crust. Extrusive ones result from the solidification of lavas on the earth's surface. Pyroclastic (or volcanoclastic) rocks are a special type of extrusive igneous rock; lava and gases are ejected upwards from a volcano so that the airborne fragments partially solidify in the air and fall back down like rain.

Most magmas have a silicate composition and the rocks derived from them contain silicate minerals whose chemistry reflects that of the original melts. The only exception is carbonatites which derive from magmas of carbonate composition. Magmas show a range of silica (SiO_2) contents, where silica is the total amount of elementary silicon in the rock expressed as an oxide. This compositional variability is described in terms of acidity. On this basis magmas are classified as:

acid	>63% silica
intermediate	52-63% silica
basic	45-52% silica
ultrabasic	<45% silica

chain like either the Andes or the Alps is generated during what is known as an orogenic event.

On the earth's surface, mountains are prominent features and therefore are subject to erosion by exogenous agents that attack the rocks by physical and chemical means. The eroded fragments are transported downstream by rivers and are deposited in a variety of environments. Thus, there are fluvial sediments along rivers, lacustrine ones in lakes, and marine ones when some of the fragments reach the ocean. In the subduction zones where a lithospheric plate with oceanic crust sinks under one with continental crust, deep trenches are formed parallel to the volcanic mountain range and are filled with notable depths of sediment.

All of the sediments described above are defined as detrital. There is also another type which we shall discuss later.

During subduction some of the trench sediments sink along with the down-going plate. As the sediments are subducted, they are subjected to increasing temperatures and pressures such that they are transformed into metamorphic rocks. Within one area of the subduction zone conditions are favorable to melt some of the rocks, and the resulting magma moves buoyantly up through the upper mantle and crust. Some of the magma may reach the earth's surface to erupt as lava and produce a volcanic edifice. The lava cools quickly giving rise to extrusive (or volcanic) rocks. The magma remaining in the inte-

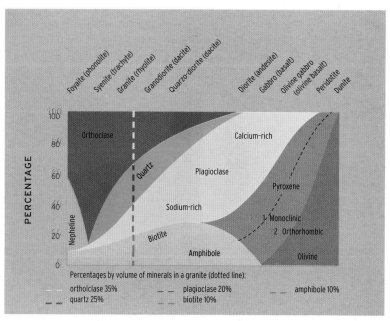

Foyaite (phonolite) · Syenite (trachyte) · Granite (rhyolite) · Granodiorite (dacite) · Quarzo-diorite (dacite) · Diorite (andesite) · Gabbro (basalt) · Olivine gabbro (olivine basalt) · Peridotite · Dunite

PERCENTAGE

100 · 80 · 60 · 40 · 20 · 0

Orthoclase

Calcium-rich

Quartz

Plagioclase

Pyroxene

Nepheline

Sodium-rich

1 Monoclinic

2 Orthorhombic

Biotite

Amphibole

Olivine

Percentages by volume of minerals in a granite (dotted line):

orthoclase 35% · plagioclase 20% · amphibole 10%
quartz 25% · biotite 10%

The degree of acidity, as we have pointed out, is also associated with the occurrence of certain minerals. Therefore some, like quartz, orthoclase, and albite consistently occur in acid and intermediate rocks. Others, like olivine, leucite, and nepheline are consistent with basic and ultrabasic rocks. Rocks containing quartz are called oversaturated; those which lack quartz are either saturated or undersaturated. Saturated rocks, in contrast to the undersaturated ones, do not contain feldspathoids.

Laboratory experiments have made it clear that rocks melt at high temperatures, generally over 1832° F (1000° C). If the experiment is performed in the presence of a gas, the melting point is lowered. From this it has been deduced that magmas that contain volatile components (gas) have lower melting points, with a minimum possible melting point of about 1292° F (700° C). The most abundant volatile component of magmas is commonly water. Chlorine, fluorine, hydrogen, hydrochloric acid, hydrofluoric acid, and carbon dioxide, might also be present. These contribute to making the magma much less viscous and, as it crystallizes, the gases tend to be concentrated in the residual melt.

INTRUSIVE ROCKS

Rocks resulting from the crystallization of magmas within the crust are particularly common in an orogenic environment. In this environment are also found a preponderance of acid rocks. In rift zones, however, intrusive rocks are basic and ultrabasic in composition.

The formation of intrusive rocks in orogenic belts is very complex. Let us consider, for example, the case of colli-

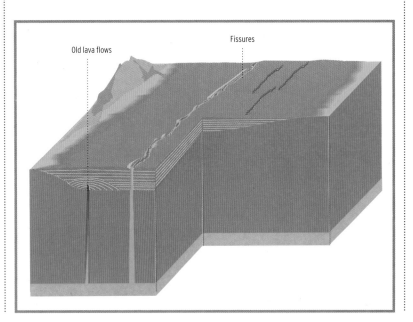

Old lava flows

Fissures

sion between oceanic crust and continental crust which occurs west of the Andes mountain chain. Oceanic crust, being made up of denser basic rocks, tends to sink below the continental one. At depth the sinking lithospheric plate partially melts and assimilates rocks from the overlying plate.

The magmas formed in these areas show a range in composition from basic to acid, and the corresponding intrusive rocks (in order of increasing acidity) are gabbros, diorites, granodiorites, and granites. It is not very clear if a basic to intermediate magma is derived from the melting of only the oceanic crust or if it also involves part of the continental one. It is evident, however, that in order for an acid magma to evolve in a subduction zone, the melting of continental crust rocks and a mixing of this melt with the basic magma derived from the oceanic crust must take place.

Intrusive rocks commonly have a granular texture. That is, they are composed of fairly large interlocking crystals of approximately the same size which can be seen by the naked eye or with the aid of a hand lens. On the

basis of the rocks that we observe in the earth's great orogenic belts, it can be easily deduced that a large mass of magma solidifies to become intrusive rocks. Only a relatively small amount actually reaches the surface through volcanoes.

EXTRUSIVE ROCKS

When a magma reaches the earth's surface, it is called lava and, depending on its acidity, may or may not build a volcanic cone. Normally, basic and ultrabasic lavas, due to their greater fluidity, tend to flow along the surface without forming volcanic cones. The great basaltic expanses of the Great Karroo, Paraná, Ethiopia, the Deccan Plateau, and others were formed in this way. On the other hand, intermediate and acid lavas, because they are more viscous, solidify more rapidly and accumulate layer upon layer to build a cone where they erupt. This type of lava is generally explosive and the resulting flows commonly alternate with pyroclastic deposits.

Unlike the intrusive rocks, the extrusives generally have a very fine-grained texture. Sometimes they contain glass as well. In some cases, visible crystals (phenocrysts) are scattered in the fine-grained groundmass. These crystals are formed in subcrustal conditions prior to eruption and are said to have a prophyritic texture. This texture is fairly common in rocks of intermediate and acid composition.

PYROCLASTIC ROCKS

Pyroclastic rocks are produced by a volcanic eruption that explosively ejects more or less solidified fragments which then fall back to earth like rain. These rocks generally have an intermediate or acid composition. Only in rare instances do basic rocks erupt explosively, and those that do tend to have an excess of potassium.

In general, a magma erupts explosively when gases are trapped in small bubbles and pockets in the magma and cause a build-up of pressure within the volcanic chimney. The passage of gases may also be blocked when a partial solidification of the magma in the uppermost portion of the chimney acts like a cork.

In any case, sufficient pressure can accumulate in a volcanic conduit to

Top: current view of Pompeii, buried under great quantities of volcanic material erupted from Mount Vesuvius in 79 A.D.

Bottom: a cloud of lethal gas and incandescent ash rolls down a side of Mount Pelée on December 16, 1902 during an eruption similar to that which had destroyed Saint-Pierre seven years earlier. This exceptional series of photographs was taken over a period of five minutes while the *nuée ardente* expanded until it reached a height of 13,123 feet (4,000 meters).

prime the explosion and the magma is ejected like uncorked champagne. From studies conducted on the erupted rocks, it is thought that the destruction of Pompeii and Herculaneum in 79 A.D. was the result of this type of eruption spouting from Mount Vesuvius.

During an explosion, the material is generally partly molten and partly solid, and it is mixed with gases. Some explosions shoot out material of various shapes and sizes launching them up to several thousand feet into the atmosphere. In other eruptions, hot clouds of gas and ash are generated that reach heights of at most a few thousand feet and then flow rapidly down the sides of the volcano. In general, volcanic eruptions with the most destructive power are those where ground water or surface water is intercepted by the rising magma (phreatomagmatic eruptions). Sometimes, however, less destructive eruptions can become more so for ancillary reasons. For example, the high temperatures associated with the eruption may cause a cover of snow to liquefy thereby suddenly creating an avalanche of debris-laden mud; the destructive power can be very great. For humans, living in proximity to a volcano often constitutes a fairly high risk, regardless of the composition of the magma. Therefore, we can consider only those volcanoes situated in uninhabited areas to be relatively innocuous.

METAMORPHIC ROCKS

As we have already indicated, metamorphic rocks are the product of changes in other rocks, be they sedimentary,

THE VOLCANO SPROUTED IN A FIELD OF CORN

On the morning of February 20, 1943, the frightened Mexican farmer, Dionisio Pulido discovered that during the night an 82 ft (25 m) long fracture had opened up in the ground. As Pulido and his family intently looked on in amazement, incandescent ashes and stones were flung from the crevice. They were witnesses to an extraordinary geological event: the birth of a volcano. Pulido raced to the nearby village of Paricutín to tell of the strange affair and very soon dozens of geologists rushed there. The volcano, which became known as Paricutín, grew with stupefying speed. The next day it was 33 ft (10 m) high, after a week it reached 541 ft (165 m), and after a year over 984 ft (300 m). Large quantities of ash had covered the surrounding countryside for a radius of almost 19 miles (30 km). Then its growth slowed, reaching a maximum height of about 1345 ft (410 m), while continuing to emit great amounts of lava. Nine years and 12 days after its spectacular birth, Paricutín became dormant. It had emitted 3.6 million tons of eruptive projectiles and lava, and scientists had been able to observe a rare event: the growth of a volcano.

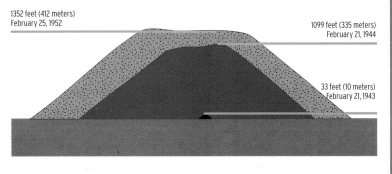

1352 feet (412 meters)
February 25, 1952

1099 feet (335 meters)
February 21, 1944

33 feet (10 meters)
February 21, 1943

igneous, or even those subjected to a previous metamorphic event. Besides the two variables of temperature and pressure, the amount and composition of any fluids present and the mineralogical composition of the original rocks result in many possible types of metamorphic rocks. Their classification is based on the presence of certain characteristic minerals that are interpreted to indicate ranges of temperature and pressure. There are, therefore, metamorphic rocks produced at either low or high temperatures and at low or high pressures.

The rocks are formed in a variety of environments. To simplify these concepts, we can subdivide metamorphism into two types: regional and contact. Regional metamorphism involves an enormous mass of rocks at great depth and is generally associated with an orogenic or mountain-building event. Contact metamorphism involves a relatively restricted area of rocks adjacent to a magmatic body.

During an orogenic event, the rocks at depth within the crust are subjected to both very intense temperature and pressure. As a rule, the intensity of metamorphism increases with depth; from not very metamorphosed shallow rocks to deeper rocks in which the minerals of the original rock have been completely replaced by a new assemblage of minerals. In some instances, at particularly high temperatures and pressures, partial melting of the rocks occurs. The magma generated in this way recrystallizes, giving birth to a granitic rock.

In contact metamorphism, on the other hand, the rocks are subjected to high temperatures caused by proximity to an igneous intrusion at relatively low pressure. Under these conditions, metamorphism generally affects a restricted area of rocks, since temperature decreases steeply with distance from the source of heat.

SEDIMENTARY ROCKS

Sedimentary rocks are formed by the deposition and accumulation of rock fragments or by chemical or biochemical precipitation onto the bottom of a basin. These rocks can be classified into the following types: detrital, organic, chemical, and residual. By far the most abundant are those of detrital origin,

Upper left: slate fragments in which schistosity or cleavage planes (vertical lines) can be seen, as well as relics of the original stratification (horizontal layers) represented by quartz-rich bands. The enlargement shows the schistosity planes as vertical lines, while the original strata or layers in the slate are folded into "s" shapes.

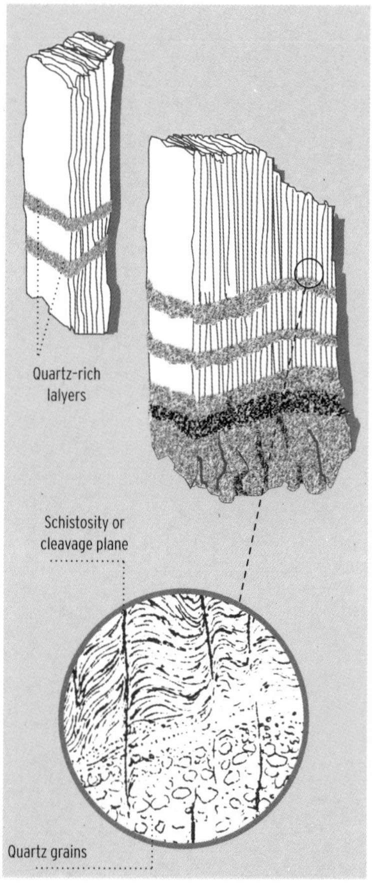

Right: diagram showing temperature and pressure conditions inside the earth's crust.

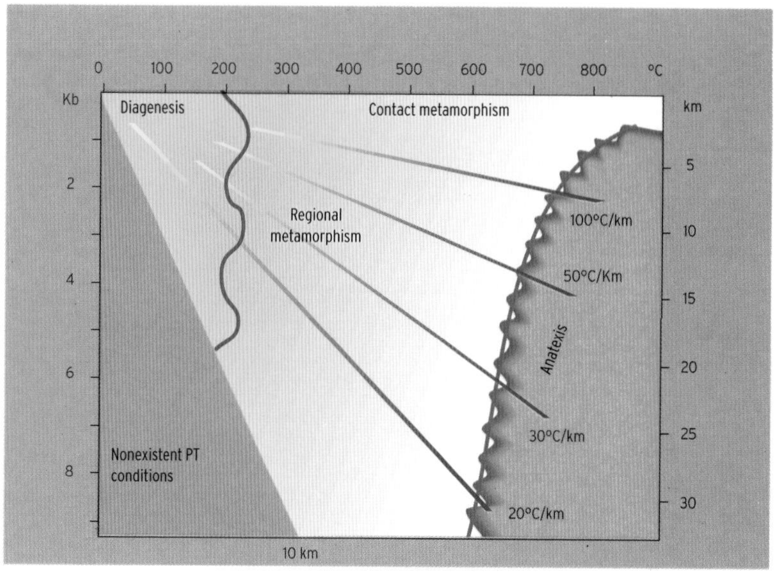

Bottom: sandstone strata of the Carboniferous Period, folded into zigzags by the internal movements of the earth (Millok Haven, Cornwall).

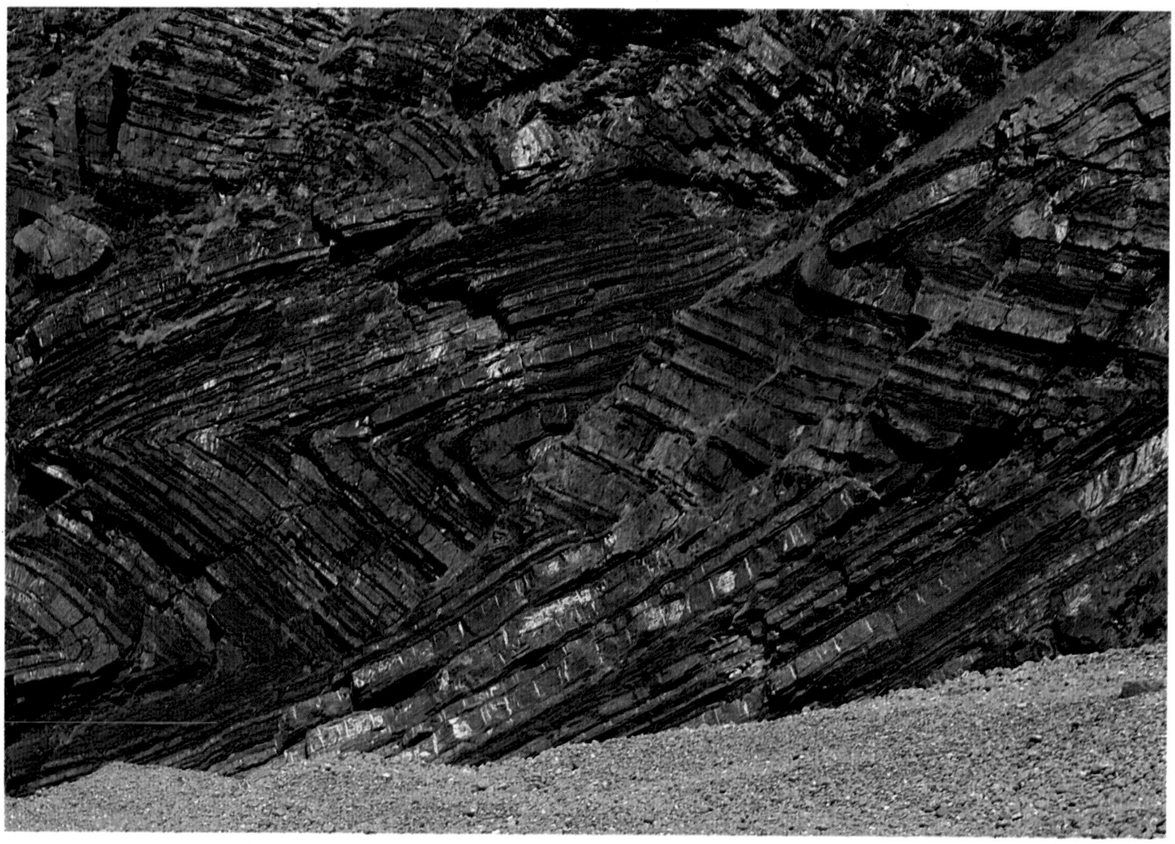

followed by the organic ones. The others are less common. Detrital rocks result from the erosion, transportation, and deposition of rock fragments in either a continental environment (fluvial, lacustrine, glacial, aeolian) or a marine one.

Sediments can be defined on the basis of the size of the particles which make them up. Many reference scales have been proposed, but the most commonly used in the field of geology is the following:

gravels	> 2 mm.
sands	0.063-2 mm.
muds	0.002-0.063 mm.
clays	< 0.002mm.

Unlike igneous and metamorphic rocks in which a chemical balance exists among the minerals that compose them, sedimentary rocks lack this balance since the individual particles can be derived from any kind of rock as long as it was present in the area of erosion. In order to better understand this concept, let us examine a gravelly sediment. We can observe the individual pebbles in it and see the heterogeneity of types.

Organic rocks are those produced by the shells or skeletons that organisms build by using the salts dissolved in the water. These rocks belong exclusively to marine environments. The organisms are incredibly varied: the corals and madrepores of the coral barrier reefs; the shells of mollusks and gastropods; and the single-celled microorganisms that make up plankton. Therefore, they give rise to sediments that are very different from one another.

Rocks of chemical origin are those which are formed by the crystallization of minerals directly from a solution rich in dissolved salts (precipitation). There are various environments in which this phenomenon may occur. They can, however, be classified essentially into two environments: marine and continental.

Marine sediments of chemical origin are the most common and are called evaporites. They are formed in the shallow water of relatively small basins in hot climates which promote evaporation of the water.

Continental sediments are subdivided into two types: those linked to a warm climate and, therefore, of similar origin to marine sediments, and those that are formed near hot springs. The first type include the endorheic or closed basins in which accumulates water relatively rich in a variety of dissolved salts. The salts are further concentrated by evaporation in a hot climate as, for example, occurs at present in Lake Chad (Central Africa). The precipitated sediment from springs is commonly calcium carbonate ($CaCO_3$), generally in the form of calcite, more rarely as aragonite. By this process, notable amounts of a rock called travertine is produced. Less common are the hot springs from which the water flow is either steady or intermittent (geysers), which precipitate other minerals such as the borates, the sulfates, silica, and sulfur.

Finally, the residual rocks are what remains after other rocks have been dis-

Top: sandy beach currently in a stage of deposition on a rocky substrate (Sardinia).

Bottom left: view of the Grand Canyon.

Bottom right: an accumulation of mollusks, an example of rocks formed from the solid parts of organisms after their death.

solved and weathered by water. These are fairly common rocks, but almost always occur as relatively small deposits. The bauxite deposits are of interest because of their high aluminum content.

Sedimentary rocks often display a stratified or layered appearance. The width of the layers is very variable, from less than a centimeter to several meters. Successions of strata are made up either of the same lithology or of differing rock types, as occurs, for example, in the alternation of sandstones with shales in some formations. Sometimes the stratification may be barely visible as in some clay-rich successions, or it may be totally lacking as in the coral reef barriers in which the organisms build an almost continuous structure.

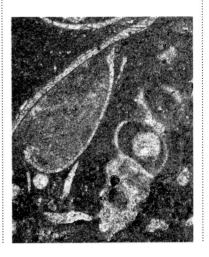

TRANSPARENT MINERALS

TRANSPARENT MINERALS

Top: the drawing depicts a transmitted-light microscope in which the route of the light is highlighted.
Center: Depiction of the link that, in the past was thought to exist between the signs of the zodiac and precious stones.
Bottom: Some specimens of transparent minerals.

Transparent minerals are those which allow light to pass through them. They represent the majority of all known minerals. Among the physical properties of minerals, those linked to the reflection and transmission of light are, without a doubt, the most readily noticed. We are attracted to colored and brilliant transparent crystals, but less so to the opaque and gray ones which, at first sight, appear indistinguishable from one another. Because of the importance of the characteristic ways in which light interacts with crystals, the polarizing transmitted-light microscope has an important role in studying minerals.

When studying transparent minerals, the microscope is set up so that light coming from below passes through a thin section of a mineral and reaches the upper part of the instrument where an eyepiece for observation is positioned.

A "thin section" consists of a slice of a rock or mineral (about 30 microns) mounted on a glass slide or pressed between two thin pieces of glass. The microscope is equipped with a number of accessories which allow one to observe the distinctive optical properties of minerals.

As we have seen, these characteristics are tied to the structure of the crystalline network and its symmetry. Among it many uses, the microscope allows one to estimate the index (or indices) of refraction of minerals (a very useful mark of identification as will be more easily understood through the examples given in the course of the text) and to observe their form. These two properties, considered jointly, are sufficient for recognizing the crystalline group (monometric, dimetric, and trimetric) and the system to which a mineral belongs.

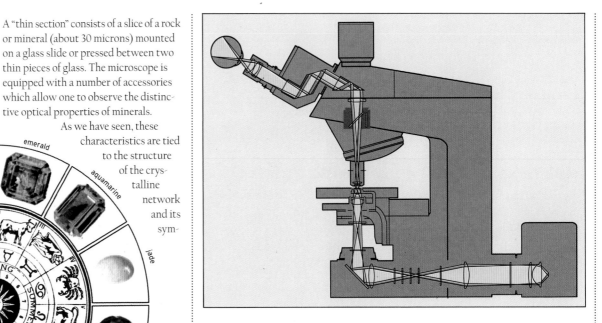

While examining the phenomenon of double refraction, we pointed out the fact that light passing through a calcite crystal splits into two rays of unequal velocity, but all light passing through a halite (rock salt) crystal has the same velocity. This behavior is tied to the system in which minerals crystallize. Halite belongs to the cubic system (monometric group), and light passing through any cubic transparent mineral has the same velocity in all directions. It is as if the light were transmitted through air or glass. However, a light ray that passes through crystals belonging to the dimetric and trimetric groups (e.g., calcite) undergoes the phenomenon of double refraction. That is, it splits into two rays and the behavior of each is independent of the other. Each of the two rays has its own velocity and consequently the mineral has two indices of refraction. The difference between the indices is called birefringence.

Determining the index or indices of refraction is very useful in identifying minerals. For example, diamond is cubic and has a single, high index of refraction. Under the microscope it is easily distinguishable from similar which are minerals such as colorless zircon, rutile, topaz, and sapphire sometimes used to imitate it because these are all birefringent. Minerals are often distinguished from each other based on the value of the index of refraction. Colorless spinel, like diamond, has only one refractive index, but its value is much lower than that of diamond. Synthetic imitations of diamonds such as cubic zirconia (cubic

Blue celestite, Madagascar.

Sulphur, Sicily.

Tiger's Eye (silicified crocodilite) Brazil.

zirconium oxide), YAG (yttrium-aluminum garnet) or GGG (gadolinium-gallium garnet) have, like diamond, high indices of refraction and are therefore more difficult to distinguish optically. In this case, the minerals can be distinguished from diamond based on other characteristics such as hardness and density.

It can be easily understood how the properties that capture people's imagination are the transparency and luster typical of many minerals and how their appearance has often encouraged research into their connection with the world of magic rather than observation of them as natural phenomena. However, over the course of history, there have been important people who, while remaining sensitive to the supernatural powers attributed to minerals, have also observed them from a practical standpoint. Certainly the most famous naturalist, who handed down so much information to us from the ancient world, was Pliny the Elder. He wrote a large compendium on observations and interpretations (often of a very fanciful nature) of natural materials and phenomena. This attentive observer of nature has related to us the most important minerals in Roman civilization: gold and silver, of course, but also other minerals from which they were able to extract commonly used metals such as lead, copper, iron, and tin.

In Roman society, as in all ancient societies, man was strictly dependent on nature to satisfy his needs and pleasures. One has only to think of the way in which these people obtained colors for dyeing cloth, and for painting houses. Some dyes were extracted from plants; others were made from mineral powders. Malachite for example, in addition to being used for ornamental objects, supplied a very beautiful, bright green pigment. Blue was obtained from azurite, brown from ferrous ocher (oxides and hydroxides of iron), and a very beautiful, intense red color from cinnabar (mercury sulfide).

Certainly gems exercised the greatest fascination because of the mysterious world they evoked. The most beautiful ones came from India and were primarily emeralds and carbuncles, the name used for all red gems such as ruby, garnet, and spinel.

Topaz, zircon, tourmaline, quartz,

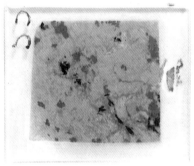

The study of transparent minerals by means of a polarizing microscope requires a preparation called a *thin section*. A thin slice of a mineral is cut, stuck on a glass support, and reduced to a thickness of 30 microns (0.03 mm) with the aid of a grindstone. Finally, it is covered with another thin piece of glass. The pictures reproduce a specimen of rock in which epidote crystals (green) are visible, and a thin section.

and other gems were also well-known, although there was some confusion about mineral names. Many stones were identified on the basis of their color and general appearance, resulting in different stones being called by the same name. Many times different colored minerals of the same kind were assigned to different species.

The transparent crystals from which gems were obtained were in demand primarily because it was believed that they were of celestial origin. Magical and medicinal properties resulting from stellar radiation were attributed to gems, according to Aristotelian theory. These ideas endured for a long time, as certain Greek works by Alexandrian authors from the first few centuries of the Christian era demonstrate. Later on, even some members of the ecclesiastical hierarchy, among them Albertus Magnus, treated minerals as substances endowed with moral and magical powers. This link to the world of the occult lasted for centuries, as can be deduced from the many works written on minerals which ignored contemporary technical and scientific advances. Considering minerals in such a way was probably also bases on the high philosophical value attributed to them in the works of Aristotle, and on the deep knowledge of natural matters handed down by Pliny the Elder. Many beliefs have in fact survived until relatively recent times. Up until the beginning of the last century, it was a common conviction that gold was formed in the equatorial zones where the sun's rays were most powerful, and that gems, believed to be of celestial origin, were associated with the signs of the zodiac. Even today, how many of us believe in a more or less conscious fashion in this link with the stars?

SULFUR

COMPOSITION: S native sulfur.

CRYSTALLINE FORM: orthorhombic system, bipyramidal class (alpha-sulfur). Appears as well developed crystals of bipyramidal form on a rhombic base, in acicular aggregates or in incrustations.

COLOR: lemon yellow, brown to black as a result of inclusions of clay or bitumen.

HARDNESS: 1.5-2.5.

SPECIFIC GRAVITY: 2.07.

LUSTER: resinous in pure crystals, greasy in those with inclusions of bitumen.

GEOLOGIC OCCURRENCE: common on top of salt domes from the conversion of sulfates (e.g. gypsum); by sublimation from fumaroles (solfataras).

ASSOCIATED MINERALS AND LOCALITIES: appears as large crystals in association with gypsum, calcite, aragonite, celestite, and barite in sedimentary deposits; occurs with realgar and other sublimates in solfataras. Beautiful specimens have been found in Italy (Sicily and Romagna), Russia, Mexico, and the United States.

USES: the chief material for the manufacture of sulfuric acid; also used in the production of explosives, fungicides, and fertilizers.

NOTES: the name is derived from the Latin *sulfur*. At present, naturally occuring deposits are of lesser interest as sulfur is an important by-product of petroleum refining. When heated quickly, sulfur fuses at about 235°F (113°C). With a slow rise in temperature, at about 203°F (95°C) it is converted into beta-sulfur (monoclinic system), typical of solfataras. Melted sulfur thrown into cold water hardens and becomes rubbery (elastic sulfur). It is a very poor conductor of heat and electricity and is soluble in chloroform and carbon sulfide.

▲ *Sulfur (ca. x 1), Belisio, Pesaro, Italy.*

DIAMOND

COMPOSITION: C native carbon.

CRYSTALLINE FORM: cubic system, hexoctahedral class. Isolated crystals with almost perfect octahedral form; other less common forms are the tetragonal trisoctahedron, the hexoctahedron, and the tetrahexahedron.

COLOR: generally it is colorless or light yellow, less commonly light green, other colors are rare.

HARDNESS: 10.

SPECIFIC GRAVITY: 3.50-3.53.

LUSTER: adamantine.

GEOLOGIC OCCURRENCE: in primary deposits diamond is most common in kimberlites, volcanic ultrabasic rocks in cylindrical or funnel-shaped deposits called "pipes." Diamond is also found in meteorites and in and around impact craters. In secondary deposits called "placers," it is present in sand and gravel.

ASSOCIATED MINERALS AND LOCALITIES: Diamond is found in South Africa, Lesotho, Namibia, Botswana, Zaire, Angola, Sierra Leone, Tanzania, Ghana, Australia, India, Brazil, Venezuela, Russia, Borneo, Canada, and rarely, in the United States.

USES: Diamond has long been the most prized of gems, and therefore of high value. Becuase it is the hardest known mineral, impure crystals which are not gem-quality are useful as abrasives ground into pastes, or included in metallic equipment to increase hardness and to reduce wear.

NOTES: The name is derived from the Greek for "indomitable." Pure stones are used as high-priced gems; colored varieties with strong color tones also fetch a good price, while weak tones considerably decrease the value of the stones; among these, light yellow (straw-yellow) is very common. Red diamonds are the most valuable gemstones per carat in the world. Carbon in its pure state crystallizes as diamond under conditions of very high pressure; but under low pressure it crystallizes in the form of graphite (hexagonal).

Octahedral diamond crystal ▶
(ca. x 0.5) Kimberley, South Africa.

SPHALERITE

COMPOSITION: ZnS zinc sulfide.

CRYSTALLINE FORM: cubic system, hextetrahedral class. Appears as crystals of generally octahedral form, sometimes the combination of two tetrahedrons simulates an octahedron; in compact masses from granular to microcrystalline.

COLOR: if pure, has a resinous yellow, brown, and reddish-brown coloration; the dark brown to black color is found in crystals in which part of the zinc has been replaced by iron.

HARDNESS: 3.5-4.

SPECIFIC GRAVITY: 3.9-4.2.

LUSTER: from adamantine to resinous, to submettalic.

GEOLOGIC OCCURRENCE: found in hydrothermal deposits of mixed sulfides characterized by abundant amounts of zinc, which form sphalerite and lead which forms galena. It can be an accessory mineral in pneumatolytic veins and pegmatites.

ASSOCIATED MINERALS AND LOCALITIES: Mixed sulfide deposits are characterized by a considerable variety of minerals, among which the most common are galena, argentite, chalcopyrite, pyrrhotite, tetrahedrite, and pyrite; also commonly associated with gangue minerals such as quartz, calcite, fluorite, and boaite. These deposits are very widespread on the earth's crust and among the most well-known are those in the United States, Canada, Mexico, Australia, Morocco, Zambia, the Balkan countries, Italy, etc.

USES: the chief mineral for the extraction of zinc and some by-products such as gallium, cadmium, and indium. Zinc is primarily used as a component of metallic alloys.

NOTES: Sphalerite is also known as blende which comes from the word for "deceitful," stemming from the fact that sphalerite commonly resembles galena, but contains no lead.

Sphalerite with ▶
galena (ca. x 1.5),
Isère, France.

WURTZITE

COMPOSITION: ZnS zinc sulfide.

CRYSTALLINE FORM: hexagonal system, dihexagonal pyramidal class. Rare as crystals, which are normally prismatic, pyramidal, and of a small size; generally appears in aggregates with a radial fibrous structure.

COLOR: from brown to black.

HARDNESS: 3.5-4.

SPECIFIC GRAVITY: 3.98-4.0.

LUSTER: from resinous to submetallic.

GEOLOGIC OCCURRENCE: in hydrothermal deposits; it may be an accessory mineral in pneumatolytic veins, in pegmatites, and in marbles.

ASSOCIATED MINERALS AND LOCALITIES: with galena, argentite, calchopyrite, pyrrhotite, fluorite, and barite. Beautiful specimens come from Bolivia, the United States, and the Czech Republic.

USES: has the same uses as sphalerite but is rarer and therefore is less commonly used.

NOTES: Wurtzite is an example of a polymorph of the ZnS compound. It is stable at a higher temperature 1875°F (1024°C) than sphalerite and contains lesser quantities of iron and greater quantities of cadmium. Wurtzite can be recognized by its radial fibrous appearance. Because it is unstable at low temperatures, below 1875°F (1024°C) wurtzite can transform into sphalerite. It fuses with difficulty, is soluble in hydrochloric acid, and is not fluorescent.

▲ *Wurtzite (ca. x 1) Lengenbach, Binn Valley, Switzerland.*

THE MOST PRECIOUS MINERAL

DIAMOND

The hardest, the most brilliant, the most valuable. Pure carbon crystallized at great depths under conditions of extremely high temperatures and high pressures, brought close to the earth's surface by volcanism and erosion. Innumerable imitations and reproductions, none of which equal its characteristic properties.

HISTORY AND LEGEND

The choice of the Greek word *indomitable* for this valuable mineral is a result of the fact that the ancients were not able to cut it, much less work with it. Some peoples believed that it could make love grow and last, and it was religiously guarded by

Rectangular scissors-cut and round-off rectangular scissors-cut.

Zircon-cut.   *Star-cut.*

married couples and lovers. Others considered it capable of exposing the adulterous liaisons of women.

Highly poisonous properties were attributed to diamond powder. Stories were handed down that this was the means of assassinating both Paracelsus, the Swiss physician and alchemist, and Benvenuto Cellini, the celebrated Florentine sculptor poisoned by Messer Durante. The naturalist Biringuccio did not believe

The Tiffany diamond.

that the powder was poisonous, but rather that, acting as would ground glass, it was the cause of lesions on the walls of the stomach. Other peoples attributed beneficent powers to the diamond, especially against serious illnesses like the plague.

INDUSTRIAL USES

The variety of diamond made up of crystals in a radial fibrous association, called a "bort," is used as an abrasive, while the variety in a microcrystalline association called a "carbonado" is incorporated into hard steels. Several years ago, in absolute secrecy, synthetic production was begun because of the difficulty of producing stones of suitable size for gemological use. Until the second half of the last century, diamonds came exclusively from secondary deposits (Indian and Brazilian placers). With

The Briolette diamond (above) and a diamond crystal (below).

the discovery of the South African kimberlite deposits, extraction took on an industrial character and the cutting system was developed.

THE MOST FAMOUS DIAMONDS

Large diamonds are very rare: the heaviest raw diamond 3,106 carats = 22 oz., about 3.4 in. x 2.4 in. (621 grams, about 10 cm. x 6 cm.) was found near Pretoria in 1905 and was

Specimens of diamonds prized for their color and cut.

cut into 9 large and 96 small stones. The largest piece, the "Cullinan I," which is 530.2 carats, adorns the scepter of England.

The famous Koh-i-Noor (mountain of light), the most coveted diamond in history, is also mounted in Queen Mary's Crown.

The 787.5 carat "Great Mogul" was found in India between 1650 and 1658 and, for a long time, was believed to be the largest diamond in the world. It was cut by the Venetian, Ortensio Borgia into a rose-form gem in the shape of half an egg, reducing its weight to only 280 carats. The Shah's anger was so great that poor Borgia was fined 10,000 rupees, a sum equal to the total of his riches. The diamond's current owner remains unknown.

CINNABAR

COMPOSITION: HgS mercury sulfide.

CRYSTALLINE FORM: trigonal system, trapezohedral class. Rhombohedral crystals are rare; more commonly, it appears in microgranular masses.

COLOR: bright red, brown-red, gray.

HARDNESS: 2–2.5.

SPECIFIC GRAVITY: 8.09.

LUSTER: adamantine in transparent red crystals, metallic in the dark ones (tending towards black).

GEOLOGIC OCCURRENCE: generally in low-temperature hydrothermal veins; as a sublimate in a fumarolic environment; and rarely in secondary deposits (placers).

ASSOCIATED MINERALS AND LOCALITIES: with calcite, stibnite, realgar; rarely with other sulfides. Principal deposits are found in Spain, China, the United States, and Mexico.

USES: principal mineral for the extraction of mercury (if pure, 86.2 weight %), as the native element is very rare. It was once used as a natural pigment (vermillion) and as a fungicide. It is the prime material in the manufacture of detonators for explosives.

NOTES: The name is probably of Indian origin. When heated in a closed tube, it evaporates at about 1076°F (580° C), depositing drops of mercury on the cold wall; insoluble in acids, it reacts with aqua regia; it is not fluorescent. Metacinnabar is a rare black variety.

▼ *Cinnabar (ca. x 1.3), Guizhou (Kweichow), China.*

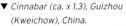

PROUSTITE

COMPOSITION: Ag_3AsS_3 silver arsenic sulfide.

CRYSTALLINE FORM: trigonal system, ditrigonal pyramidal class. The crystals have a prismatic or rhombohedral habit, are typically striated, and commonly twinned; in massive aggregates.

COLOR: dark red, black.

HARDNESS: 2-3.5.

SPECIFIC GRAVITY: 5.57-5.62.

LUSTER: adamantine in transparent red crystals, metallic in the dark ones (tending towards black).

GEOLOGIC OCCURRENCE: in low-temperature hydrothermal sulfide veins.

ASSOCIATED MINERALS AND LOCALITIES: with argentite and other sulfides and gangue minerals of similar origin. Very beautiful specimens come from Chile (Chañarcillo) and the German deposits are historically noteworthy.

USES: it is an excellent mineral for the extraction of silver (65.4%), but is fairly rare.

NOTES: named in honor of the French chemist J. L. Proust (1754-1826). Proustite constitutes an isomorphic series (group) with pyrargyrite and, therefore, thus minerals of intermediate composition and with intermediate characteristics are possible. In general, its characteristics are less metallic than those of pyrargyrite (color not as dark and more transparency). They are both called ruby silvers. It fuses easily, is soluble in nitric acid with the separation of sulfur, and is not fluorescent.

▼ Proustite (ca. x 1.5), Chañarcillo, Chile.

PYRARGYRITE

COMPOSITION: Ag_3SbS_3 silver antimony sulfide.

CRYSTALLINE FORM: trigonal system, ditrigonal pyramidal class. The crystals have a prismatic or rhombohedral habit with typical striations, usually in compact masses.

COLOR: black with dark red hues.

HARDNESS: 2.

SPECIFIC GRAVITY: 5.77-5.87.

LUSTER: from submetallic to adamantine.

GEOLOGIC OCCURRENCE: in low-temperature hydrothermal sulfide veins; as an alteration product of argentite or argentiferous galena.

ASSOCIATED MINERALS AND LOCALITIES: with argentite and other sulfides of similar origin, and gangue minerals. Very beautiful specimens come from Chile (Chañarcillo), the Czech Republic, and Canada; the German deposits (Harz) are historically noteworthy.

USES: it is an excellent mineral for the extraction of silver (about 60%), but is fairly rare.

NOTES: The name is derived from the Greek for "fire and silver", relating to the color and composition of the mineral. It constitutes an isomorphic series with proustite, and consequently minerals of intermediate composition and with intermediate characteristics are possible. In general, pyrargyrite has more strikingly metallic characteristics than proustite (darker color and less transparent). They are both called ruby silvers. It fuses easily, is soluble in nitric acid, and is not fluorescent.

▼ Pyrargyrite (ca. x 2), St. Andreasberg, Harz, Germany.

POLYBASITE

COMPOSITION: $(Ag, Cu)_{16}Sb_2S_{11}$ silver and copper antimony sulfide.

CRYSTALLINE FORM: monoclinic system, prismatic class. The crystals have a tabular, pseudohexagonal habit.

COLOR: black, or cherry-red if in thin chips.

HARDNESS: 2-3.

SPECIFIC GRAVITY: 6-6.2.

LUSTER: metallic.

GEOLOGIC OCCURRENCE: in low-temperature hydrothermal sulfide veins.

ASSOCIATED MINERALS AND LOCALITIES: with tetrahedrite, stephanite, and gangue minerals. Beautiful specimens come from Mexico, Chile, Bolivia, and the United States; the German deposits (Harz) are historically noteworthy.

USES: it may contribute locally to the production of silver (about 70%).

NOTES: similar to hematite (ferric oxide), it is softer, fuses at a lower temperature, and gives off sulfurous and antimony fumes. It is soluble in nitric acid and is not fluorescent.

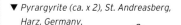

▲ Polybasite (ca. x 1.2), Mexico.

ORPIMENT

COMPOSITION: As_2S_3 arsenic sulfide.

CRYSTALLINE FORM: monoclinic system, prismatic class. The rare crystals have a flattened prismatic habit; in fibrous or lamellar aggregates; in foliated masses or in earthy ones with a pulverulent appearance.

COLOR: various shades of yellow.

HARDNESS: 1.5-2.

SPECIFIC GRAVITY: 3.48.

LUSTER: from adamantine to resinous.

GEOLOGIC OCCURRENCE: in low-temperature hydrothermal deposits; common among the sublimates of fumaroles; rarer in relation to hot springs.

ASSOCIATED MINERALS AND LOCALITIES: with realgar, cinnabar, borates. Beautiful specimens come from Turkey, France, Hungary, and the United States.

USES: in the leather goods industry, and for the extraction of arsenic (61%). In the past, especially in the Orient, it was used as a gold pigment.

NOTES: the name comes from the Latin *auripigmentum* (already mentioned by Pliny), relating to the mineral's color. It fuses easily, giving off arsenical fumes with a garlicky odor. It is soluble in nitric acid and in aqua regia with separation of sulfur. In the air and in light, it tends to disintegrate into powder. It may be derived by alteration from realgar, and it is not fluorescent.

▼ Orpiment (ca. x 0.6), Greece.

REALGAR

COMPOSITION: AsS arsenic sulfide.
CRYSTALLINE FORM: monoclinic system, prismatic class. The rare crystals have a prismatic habit, commonly acicular; in compact, granular masses and incrustations.
COLOR: ruby-red.
HARDNESS: 1.5-2.
SPECIFIC GRAVITY: 3.55.
LUSTER: resinous.
GEOLOGIC OCCURRENCE: in low temperature hydrothermal deposits; common among sublimates of fumaroles; rarer in ralation to hot springs.
ASSOCIATED MINERALS AND LOCALITIES: with borates, sphalerite, tetrahedrite. Very beautiful specimens come from Turkey, Romania, Hungary, Macedonia, and Switzerland.
USES: in the fireworks industry for the color white, in the paint industry, for the manufacture of arsenious anhydride, and for the extraction of arsenic (70.1%).
NOTES: the name is derived from the Arabic *rahj al-ghar*, powder of the mine. It fuses easily, giving off arsenic fumes with a garlicky odor; it is soluble in nitric acid and in aqua regia with separation of sulfur. In the air and in light, it tends to alter to pulverulent orpiment. It is not fluorescent, and if it is ground into powder and mixed with potassium nitrate or potassium chlorate, it deflagrates.

▼ *Realgar (ca. x 2) Corsica, France.*

KERMESITE

COMPOSITION: Sb_2S_2O antimony oxysulfide.
CRYSTALLINE FORM: monoclinic system, prismatic class. The rare crystals are acicular, needle-shaped, commonly joined together in tufts or bands.
COLOR: cherry-red.
HARDNESS: 1-1.5.
SPECIFIC GRAVITY: 4.68.
LUSTER: from submetallic to adamantine.
GEOLOGIC OCCURRENCE: alteration product of stibnite.
ASSOCIATED MINERALS AND LOCALITIES: with stibnite, klebelsbergite, quartz, calcite. Beautiful specimens come from Germany, Hungary, Japan, and Italy.
USES: use limited to research and collecting.
NOTES: Kermesite is a rare antimony mineral. It fuses easily and becomes black when heated. It is not fluorescent.

▲ *Kermesite (ca. x 1) Braunsdorf, Saxony, Germany.*

HALITE

COMPOSITION: NaCl sodium chloride.
CRYSTALLINE FORM: cubic system, hexoctahedral class. The very common crystals are cubic; hopper-shaped forms are typical, the octahedrons are more rare, in microcrystalline masses.
COLOR: colorless if pure, blue or violet from lattice defects, grayish from inclusions of clay, yellow or brown from inclusions of ferric oxides.
HARDNESS: 2.5.
SPECIFIC GRAVITY: 2.168.
LUSTER: vitreous.
GEOLOGIC OCCURRENCE: in sedimentary deposits resulting from the evaporation of sea water. Deep deposits, due to their low density with respect to the surrounding rocks and to the notable plasticity of the mineral, can slowly rise up to form salt domes. In the cap-rock gypsum, anhydrite and sometimes native sulfur are concentrated. Halite may also be found in small quantities as a sublimation product in volcanic craters.
ASSOCIATED MINERALS AND LOCALITIES: with gypsum and other halogenous evaporites. Very beautiful specimens come from Poland, Germany, England, Denmark, Austria, the United States, etc.
USES: basic food for man and for animal life, prime material in the food industry and in the chemical industry for the production of chlorine and sodium.
NOTES: The name derives from the Greek for salt. Deliquescent when it contains inclusions of calcium or magnesium chloride. Easily soluble in water or in a humid environment. An excellent conductor of heat, it fuses easily coloring the flame a bright yellow. It is fluorescent, emitting blue or brown light.

SYLVITE

COMPOSITION: KCl potassium chloride.
CRYSTALLINE FORM: cubic system, hexoctahedral class. Crystals are generally cubic, octahedrons are rare, very commonly in microcrystalline masses.
COLOR: colorless if pure; violet, reddish from inclusions of ferric oxide (hematite).
HARDNESS: 2-2.5.
SPECIFIC GRAVITY: 1.993.
LUSTER: vitreous.
GEOLOGIC OCCURRENCE: most commonly in evaporitic sedimentary deposits associated with rock salt. Rarely, it can be found as a sublimation product in volcanic craters.
ASSOCIATED MINERALS AND LOCALITIES: with rock salt, kieserite, kainite, cornallite, polyhalite, gypsum. Beautiful specimens come from Poland, Germany, Spain, Canada, the United States, Ukraine, etc.
USES: other than constituting the principal potassium mineral, it can be used directly as a fertilizer.
NOTES: less deliquescent than rock salt, it has a much more bitter taste. It is easily soluble in water and fuses easily, coloring the flame violet-red; it is not fluorescent.

▼ *Sylvite (ca. x 1.2), Stassfurt, Germany.*

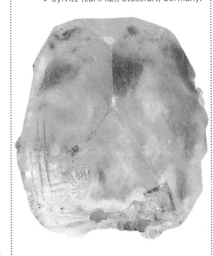

◄ *Violet rock salt (ca. x 1), Calascibetta, Enna, Italy.*

FLUORITE

COMPOSITION: CaF_2 calcium fluoride.
CRYSTALLINE FORM: cubic system, hexoctahedral class. Crystals are cubic, hexoctahedral, dodecahedral, or, very rarely, composed of more than one form; in compact, sparry, zonal masses.
COLOR: colorless when pure; red, yellow, green, violet, blue to black.
HARDNESS: 4.
SPECIFIC GRAVITY: 3.18.
LUSTER: vitreous.
GEOLOGIC OCCURRENCE: as a gangue mineral in metalliferous veins; an accessory mineral in pegmatites; more rarely as a precipitate from mineral waters of volcanic origin.
ASSOCIATED MINERALS AND LOCALITIES: with quartz, tourmaline, topaz, apatite, cassiterite, calcite, barite, etc. Very beautiful specimens come from Brazil, Sri Lanka, India, Burma, Madagascar, Namibia, Norway, etc.
USES: in metallurgy as a base flux (treatment of bauxite or as a liquefacient), and in the ceramics industry; in the preparation of hydrofluoric acid; very pure crystals are used in the making of microscopical lenses (called apochromatic lenses) prisms for spectrography.
NOTES: the name is derived from the Latin *fluere* (to flow), relating to its fluidizing property. It fuses with difficulty, coloring the flame brick-red and is soluble in concentrated sulfuric acid, giving off hydrofluoric acid fumes. It is very fluorescent, so much so that the phenomenon of fluorescence owes its name to this mineral. (See box.)

▼ *Fluorite (ca. x 1), Cumberland, Great Britain.*

CARNALLITE

COMPOSITION: $KCl \cdot MgCl_2 \cdot 6H_2O$ hydrated potassium magnesium chloride.
CRYSTALLINE FORM: orthorhombic system, hipiramidal class. The rare crystals have a pseudohexagonal form, very commonly in compact microcrystalline masses.
COLOR: colorless when pure, sometimes reddish from inclusions of minute sheets of hematite.
HARDNESS: 2.5.
SPECIFIC GRAVITY: 1.6.
LUSTER: vitreous.
GEOLOGIC OCCURRENCE: in deposits connected to evaporite rocks.
ASSOCIATED MINERALS AND LOCALITIES: with kieserite, anhydrite, sylvite, rock salt, polyhalite. Beautiful specimens come from Iran, Tunisia, China, Mali, the United States, and the Ukraine.
USES: in the fertilizer industry and for the extraction of potassium, magnesium, chlorine, and bromine and cesium which substitute for chlorine.
NOTES: the name pays homage to the German mining engineer Rudolf von Karnall. It fuses easily, coloring the flame violet (potassium), is soluble in water, deliquescent, disintegrates in the air, and is very phosphorescent.

▲ *Carnallite (ca. x 1), Stassfurt, Germany.*

LIGHT EFFECTS

FLUORITE

Fluorite is a very common mineral with numerous variations in color. It is primarily used for the production of decorative items or sculptures, especially in the Orient where true works of art that emphasize the delicacy and beauty of the material are created.

COLORS AND PROPERTIES
Colorless, yellow, violet to blue and black fluorite (see this page) crystals are common; the blue, green, pink, and sometimes even zoned ones are more rare. The reason for this chro-

Drawing of a cubic crystal of fluorite with cleavage marks which produce an octahedral crystal.

matic variability is not yet known: it is believed that it is essentially due to impurities and perhaps partly to lattice defects which result from impurities. In some cases a connection with the presence of radioactive elements has also been hypothesized. In fact, europium, an element in the rare-earth group, has been found in the green variet. In many fluorites the phenomenon of fluorescence is remarkable. This characteristic appears to be linked to the environments in which crystallization takes place. When exposed to ultra-violet rays, colorless fluorites exhibit a blue fluorescence, the green ones a bright green response tending toward violet. Exposed to X-rays, the first emit an azure light, the red ones emit blue-green light followed by phosphorescence, the violet ones emit milky-blue light followed by very strong white-light blue

phosphorescence which slowly turns dull violet and remains active for many hours. In these conditions, it may also make an impression on a shaded photographic plate. A rare yellow variety, called chlorophane, emits an intense green phosphorescence simply upon heating a phenomenon which in the past

Oriental fluorite vase with prized features.

encouraged bizarre tales about the occult properties of this mineral.

USES:
Fluorite is characterized by easily cleaved octahedra, so much so that geometrically perfect octahedrons can be obtained from cubic crystals or from shapeless masses. Because of its low hardness (4 on the Mohs scale), it is not often used as a gem. However, sometimes the green variety is emerald-cut and carefully mounted so as to protect it from shocks. They are, in any case, low cost stones, appreciated by collectors only in speciments of over 8–10 carats.

Octahedron from cleavage of fluorite (Great Britain).

CRYOLITE

COMPOSITION: Na_3AlF_6 sodium aluminum fluoride.

CRYSTALLINE FORM: monoclinic system, prismatic class. Crystals have a pseudo-cubic form, very commonly in compact microcrystalline masses.

COLOR: colorless when pure, white, sometimes reddish brown.

HARDNESS: 2.5-3.

SPECIFIC GRAVITY: 2.97.

LUSTER: from vitreous to silky.

GEOLOGIC OCCURRENCE: a precipitate from fluorine-rich fluids in pegmatitic and in pneumatatolytic environments.

ASSOCIATED MINERALS AND LOCALITIES: with pegmatitic minerals. Beautiful specimens come from Greenland, the United States, and Nigeria.

USES: as a flux in enamel factories and for the manufacture of translucent glass and china; in the past, it was used in the elec-trometallurgy of aluminum, but today it has been replaced by synthetic products.

NOTES: the name is derived from the Greek for "cold and stone", in connection with its ice-like appearance resulting from the low value of the indexes of refraction; very close to that of water, so that when it is immersed, it disappears from sight. It fuses easily, coloring the flame yellow (sodium), is soluble in sulfuric acid with the production of hydrochloric acid, and is not fluorescent.

▼ *Cryolite (ca. x 1.5), Ivigtut, Greenland.*

ATACAMITE

COMPOSITION: $Cu_2Cl(OH)_3$ hydrated copper chloride.

CRYSTALLINE FORM: orthorhombic system, bipyramidal class. Crystals have a pris-matic form and are striated and elongat-ed; in acicular, lamellar aggregates. Very commonly in microcrystalline masses.

COLOR: intense green.

HARDNESS: 3-3.5.

SPECIFIC GRAVITY: 3.75-3.77.

LUSTER: vitreous.

GEOLOGIC OCCURRENCE: product of alter-ation in copper sulfides and of sublima-tion in patinas from volcanic fumes.

ASSOCIATED MINERALS AND LOCALITIES: alteration minerals of copper deposits. Beautiful specimens come from Chile, Peru, Bolivia, the United States, Australia, and Namibia.

USES: a secondary mineral for the extrac-tion of copper when in high concentra-tions such as those in the Chilean mines.

NOTES: the name is derived from the Ataca-ma Desert in Chile. It fuses easily, coloring the flame first blue, then green; it decrepi-tates during heating and is soluble in hydrochloric and nitric acids without effervescing; it is not fluorescent.

▼ *Atacamite (ca. x 1.5), Copiapo, Chile.*

CUPRITE

COMPOSITION: Cu_2O Copper oxide.

CRYSTALLINE FORM: cubic system, hexoc-tahedral class. Crystals have octahedral, rhombic dodecahedral, and cubic forms.

COLOR: copper-red or blackish, sometimes with a greenish (malachite) alteration patina.

HARDNESS: 3.5-4.

SPECIFIC GRAVITY: 6.14.

LUSTER: from adamantine to submetallic.

GEOLOGIC OCCURRENCE: especially in alteration zones of copper deposits.

ASSOCIATED MINERALS AND LOCALITIES: withcopper, azurite, malachite, tenorite. Very beautiful specimens come from France, Chile, Bolivia, the United States, Namibia, and South Africa.

USES: when abundant, it constitutes an excellent mineral for the extraction of copper.

NOTES: the name is derived from the Latin *cuprum* (copper), relating to its composi-tion. It fuses easily, coloring the flame green; is soluble in concentrated hydro-chloric acid and in dilute nitric acid; it is not fluorescent, and when exposed to the air it becomes semiopaque.

▲ *Cuprite (ca. x 1.5), Cornwall, Great Britain.*

SPINEL

COMPOSITION: $MgAl_2O_4$ magnesium aluminum oxide.

CRYSTALLINE FORM: cubic system, hexoc-tahedral class. Small, well-formed octahe-dral crystals, commonly twinned; in rounded granules.

COLOR: colorless if pure (very rare); with inclusions it appears red, pink, blue, azure, blackish brown in the ferriferous variety (pleonaste), yellow-brown from the presence of chromium (picotite).

HARDNESS: 7.5-8.

SPECIFIC GRAVITY: 3.5-4.1.

LUSTER: vitreous.

GEOLOGIC OCCURRENCE: typical of lime-stones and dolomites subjected to con-tact metamorphism; in ultrabasic magmatic rocks; concentrates in alluvial and marine deposits (placers).

ASSOCIATED MINERALS AND LOCALITIES: with pargasite, phologopite, chondrodite, calcite. Beautiful specimens come from India, Thailand, Madagascar, Sri Lanka, and Afghanistan.

USES: clear and transparent varieties of var-ious colors are used as gems; currently also produced synthetically.

NOTES: the term is of uncertain origin, prob-ably from the Greek for "spark"; it gives its name to a group of minerals divisible into three series, according to whether the trivalent ion is aluminum (spinel series), iron (magnetite series), or chromi-um (chromite series). Infusible, soluble only in concentrated sulfuric acid; in ultra-violet light, it has an orange to ruby-red fluorescence. (See box, p. 40.)

▼ *Spinel, pleonaste variety (ca. x 1.5), Pamir, Jakutien, Tajikistan, Russia.*

CHRYSOBERYL

COMPOSITION: $BeAl_2O_4$ beryllium aluminum oxide.

CRYSTALLINE FORM: orthorhombic system, bipyramidal class. Crystals are prismatic, tabular, commmonly twinned with typical V-shaped striations.

COLOR: varied: greenish yellow (common), emerald-green, red-violet, gray, brown, colorless.

HARDNESS: 8.5.

SPECIFIC GRAVITY: 3.65-3.85.

LUSTER: vitreous.

GEOLOGIC OCCURRENCE: in pegmatites, commonly in large semi opaque crystals: included in mica schists and alluvial deposits.

ASSOCIATED MINERALS AND LOCALITIES: with quartz, topaz, corundum, feldspars. Very beautiful specimens come from Brazil, Sri Lanka, the United States, and from the Urals region.

USES: the clearest and most colorful varieties are used as gems because of their hardness. The alexandrite variety (green in natural light, red in artificial), in particular, is highly valued while cymophane (noble or oriental cat's eye) has a slightly lower value. The latter has a silky luster, is very chatoyant, due to needle-shaped mineral inclusions, and is cabochon-cut.

NOTES: the name is derived from the Greek for "color" and "beryl," emphasizing its more intense yellow color than that of beryl. Alexandrite owes its name to Alexander II, as it was found in the Urals on his birthday in 1830. Its colors (dark green and red) correspond to the Imperial Russian colors. Cymophane takes its name from the Greek for wave and bright because of its very beautiful chatoyant effect. This highly prized variety may be confused with the very common cat's eye quartz. It is infusible, is unaffected by acids, and is not fluorescent.

▼ *Chrysoberyl (ca. x 1), Brazil.*

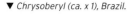

CORUNDUM

COMPOSITION: Al_2O_3 aluminum oxide.

CRYSTALLINE FORM: trigonal system, ditrigonal scalenohedral class. Crystals have a prismatic, tabular, or even bipyramidal form.

COLOR: varied: colorless, blue (sapphire), red (ruby), gray, black, yellow; more rarely green, pink, violet.

HARDNESS: 9.

SPECIFIC GRAVITY: 4-4.1.

LUSTER: from adamantine to vitreous.

GEOLOGIC OCCURRENCE: typical accessory mineral of metamorphic rocks rich in aluminum (marbles, mica schists), also present in magmatic rocks undersaturated with silica (syenites, nepheline syenites) and in pegmatites. Because of its high resistance to alteration, concentrations in alluvial or marine sands are common.

ASSOCIATED MINERALS AND LOCALITIES: with zoisite, calcite, basic feldspars, scapolite. Very beautiful specimens come from Burma, Thailand, Sri Lanka, Cambodia, Australia, Afghanistan, Greece, Tanzania, the United States, etc.

USES: the semiopaque varieties (emery) are used as abrasives; while clear, transparent varieties are cut as gems, some of which are very precious (e.g., ruby, sapphire).

NOTES: the name is derived from the Sanskrit *corind*, of uncertain meaning. Some highly prized varieties show a characteristic internal luminosity, similar to a six-rayed star, called an asterism. Others are fluorescent in ultraviolet light, with emission of yellow light. It is infusible and insoluble. (See box, p. 42.)

▼ *Corundum (ca. x 1.5), Burma.*

RED . . . SPINEL

Its wide chromatic variability and elevated hardness have made spinel a stone that has been traditionally favored by jewellers. The most well-known variety, the red, was in the past often mistaken for the ruby, and both of them were formerly indiscriminately called "carbuncles."

HISTORY AND LEGEND

The classic spinel (see p. 39) is the red one which in the past, was regarded as a symbol of war and victory. It was also used as an antidote for poisons and against the

Blue spinel and pink spinel, 1.38 c. (Sri Lanka).

plague, and was thought to dispel melancholy and overcome lust. It was believed that, as disaster approached, this stone took on a darker color and then returned to its natural one as the difficulty was overcome.

FAMOUS STONES

The most famous spinel is the one that Pedro the Cruel, King of Castille, gave to his ally the Black Prince, son of Edward III of England, in 1367 after the victory over the Moors at Najera. This was a 170 carat balas ruby (spinel) now mounted in the British Imperial crown. Also belonging to the crown treasury are two large, very bright red, flawless spinels of 197 and 102 carats which after cutting were reduced to 81 and 72 carats respectively. The treasury is also enriched by a truly unique 361 carat spinel called the Timur-Ruby which belonged first to the

Mongol conqueror Tamerlane, then, to the Great Mogol of New Delhi, to the Shah of Persia, and to the king of Lahore. Probably the most beautiful spinel is the 400 carat stone in Russia, which belonged to the czar who bought it in Beijing and had it mounted in the czarina's crown. Other important spinels are in various collections in the world: in the British Museum in London there is a magnificent blue spinel as well as other very large red ones; the Crown Treasury of Iran (not on exhibition) includes two exceptional raw red spinels of 500 and 270 carats, in the shape of rounded pebbles.

COLORS

Numerous varieties of spinel are found in nature because a variety of chemical elements can enter into its crystalline structure. This fact allows for a multiplicity of colors shown by other minerals such as corundum and tourmaline. The noble spinel is the red variety and it is also the most sought-after because of its strong similarity to the ruby, even though it is of lesser value.

The balas spinel (from Badakhstan, region of India) has a color from violet-red to violet. Picotite (from Picot de la Peyrouse) is a chromium spinel of brown or dark green color. Pleonaste is characterized by highly modified crystals with small crystalline faces, and it is black. Rubicelle is orange-red. Blue spinel has a velvety appearance and a color similar to the sapphire. Finally, gahnite comes in various colors, from reddish to dark green to black.

Spinel earring surrounded by gold and diamond pave.

ARSENOLITE

COMPOSITION: As_2O_3 arsenic oxide.

CRYSTALLINE FORM: cubic system, hexoctahedral class. The rare crystals have an octahedral form; commonly in pulverulent patinas and crusts, in earthy masses.

COLOR: white, blue, yellow.

HARDNESS: 1.5.

SPECIFIC GRAVITY: 3.88.

LUSTER: vitreous.

GEOLOGIC OCCURRENCE: alteration product in mixed sulfide deposits; in coal fields in the form of efflorescence, by metamorphic segregation.

ASSOCIATED MINERALS AND LOCALITIES: with arsenopyrite, tetrahedrite, realgar and other arsenic minerals, native arsenic. Beautiful specimens come from Germany, Hungary, the Czech Republic, and the United States.

USES: no industrial importance, interesting to researchers and collectors.

NOTES: the compound As_2O_3 is dimorphous with claudetite (monoclinic). It is infusible, sublimes in even a weak flame, giving off arsenical fumes with a garlicky odor; soluble in acids and warm water; it is not fluorescent and is poisonous if ingested.

▲ Arsenolite with arsenic (ca. x 1), Sondalo, Sondrio, Italy.

RUTILE

COMPOSITION: TiO_2 titanium oxide.

CRYSTALLINE FORM: tetragonal system, ditetragonal bipyramidal class. The crystals have a prismatic form, sometimes very elongated and striated; knee-shaped twins, in which two crystals are connected at an acute angle of about 50 degrees, are typical.

COLOR: generally red or metallic gray with red, brown, blue, violet, black, and in the sagenite variety, golden yellow reflections.

HARDNESS: 6-6.5.

SPECIFIC GRAVITY: 4.25.

LUSTER: from adamantine in transparent crystals to metallic in those tending towards black.

GEOLOGIC OCCURRENCE: very common accessory mineral in intrusive magmatic and metamorphic rocks; present in pegmatites; due to its high resistance to alteration, concentrations in marine sands are frequent.

ASSOCIATED MINERALS AND LOCALITIES: it is found with quartz, pyrophyllite, hematite, hornblende, feldspars, apatite, and other accessory minerals. Very beautiful specimens come from Brazil, Madagascar, the United States, Austria, and Norway.

USES: useful mineral of titanium, used in the production of special steels.

NOTES: the name is derived from the Latin rutilus (reddish). The acicular crystals of the sagenite variety are commonly included in clear quartz crystals, giving rise to a beautiful stone called "Venus' hair." Octahedrite (or anatase) and brookite have the same composition, but have different forms. It is infusible and unaffected by acids.

▼ Sagenite rutile (ca. x 1), Ticino, Switzerland.

CASSITERITE

COMPOSITION: SnO_2 tin oxide.

CRYSTALLINE FORM: tetragonal system, ditetragonal bipyramidal class. Prismatic, short crystals, commonly twinned in "visor tin" shape; in fibrous or granular masses and in dispersed granules.

COLOR: black, iron-gray, brown, hazel, pink.

HARDNESS: 6-7.

SPECIFIC GRAVITY: 6.9.

LUSTER: from submetallic to adamantine to resinous.

GEOLOGIC OCCURRENCE: it is found in acid intrusive rocks and in pegmatites; in metasomatic deposits such as skarn; in alluvial and marine deposits (placers).

ASSOCIATED MINERALS AND LOCALITIES: with wolframite, quartz, tourmaline, topaz, lepidolite, and fluorite. Beautiful specimens come from Malaysia, China, Bolivia, Germany, and Russia.

USES: it is the principal mineral for the extraction of tin (78.5 wt %); it is used in the preparation of alloys (bronze and brass), as an antioxidant (tinning), and for ceramic pigments.

NOTES: the name is derived from the Greek for tin. Isomorphous with rutile and pyrolusite; a variety of cassiterite is wood tin which forms in the oxidation zones of "tin" deposits. Infusible; in the reducing flame, with the addition of sodium carbonate, gives particles of metallic tin. It is insoluble in acids and is not fluorescent.

▲ Cassiterite (ca. x 2), La Paz, Viloco M.ra, Bolivia.

STIBICONITE

COMPOSITION: $Sb_3O_6(OH)$ hydrous antimony oxide.

CRYSTALLINE FORM: cubic system, hexoctahedral class. The rare crystals are always very small; commonly occurs in earthy masses.

COLOR: yellow-white, yellowish, reddish.

HARDNESS: 4.5-5.

SPECIFIC GRAVITY: 5.58.

LUSTER: from vitreous to pearly.

GEOLOGIC OCCURRENCE: typical mineral in the alteration zones of antimony mineral deposits.

ASSOCIATED MINERALS AND LOCALITIES: with stibnite, cervantite, quartz, calcite. Beautiful specimens come from Japan, Hungary, Romania, the United States, Mexico, Peru, and Bolivia.

USES: useful mineral for the extraction of antimony, if present in high concentrations.

NOTES: commonly forms beautiful crystalline aggregates, pseudomorphic after stibnite; it fuses with difficulty and is insoluble in hydrochloric acid; in a closed tube yields water; it is not fluorescent; also known in the past under the name antimony ochre.

▼ Stibiconite (ca. x 1), Valle Serra, Trento, Italy.

CORUNDUM

Corundum is frequently found as very small crystals. Large crystals are commonly gray, brown, or opaque as a result of abundant inclusions. Transparent corundum (see p. 40) crystals are rare, occur in a variety of colors, and are therefore valuable gemstones. Both the value and the name of corundum gems are based on color.

THE RUBY

In the past, the ruby (from the Latin *rubrum*, red) was called *carbunculus* (carbuncle, vivid charcoal) and was considered a strictly masculine stone, prized by kings, emperors, and maharajahs. Numerous legends elucidated the powers of this stone. It was said that it could allow the possessor to conquer realms and dominions, stop water from boiling, soothe cholera, and safeguard against temptation. It was considered to be the symbol of victory, charity, love, faith, and, according to Mandeville, could assume evil properties when worn on the right.

Because it is so hard, any variety of corundum is suitable for various types of cutting.

Depending on its color, size, and other characteristics, a ruby can have a higher value than a diamond. The most beautiful crystals, those most markedly vermillion, come from

Gray-blue asteria (star sapphire), 1.50c.

Colorless sapphire, 1.50c.

Burma (Mogok region), although splendid stones are also found throughout Southeast Asia. One distinctive property of rubies is that they brighten in color when subjected to intense light. A silky effect is also typical, though may not be present in every ruby.

Stones of similar color can be dis-

RUBY: various shades of red
SAPPHIRE: from azure to various shades of blue
PADPARADSCHA (OR ROYAL TOPAZ): yellow-orange, tangerine
ASTERIA (STAR-STONE): red, blue, gray
Yellow sapphire (or yellow corundum): light yellow, lemon-yellow
LEUCOSAPPHIRE: colorless

SUNflOWER RUBY (OR CAT'S EYE RUBY): chatoyant red
PINK SAPPHIRE (OR PINK CORUNDUM): pink, pink-violet
SUNflOWER SAPPHIRE (OR CAT'S EYE SAPPHIRE): chatoyant dark blue
OPALESCENT SAPPHIRE: milky gray or bluish opalescent

Classic ring with an oriental ruby surrounded by ten diamonds.

tinguished from ruby on the basis of several properties. Red spinel lacks both the silky effect and pleochroism typical of rubies; the garnet lacks both pleochrosim and color augmentation in intense light. Red tourmaline also fails to brighten in color under intense light. These minerals can also be distinguished from ruby on the basis of hardness, specific gravity, and refractive index. As a result of the high level of current technologies in gem synthesis, it is very difficult to distinguish real rubies from synthetic ones.

Among the most famous rubies are the 17.5 carat stone described by Tavernier and owned by the Maharajah of Visapour and the 50.75 carat Banarous ruby owned by an Indian merchant.

THE SAPPHIRE

In the past, the sapphire (from the Greek, perhaps of early Sanskrit origin) was called the "gem of gems." As with all precious stones, medical powers were attributed to it. It was also held to be the symbol of generosity, goodness of heart, and loyalty. According to Galen, sapphire, when held in the mouth, safeguarded against scorpion stings and, according to Dioscorides, against intestinal sores. It was thought to placate amorous

desires and was much used by priests.

Although sapphire is more abundant than ruby, it is one of the most valuable precious stones. The finest crystals come from Southeast Asia, Sri Lanka, the United States, and Australia. The peculiar property of this stone is its pleochroism when it has an intense color. If observed from above, it appears azure-blue, while from the side, it exhibits a blue color with an obvious dark green tone. Stones of similar color are the tanzanite (with an obvious violet hue), cordierite (blue-violet or grayish-blue), indicolite (greenish-blue tourmaline), blue zircon (electric sky-blue), and the blue spinel. Once again, sapphire can be distinguished from other similar minerals based on physical properties.

REAL AND IMITATION

At present, very carefully worked imitations are on the market. When natural sapphires with light coloring and inclusions of rutile (titanium oxide, which produces the silky effect), are heated to 2732–2912° F (1500–1600° C) in a reducing environment, they take on an intense blue color due to the "redistribution" of titanium in their interi-

or. Another procedure consists in covering the gems with a paste containing abundant amounts of titanium and iron and subjecting them to temperatures of about 3092° F (1700° C) for a few days. In this second case, the titanium and iron oxides that are formed

Raw corundums of various colors.

penetrate about 1 millimeter into the interior of the stones giving them the typical blue-sapphire coloration.

Among the most famous sapphires, the largest (951 carats) is from India. In the mineralogical museum of Paris, a 136 carat stone, called the "Ruspoli," cut in the shape of an oblique prism, is exhibited. In Russia a specimen of 249 carats was kept in the czar's treasury.

Sapphires of various shades of blue.

BRUCITE

COMPOSITION: $Mg(OH)_2$ magnesium hydroxide.

CRYSTALLINE FORM: trigonal system, ditrigonal scalenohedral class. The rare crystals have a pseudohexagonal, tabular form; commonly in scaly, microcrystalline masses.

COLOR: colorless, green, blue, yellow-pink, brown in the manganous varieties.

HARDNESS: 2.5.

SPECIFIC GRAVITY: 2.4.

LUSTER: silky.

GEOLOGIC OCCURRENCE: in low temperature hydrothermal veins within serpentine rocks; alteration product of periclase (MgO) in contact metamorphism of dolomites.

ASSOCIATED MINERALS AND LOCALITIES: with magnesium carbonates, aragonite, chromite, chrysotile. Beautiful specimens come from Canada, the United States, and Scotland.

USES: for the extraction of metallic magnesium and for the manufacturing of refractories.

NOTES: the name pays homage to the American mineralogist Archibald Bruce. It is infusible, easily soluble in cold dilute acids, and, unlike some magnesium carbonates, does not efferevesce (even after heating). It is not fluorescent.

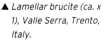

▲ Lamellar brucite (ca. x 1), Valle Serra, Trento, Italy.

CALCITE

COMPOSITION: $CaCO_3$ calcium carbonate.

CRYSTALLINE FORM: trigonal system, ditrigonal scalenohedral class. These very common crystals have a very varied form, from acicular to prismatic, scalenohedral, rhombohedral, twins are frequent; also found as zoned concretions (stalactites, stalagmites).

COLOR: colorless, white, yellow, pink, green, brown.

HARDNESS: 3.

SPECIFIC GRAVITY: 2.71.

LUSTER: vitreous to pearly.

GEOLOGIC OCCURRENCE: in sedimentary rocks (limestones of various kinds), metamorphic rocks (marbles), rarely in igneous rocks (carbonatites); as a gangue mineral in numerous types of mining deposits.

ASSOCIATED MINERALS AND LOCALITIES: because of its wide range of occurrence, it is associated with many minerals. Very beautiful specimens come from England, Germany, Canada, the United States, Mexico, etc.; the famous spar is typical of Iceland but also common in other places.

USES: in the building industry, in metallurgy as a flux and to remove slag, in the chemical industry, in fertilizers, as an ornamental stone, etc.

NOTES: Calcite is a very common mineral and its name is derived from the Latin *calx* which was the word for quicklime (CaO), obtained by roasting limestone. It is the fundamental component of shells and animal skeletons. Iceland spar is unusual for its ability to split images seen through it (a result of double refraction). Calcite is one in a group of isomorphic carbonate minerals in which calcium is replaced by cadmium, manganese, iron, cobalt, zinc, and magnesium. (See box, p. 45.)

Calcite (ca. x .09) England, ▶ Great Britain.

MAGNESITE

COMPOSITION: $MgCO_3$ magnesium carbonate.

CRYSTALLINE FORM: trigonal system, ditrigonal scalenohedral class. The rare crystals are prismatic, rhombohedral, scalenohedral; common in compact microcrystalline, saccharoidal or, sometimes, concreted masses.

COLOR: colorless, white, yellow, brown.

HARDNESS: 4-4.5.

SPECIFIC GRAVITY: 3.

LUSTER: vitreous to pearly.

GEOLOGIC OCCURRENCE: secondary mineral in rocks rich in magnesium minerals (serpentine, olivine); of metasomatic origin in limestones and dolomites; rarely as a gangue mineral in metalliferous veins.

ASSOCIATED MINERALS AND LOCALITIES: with talc, chlorite, calcite, and metallic minerals. Beautiful specimens come from India, Manchuria, Brazil, the United States, Greece, Poland, etc.

USES: for the extraction of magnesium, in the iron and steel and pharmaceutical industries, for refractories and special cements.

NOTES: Magnesite is a member of the isomorphous calcite group; it is soluble with effervescence in hot hydrochloric acid, but, unlike calcite which dissolves quickly in cold and dolomite which dissolves slowly, it is insoluble in cold. Infusible, it decrepitates during heating. It is fluorescent, emitting green or bluish light.

▼ *Magnesite, breunnerite variety (ca. x 0.4), Val Malenco, Sondrio, Italy.*

SIDERITE

COMPOSITION: $FeCO_3$ iron carbonate.

CRYSTALLINE FORM: trigonal system, ditrigonal scalenohedral class. Crystals are rhombohedral or scalenohedral, commonly with curved and striated faces; common in radial fibrous and botryoidal masses.

COLOR: from light yellow to brown from oxidation of iron.

HARDNESS: 4-4.5.

SPECIFIC GRAVITY: 3.89.

LUSTER: vitreous.

GEOLOGIC OCCURRENCE: in low temperature hydrothermal veins and in sedimentary deposits of chemical origin; as a gangue mineral in metalliferous deposits.

ASSOCIATED MINERALS AND LOCALITIES: found with metallic minerals of various kinds. Beautiful specimens come from Austria, Germany, France, England, Spain, Tunisia, Algeria, the United States, Greenland, Brazil, etc.

USES: if in good concentration, it is an excellent mineral for the extraction of iron.

NOTES: Siderite is a member of the isomorphous calcite group. It is slowly soluble in cold hydrochloric acid, while when heated in the acid, it effervesces; infusible, but decomposes at moderate temperatures, darkens, and becomes magnetic. It alters easily to ferric oxides (limonite). It is not fluorescent.

▲ *Siderite (ca. x 1.5), St. Hilaire, Québec, Canada.*

RHODOCHROSITE

COMPOSITION: $MnCO_3$ manganese carbonate.

CRYSTALLINE FORM: trigonal system, ditrigonal scalenohedral class. The rare crystals are rhombohedral or scalenohedral; most commonly in granular, sparry, radial fibrous, and botryoidal masses.

COLOR: more or less intense pink, flesh pink if the mineral is fresh; alters to black.

HARDNESS: 4.

SPECIFIC GRAVITY: 3.7.

LUSTER: vitreous.

GEOLOGIC OCCURRENCE: in low-temperature hydrothermal veins; as a gangue mineral in silver, copper, lead, and zinc mineral deposits.

ASSOCIATED MINERALS AND LOCALITIES: with calcite, quartz, and metallic minerals of various kinds. Beautiful specimens come from Romania, Spain, France, Germany, the United States, Argentina, South Africa, etc.

USES: for the extraction of manganese; in the metallurgic industry for special steels; for jewelry and furnishings.

NOTES: The name is derived from the Greek for "rose" and "color". It is a member of the isomorphous calcite group. It is insoluble in cold hydrochloric acid, but is soluble when hot; is infusible and blackens during heating. It is distinguishable from rhodonite and thulide, of similar color, by its lower hardness. It is sometimes fluorescent in ultraviolet light, emitting dark red light.

▼ Rhombohedral rhodochrosite (ca. x 1), Colorado, the United States.

SMITHSONITE

COMPOSITION: $ZnCO_3$ zinc carbonate.

CRYSTALLINE FORM: trigonal system, ditrigonal scalenohedral class. The rare crystals are rhombohedral or scalenohedral; most commonly in radial fibrous and botryoidal masses.

COLOR: colorless, white, blue and green from inclusions of copper, brown from iron hydroxides, yellow from the presence of cadmium, pink from cobalt.

HARDNESS: 4.5-5.

SPECIFIC GRAVITY: 4-4.5.

LUSTER: vitreous.

GEOLOGIC OCCURRENCE: mineral of metasomatic origin in the oxidation zones of zinc mineral deposits.

ASSOCIATED MINERALS AND LOCALITIES: with hemimorphite, cerussite, malachite, anglesite, pyromorphite. Beautiful specimens come from Namibia, Zambia, the United States, Turkey, and Greece.

USES: for the extraction of zinc; the banded fibrous masses are used as an ornamental stone.

NOTES: its name pays homage to the English chemist and mineralogist, James Smithson. It is a member of the isomorphous calcite group. It is soluble in hydrochloric acid with effervescence and infusible. Some specimens are fluorescent, emitting pink light.

▼ Smithsonite (ca. x 0.5), Namibia.

SPHAEROCOBALTITE

COMPOSITION: $CoCO_3$ cobalt carbonate.

CRYSTALLINE FORM: trigonal system, ditrigonal scalenohedral class. The rare crystals are small rhombohedrons or scalenohedrons; most commonly in radial fibrous and botryoidal masses, commonly with a spherical appearance.

COLOR: red, pink, alters to black.

HARDNESS: 4.

SPECIFIC GRAVITY: 4.13.

LUSTER: vitreous.

GEOLOGIC OCCURRENCE: present, rarely, in pyrite deposits and in Co-Ni veins.

ASSOCIATED MINERALS AND LOCALITIES: found with pyrite, calcite, quartz. Beautiful specimens come from the Congo, the United States, Italy, and Germany.

USES: interest limited to researchers and collectors.

NOTES: Sphaerocobaltite is also known as cobaltocalcite. It belongs to the isomorphous calcite group and is soluble in cold hydrochloric acid with effervescence. It is infusible and is not fluorescent.

▼ Sphaerocobaltite (ca. x 0.8), Likasi, Kakanda Shaba, Zaire.

DOLOMITE

COMPOSITION: $CaMg(CO_3)_2$ calcium magnesium carbonate.

CRYSTALLINE FORM: trigonal system, rhombohedral class. Appears as rhombohedral crystals and as clusters of crystals with curved faces (saddle-shaped dolomite); in sparry or saccharoidal masses (dolomite rocks).

COLOR: colorless, white, gray, yellow, brown, blue.

HARDNESS: 3.5-4.

SPECIFIC GRAVITY: 2.87.

LUSTER: vitreous to pearly.

GEOLOGIC OCCURRENCE: the principal constituent of dolomitic rocks (dolomites), as a replacement of calcite or, rarely, as a precipitate from saline solutions in an evaporative environment (primary dolomite); in metamorphic rocks (dolomitic marbles).

ASSOCIATED MINERALS AND LOCALITIES: commonly associated with calcite, fluorite, barite, siderite, quartz, magnesite. It is a widely dispersed mineral and beautiful specimens come from Brazil, Mexico, the United States, Italy, etc.

USES: dolomite deposits are dispersed in carbonate environments. It is used in the building trade as a hydraulic binder and in the metallugic industry for the preparation of refractories and for the extraction of metallic magnesium. It can also be used as an ornamental stone.

NOTES: The name pays homage to the French chemist, Dolomieu (1750-1800), who distinguished it from calcite. It is infusible and soluble in hot hydrochloric acid with effervescence. It is not fluorescent.

▼ Dolomite (ca. x 1), Spain.

ARAGONITE

COMPOSITION: $CaCO_3$ calcium carbonate.

CRYSTALLINE FORM: orthorhombic system, bipyramidal class. The crystals are elongated prisms commonly twinned along the faces of the vertical prism (threelings which simulate a pseudohexagonal symmetry); in acicular aggregates with a radial fibrous structure and in pisolitic, stalactite, and coralloid (flos ferri) concretions; makes up the shell of many mollusks, and together with conchiolin, mother-of-pearl.

COLOR: colorless, white, sky-blue, yellow, pink, gray.

HARDNESS: 3.5-4.

SPECIFIC GRAVITY: 2.93.

LUSTER: vitreous.

GEOLOGIC OCCURRENCE: in evaporative sedimentary deposits (associated with sulfur); as a precipitate from hot springs (travertine); in geodes in volcanic rocks (andesites and basalts); in alteration zones of sulfide deposits; in high-pressure metamorphic rocks (glaucophane schists).

ASSOCIATED MINERALS AND LOCALITIES: with sulfur, celestite, barite, calcite, zeolites, dolomite, glaucophane, limonite, malachite. Beautiful specimens come from Mexico, the United States, Spain, Italy, etc.

USES: interest limited to researchers and collectors.

NOTES: The name is derived from the Spanish region, Aragon, where it was first recognized. It is infusible and when heated, it transforms into calcite; soluble in cold, dilute hydrochloric acid with strong effervescence; some varieties are fluorescent, emitting azure, pink, yellow light. Polymorphous with calcite, it is stable at high pressure. (See box.)

▲ *Aragonite (ca. x 0.4), Sicily, Italy.*

CALCITE AND ARAGONITE

The most common carbonate in nature is that of calcium. Calcium carbonate crystallizes in two different forms: as calcite at low pressure and as aragonite at high pressure. Aragonite is the principal component of pearls.

HISTORY AND LEGEND

The oldest jewelry is a parure from the fifth century B.C. held in the Louvre. In ancient China, because of their white color and general desirability, pearls were considered one of the eight delights of the world. They were associated with the notion of tears, so much so that the same

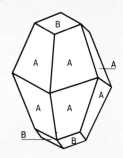

The typical morphology of a calcite crystal is made up of a combination of scalenohedron (A) and rhombohedron (B) (above, the pattern; below, a crystal).

ideogram was used to designate. The popularity and value of pearls for all ancient peoples has been chronicled throughout history and there are many legends about them. In the Orient, it was thought that Adam and Eve, shedding tears for the death of Abel, formed a lake in which the most beautiful pearls, born of love and grief, were found. Pearls were thought to have been born during the

Detail of a precious pearl necklace.

full moon when pearl oysters, rising to the surface of the sea, opened to the moon and became fertile from its beams. The spread of pearls in the West dates back to the time of the Crusades, and since then they have been prized by queens. The largest pearl "La Pellegrina" is held in the Moscow Museum. The most extraordinary is certainly the "Southern Cross" formed by nine pearls joined together in the shape of a cross. It was fished out in Australia in 1886.

THE MINERALS

Calcite (see page 43) is one of the most widespread minerals on the earth's crust. It is a fundamental component of calcareous sedimentary rocks, of metamorphic marbles, and of interesting though rare igneous rocks called carbonatites.

Calcite and aragonite (see this page) are also amply present in the "living" world: they are, in fact, among the principal constituents of many hard parts such as the shells of plankton and mollusks, and the skeletons of fish and mammals, humans included.

Calcite occurs in many colors and crystal forms. The most common forms are the rhombohedron and the scalenohedron, which can coexist in the same crystal. Less common are prismatic crystals which can be mistaken for those of aragonite and other minerals. Among the numerous forms in which this crystal occurs, the stalactite and stalagmite are among the most fascinating. These are typical of caves and karst terrain, and are formed by slow precipitation of calcite from water. Iceland spar is a very transparent type of calcite which highlights the phenomenon of double refraction: when viewed through a piece of Iceland spar, a single image appears as two.

Aragonite commonly crystallizes in hexagonal prisms, a form resulting from twinning. It is a fundamental component of mollusk shells and of pearls. The latter are considered to be real gems and, in addition to their characteristic white color, they can also be pink, golden, white-yellow, brown, and black colored.

Pearls of various shades of color.

STRONTIANITE

COMPOSITION: $SrCO_3$ strontium carbonate.

CRYSTALLINE FORM: orthorhombic system, bipyramidal class. Well developed crystals are rare; more commonly in bacillary aggregates with a radial structure and in compact microgranular masses.

COLOR: colorless, white, gray, green, pink.

HARDNESS: 3.5-4.

SPECIFIC GRAVITY: 3.76.

LUSTER: vitreous.

GEOLOGIC OCCURRENCE: gangue mineral in metalliferous hydrothermal veins; in oxidation zones of metalliferous deposits.

ASSOCIATED MINERALS AND LOCALITIES: celestite, witherite, barite, and various sulfides. Beautiful specimens come from Mexico, the United States, Great Britain, Spain, Austria, etc.

USES: useful mineral for the extraction of strontium and its salts; used in the fireworks industry (colors the flame purple red), in the sugar industry, and the special glass industry.

NOTES: The name is derived from the city of Strontian (Scotland) where it was first recognized. Belongs to the isomorphous aragonite group, with cerussite and witherite; it fuses at high temperature (about 2727°F or 1497° C) and is soluble in dilute hydrochloric acid with effervescence. It is fluorescent when exposed to ultraviolet light, emitting blue light.

▼ *Strontianite (ca. x 1.5), Westphalia, Germany.*

WITHERITE

COMPOSITION: $BaCO_3$ barium carbonate.

CRYSTALLINE FORM: orthorhombic system, bipyramidal class; in crystals, commonly threelings which simulate a pseudohexagonal bipyramidal symmetry that can take on a globular or prismatic appearance with horizontal striations; in fibrous, granular masses.

COLOR: colorless, white or gray.

HARDNESS: 3-3.5.

SPECIFIC GRAVITY: 4.29.

LUSTER: vitreous to resinous.

GEOLOGIC OCCURRENCE: gangue mineral in metalliferous hydrothermal veins; in oxidation zones of lead sulfides deposits. It can be formed by metasomatism of carbonate rocks.

ASSOCIATED MINERALS AND LOCALITIES: with barite, galena, anglesite, calcite, and dolomite. Beautiful crystals come from the United States, Canada, Great Britain, South Africa, etc.

USES: for the extraction of barium, and in the production of special glasses (barium glasses).

NOTES: Witherite takes its name from the Englishman, D. J. Withering (1741-1799), who first studied this mineral. It belongs to the isomorphous aragonite group, fuses with relative ease, and colors the flame yellowish green; soluble in dilute hydrochloric acid with effervescence; is fluorescent emitting blue light. It is poisonous, as are all soluble barium compounds.

▲ *Witherite (ca. x 1.5), Cumberland, Alston Moor, England, Great Britain.*

CERUSSITE

COMPOSITION: $PbCO_3$ lead carbonate.

CRYSTALLINE FORM: orthorhombic system, bipyramidal class. It appears as prismatic and bipyramidal crystals, elongated or tabular; there are common threelings, which simulate a pseudohexagonal symmetry with star-shaped branches. In compact, fibrous masses, and also in stalactite concretions.

COLOR: colorless, gray from inclusions of galena.

HARDNESS: 3-3.5.

SPECIFIC GRAVITY: 6.55.

LUSTER: adamantine.

GEOLOGIC OCCURRENCE: secondary mineral in oxidation zones of lead deposits.

ASSOCIATED MINERALS AND LOCALITIES: commonly associated with galena, sphalerite, anglesite, smithsonite, malachite. Beautiful crystals come from Namibia, the Congo, Tunisia, Australia, the United States, and Germany.

USES: useful mineral for the extraction of lead and, secondarily, of silver, present in subordinate quantities.

NOTES: Cerussite was named by the German mineralogist Wilhelm von Haidinger (1785-1871) from the Latin word *cerussa*, because of its appearance similar to that of wax. It fuses easily, yielding a globule of metallic lead; is soluble in nitric acid with strong effervescence (distinguishes it from anglesite); when exposed to ultraviolet light, is fluorescent emitting greenish blue light. Belongs to the isomorphous aragonite group.

▼ *Cerussite (ca. x 1.5), Morocco.*

AZURITE

COMPOSITION: $Cu_3(CO_3)_2(OH)_2$ hydrous copper carbonate.

CRYSTALLINE FORM: monoclinic system, prismatic class. Appears as elongated prismatic or tabular crystals, highly modified and sometimes striated; in nodular aggregates with a radial fibrous structure with a mammillary surface; in compact masses and earthy patinas.

COLOR: intense azure.

HARDNESS: 3.5-4.

SPECIFIC GRAVITY: 3.7-3.8.

LUSTER: vitreous to adamantine.

GEOLOGIC OCCURRENCE: secondary mineral in oxidation zones of copper deposits.

ASSOCIATED MINERALS AND LOCALITIES: associated with other copper minerals (malachite, cuprite, covellite, atacamite, etc.), native copper, limonite, and calcite. Very beautiful crystals come from Namibia, Mexico, the United States, and France. Those from Sardinia are famous.

USES: secondary mineral for the extraction of copper. It is used for prized ornamental objects. It can be cut into gems but because of its fragility and low hardness it is not very valuable. In the past, it was ground into powder (mountain blue or Armenia stone) and employed as a pigment for an oil paint that, over time, altering to malachite, took on a greenish color.

NOTES: Azurite fuses easily and colors the flame green from the presence of copper; during heating it turns black due to the loss of water. It is soluble in ammonia, while in dilute acids it reacts with effervescence. It is not fluorescent. (See box, page 48.)

▼ *Azurite (ca. x 0.5), Alghero, Sassari, Italy.*

MALACHITE

COMPOSITION: $Cu_2CO_3(OH)_2$ hydrous copper carbonate.

CRYSTALLINE FORM: monoclinic system, prismatic class. Rare as well-formed single crystals, common as acicular crystals in radial fibrous aggregates; generally occurs in compact masses with a radial and banded structure and a reniform or botryoidal appearance; pseudomorphous substitutions after azurite and cuprite are common.

COLOR: intense green, mauve.

HARDNESS: 3.5-4.

SPECIFIC GRAVITY: 4.05.

LUSTER: vitreous to silky in fibrous varieties.

GEOLOGIC OCCURRENCE: secondary mineral produced by alteration in oxidation zones of copper deposits.

ASSOCIATED MINERALS AND LOCALITIES: with azurite, native copper, cuprite, limonite, calcite. The most valuable masses come from Zaire, Zambia, Chile, Australia, Namibia, Zimbabwe, and the Urals.

USES: secondary mineral of copper; used as an ornamental stone of some value, especially the banded variety. In the past, it was ground into powder (mountain green), and it was also used as a pigment for oil paints.

NOTES: the name is derived from the Greek word for "mallow". It fuses easily and colors the flame green from the presence of copper; during heating, turns black due to the loss of water. It is soluble in hydrochloric acid and reacts with effervescence; is not fluorescent. (See box, page 48.)

▼ *Malachite (ca. x 0.6), Touissit, Morocco.*

AURICHALCITE

COMPOSITION: $(Zn, Cu)_5(CO_3)_2(OH)_6$ hydrous copper zinc carbonate.

CRYSTALLINE FORM: orthorhombic system, sphenoidal class. Rare as isolated crystals, normally appears in effluorescences or incrustations made up of bands of small, acicular crystals.

COLOR: sky-blue.

HARDNESS: 2.

SPECIFIC GRAVITY: 3.5-4.

LUSTER: silky.

GEOLOGIC OCCURRENCE: secondary mineral produced by alteration in the oxidation zones of zinc and copper deposits.

ASSOCIATED MINERALS AND LOCALITIES: generally associated with various sulfides. Beautiful specimens come from Namibia, Mexico, the United States, France, Greece, etc.

USES: interest limited to researchers and collectors.

NOTES: The name is probably derived from the Greek for "brass flowers". Infusible, when heated, it colors the flame green due to the presence of copper. It is soluble in dilute hydrochloric acid and reacts with effervescence. It is not fluorescent.

▲ *Aurichalcite (ca. x 1), Durango, Mexico.*

PHOSGENITE

COMPOSITION: $Pb_2(CO_3)Cl_2$ lead chlorocarbonate.

CRYSTALLINE FORM: tetragonal system, trapezohedral class. Appears as prismatic, tabular crystals, striated in a lengthwise direction, sometimes with telescopic terminations.

COLOR: form colorless to light yellow, brown.

HARDNESS: 2-3.

SPECIFIC GRAVITY: 6.5.

LUSTER: adamantine.

GEOLOGIC OCCURRENCE: secondary mineral produced by alteration in oxidation zones of lead deposits.

ASSOCIATED MINERALS AND LOCALITIES: found with galena and gangue minerals. Very beautiful crystals come from Namibia, the United States, Italy, and Greece.

USES: interest limited to researchers and collectors.

NOTES: Phosgenite was originally considered to be a derivative of phosgene ($COCl_2$). It fuses easily and is soluble in cold distilled water. In dilute nitric acid it reacts with effervescence, and when exposed to ultraviolet light it is fluorescent, emitting yellow light.

▼ *Phosgenite (ca. x 0.5), Monteponi, Cagliari, Sardinia, Italy.*

LEADHILLITE

COMPOSITION: $Pb_4SO_4(CO_3)_2(OH)_2$ lead carbonate sulfate.

CRYSTALLINE FORM: monoclinic system, prismatic class. Appears as small tabular, pseudohexagonal crystals sometimes, with very highly modified faces.

COLOR: colorless, white, green, blue, gray.

HARDNESS: 2.5-3.

SPECIFIC GRAVITY: 6.55.

LUSTER: resinous to adamantine.

GEOLOGIC OCCURRENCE: secondary mineral produced by alteration in oxidation zones of lead.

ASSOCIATED MINERALS AND LOCALITIES: with cerussite, phosgenite, anglesite and various sulfides. Beautiful crystals come from Namibia, the United States, Tasmania, Japan, Tunisia, and Austria.

USES: interest limited to researchers and collectors.

NOTES: The name is derived from the Scottish locality Leadhills. It fuses easily, coloring the flame yellow. It is soluble in nitric acid and reacts with effervescence, forming an insoluble residuum. When exposed to ultraviolet light, it is fluorescent, emitting yellow light.

▲ *Leadhillite (ca. x 2), Malacalsetta, Cagliari, Italy.*

ARTINITE

COMPOSITION: $Mg_2CO_3(OH)_2 \cdot 3H_2O$ hydrated magnesium carbonate.

CRYSTALLINE FORM: monoclinic system, sphenoidal class. Appears in the form of very slender needles in spherical aggregates or in incrustations or small veins of radial fibrous structure.

COLOR: grayish white.

HARDNESS: 2.

SPECIFIC GRAVITY: 2.03.

LUSTER: silky.

GEOLOGIC OCCURRENCE: found in low-temperature hydrothermal veins in serpentine rocks.

ASSOCIATED MINERALS AND LOCALITIES: very rare, found with amianthus brucite, hydromagnesite. Specimens come from Austria, the United States, and Italy.

USES: interest limited to researchers and collectors.

NOTES: Artinite was discovered in 1902 in Val Malenco, Italy by Luigi Brugnatelli who called it this in honor of the Italian researcher Ettore Artini. It is infusible and when heated, loses molecules of water and carbon dioxide and becomes milky-white in color. It is soluble in hydrochloric acid, reacting with lively effervescence, and it is not fluorescent.

▼ *Artinite (ca. x 1.5), Val Malenco, Sondrio, Italy.*

NITER

COMPOSITION: KNO_3 potassium nitrate.

CRYSTALLINE FORM: orthorhombic system, bipyramidal class. Appears in aggregates of slender acicular crystals, in crusts, efflorescences, and sometimes in compact microgranular or earthy masses.

COLOR: colorless, white, gray.

HARDNESS: 2.

SPECIFIC GRAVITY: 2.109.

LUSTER: vitreous.

GEOLOGIC OCCURRENCE: is found in superficial accumulations in arid regions, resulting from the action of bacteria on nitrogenous organic substances.

ASSOCIATED MINERALS AND LOCALITIES: occurs with soda niter, the most well-known accumulations are in Chile, Arabia, India, Iran, and the United States.

USES: of little practical use. It was used in the United States during the Civil War for the manufacture of gunpowder.

NOTES: Niter is very similar in appearance to soda niter ($NaNO_3$), which, however, crystallizes in the scalenohedral class of the trigonal system and has greater solubility in water. Isotypous with aragonite, niter is not hydroscopic and detonates when mixed with combustible substances. Also known as saltpeter. It fuses easily and is not fluorescent.

▲ *Niter (ca. x 2.8), La Maiella, Abruzzo, Italy.*

FROM AZURE TO GREEN

AZURITE AND MALACHITE

The chemical composition of these two minerals is very similar, so much so that malachite is commonly derived from alteration of azurite. In this transformation, the color passes from an intense azure to brilliant green.

Azurite (see page 46) and malachite (see page 47) are hydrous carbonates of copper. Both form in the oxidation zones of copper sulfide deposits. The former is metastable and, with time, alters to malachite. Because of this tendency, azurite is not as well-known and is not used as much as malachite. It appears as very beautiful intense azure crystals (malachite is commonly microcrystalline and needle-shaped), and in pseudomorph after

Small fans in mixed malachite and azurite.

acicular azurite (Sassari, Italy).

acicular malachite (Germany).

azurite, with a green color that calls to mind that of the mallow plant. Radial fibrous masses are common with concentric colors that range from mauve green to intense bottle green. These masses are much sought after and from them prized ornamental objects can be created.

HISTORY AND LEGEND

The Egyptians, Greeks, and Romans attributed therapeutic powers to malachite. In fact, because of its copper content, it acted as a purgative and was used for intestinal colic. In the Middle Ages, it was thought to have power over evil spirits and was cherished by wayfarers who believed it would protect them from lightning, from dangers in the night, and from fears in general. In the Orient, malachite was a part of sacred symbology and was used to decorate places of worship.

USES

The most famous malachite deposits are in the Urals and Russia. This mineral was used both in the production of ornamental objects and as a casing stone. Because of its intense color, malachite was for a long time used as a pigment in the preparation of oil paints. The collection in the Hermitage Museum is a particularly rich one.

At one time, azurite was also used as a pigment for oil paints, but it frequently altered to malachite and after a short time turned the pigment green. This phenomenon may be observed in many Renaissance frescoes.

Microscrystalline aggregate of malachite.

SASSOLITE

COMPOSITION: $B(OH)_3$ (boric acid), nesoborate.

CRYSTALLINE FORM: triclinic system, pinacoidal class. Appears in tabular crystals with a pseudohexagonal outline and in scaly aggregates.

COLOR: white, gray, yellow, blue.

HARDNESS: 1.

SPECIFIC GRAVITY: 1.48.

LUSTER: pearly.

GEOLOGIC OCCURRENCE: as a chemical precipitate from jets of hot steam of endogenous origin (the boric fumaroles of Tuscany are famous); rare as a sublimate in a fumarolic environment.

ASSOCIATED MINERALS AND LOCALITIES: found with borates and sulfates of the same origin. Beautiful specimens come from Italy, the United States, and Turkey.

USES: used in the pharmaceutical industry and for the production of borax.

NOTES: The name is derived from the Tuscan locality Sasso, Italy. It fuses easily and colors the flame green (boron). It is soluble in hot water and in alcohol, and it is not fluorescent.

▲ *Sassolite (aggregate of microscopic crystals), Larderello, Pisa.*

BORAX

COMPOSITION: $Na_2B_4O_5(OH)_4 \cdot 8H_2O$ sodium soroborate.

CRYSTALLINE FORM: monoclinic system, prismatic class. Appears as slightly elongated prismatic or tabular crystals, commonly with vertical striations; in earthy, compact masses and in efflorescences.

COLOR: colorless, white, gray, blue, green.

HARDNESS: 2-2.5.

SPECIFIC GRAVITY: 1.73.

LUSTER: vitreous to resinous.

GEOLOGIC OCCURRENCE: found as a chemical precipitate in evaporative deposits; rarely, in a fumarolic environment or in boric fumaroles.

ASSOCIATED MINERALS AND LOCALITIES: with rock salt, trona, ulexite, soda niter, gypsum, glauberite. Historically, the Tibetan deposits were well-known, but are now exhausted. Today it is found in the United States and Iran.

USES: principal mineral for the extraction of boric acid used in the metallurgical and pharmaceutical industries.

NOTES: Borax is also known as tincal. It fuses easily with swelling and colors the flame yellow from the presence of sodium. It is soluble in water, and when exposed to the air, it alters due to the loss of water molecules and is covered with an opaque patina of tincalconite. It is not fluorescent.

Borax (ca. x 1), ▶
Turkey.

ULEXITE

COMPOSITION: $NaCa(B_5O_6)(OH)_6 \cdot 5H_2O$ hydrated sodium calcium soroborate.

CRYSTALLINE FORM: triclinic system, pinacoidal class. Rare as well-developed crystals, it normally appears as acicular aggregates with a radial fibrous structure and a rounded shape, called cotton balls; in fibrous crusts with botryoidal surface.

COLOR: white.

HARDNESS: 2.5.

SPECIFIC GRAVITY: 2.

LUSTER: silky.

GEOLOGIC OCCURRENCE: found in evaporative sedimentary deposits; rarely in fumaroles and in geysers.

ASSOCIATED MINERALS AND LOCALITIES: with borax, saltpeter, and other sulfates and borates of the same origin. The principal deposits are in Chile, Argentina, Peru, the United States, and Canada.

USES: useful mineral for the extraction of boron.

NOTES: The name honors the German chemist G. L. Ulex who first analyzed the mineral. It fuses easily and during heating it swells, coloring the flame yellow (sodium). It is soluble in warm water and when exposed to ultraviolet light, it is fluorescent, emitting blue, green light; it is commonly phosphorescent. Also known as "television stone:" the crystals, cut along the fibrous grain and the cut surfaces smoothed, when rested on printed pages, project them on the upper surface as on a screen.

Ulexite (ca. x 1), ▶
Turkey.

COLEMANITE

COMPOSITION: $Ca_2B_6O_{11} \cdot 5H_2O$ hydated calcium inoborate.

CRYSTALLINE FORM: monoclinic system, prismatic class. It appears as beautiful prismatic crystals, generally equidimensional, sometimes similar to distorted rhombohedrons or octahedrons and highly modified; sometimes it forms druses or carpets geodes; in compact or fibrous masses.

COLOR: colorless, white.

HARDNESS: 4.5.

SPECIFIC GRAVITY: 2.4.

LUSTER: vitreous to adamantine.

GEOLOGIC OCCURRENCE: found in evaporative sedimentary deposits.

ASSOCIATED MINERALS AND LOCALITIES: with borax, gypsum, calcite, celestite, ulexite, realgar. Beautiful specimens come from the United States, Turkey, Chile, and Argentina.

USES: useful mineral for the extraction of boron; it is used in the pharmaceutical and cosmetic industries; for the preparation of heat resistant glassware (pyrex) and fluxes for soldering.

NOTES: The name is derived from W. T. Coleman, the owner of the mine in Death Valley California where the mineral was first recognized. It fuses easily and during heating, it decrepitates, coloring the flame green (boron). It is soluble in hydrochloric acid; on lowering the temperature, small needles of boric acid precipitate from solution. When exposed to ultraviolet light, shows fluorescence, emitting green light.

▼ *Colemanite (ca. x 0.5), Bigadic, Turkey.*

BORACITE

COMPOSITION: $Mg_3B_7O_{13}Cl$ magnesium tectoborate.

CRYSTALLINE FORM: orthorhombic system, pyramidal class. The crystals are cubic or octahedral through paramorphosis; in fibrous masses, has the name stassfurtite.

COLOR: white, gray, yellow, blue, various shades of green from the presence of iron.

HARDNESS: 7-7.5.

SPECIFIC GRAVITY: 2.9-3.10.

LUSTER: vitreous.

GEOLOGIC OCCURRENCE: in saline deposits.

ASSOCIATED MINERALS AND LOCALITIES: with gypsum, anhydrite, and various halogen salts. Beautiful crystals come from Germany, France, England, and the United States.

USES: interest limited to researchers and collectors.

NOTES: Boracite is dimorphous, that is, it crystallizes at low temperatures in the alpha phase (orthorhombic system) and at about 509°F (265° C) it changes to the beta phase (cubic system). The crystals, which in nature externally present a cubic morphology, are in reality made up of an interlacing of small rhombic crystals (paramorphosis). It is slowly soluble in hydrochloric acid and very slowly in water. It is not fluorescent.

▲ *Boracite (ca. x 1), Hohenfels, Selnode, Kaliwerk, Germany.*

ANHYDRITE

COMPOSITION: $Ca(SO_4)$ anhydrous calcium sulfate.

CRYSTALLINE FORM: orthorhombic system, bipyramidal class. Rare as prismatic stubby and tabular crystals; generally appears in crystalline masses or in granular, fibrous, or concreted masses.

COLOR: white, pale pink, gray to black.

HARDNESS: 3-3.5.

SPECIFIC GRAVITY: 2.96.

LUSTER: vitreous to silky.

GEOLOGIC OCCURRENCE: in evaporative sedimentary deposits; as a product of hydrothermal alteration of calcareous and dolomite rocks; rarely in metalliferous veins as a product of oxidation; rare in fumaroles.

ASSOCIATED MINERALS AND LOCALITIES: with gypsum, rock salt, calcite, dolomite, glauberite. Large masses make up the cap-rock of salt domes in Texas and Louisiana, France, Germany, Poland, India, etc.

USES: used in the production of sulfuric acid; as an ornamental stone for interior decorating.

NOTES: The name is derived from the Greek for "without water." It fuses easily, coloring the flame brick-red. It is soluble with difficulty in acids, and it is not fluorescent. When exposed to atmospheric agents, it slowly hydrates, changing into gypsum with a noteworthy increase in volume.

▼ *Anhydrite (ca. x 1.5), Hannover, Germany.*

CELESTITE

COMPOSITION: $Sr(SO_4)$ strontium sulfate.

CRYSTALLINE FORM: orthorhombic system, bipyramidal class. Appears in beautiful prismatic, sometimes tabular crystals, sometimes of a very large size; or in radial fibrous masses.

COLOR: colorless, milky-white, blue, yellowish, brown, gray.

HARDNESS: 3-3.5.

SPECIFIC GRAVITY: 3.95.

LUSTER: vitreous to resinous.

GEOLOGIC OCCURRENCE: a primary mineral in hydrothermal veins; in cracks and internal cavities of dolomitic rocks; in evaporative sedimentary deposits; rarely in metalliferous veins.

ASSOCIATED MINERALS AND LOCALITIES: with calcite, dolomite, gypsum, sulfur, sphalerite, galena, pyrite. It is very common, and very beautiful specimens come from Egypt, England, Madagascar, Tunisia, and the United States.

USES: very important mineral for the extraction of strontium, used in the fireworks and nuclear industries and in the preparation of special iridescent glasses and pottery.

NOTES: Celestite fuses easily, decrepitates and colors the flame red, forming a white pearl. It is slightly soluble in water and in acids. When impure, from inclusions of organic substances, it becomes fluorescent in ultraviolet light, emitting whitish-blue light. It is sometimes thermoluminescent.

▲ *Celestite (ca. x 0.6), Syracuse, Italy.*

BARITE

COMPOSITION: $Ba(SO_4)$ barium sulfate.

CRYSTALLINE FORM: orthorhombic system, bipyramidal class. Appears as tabular, prismatic crystals or in lamellar aggregates; in compact, sparry, concreted masses.

COLOR: colorless when pure, yellow, red, green, chestnut from inclusions of oxides and hydroxides of iron, blue from exposure to radium radiation.

HARDNESS: 3.

SPECIFIC GRAVITY: 4.47.

LUSTER: vitreous.

GEOLOGIC OCCURRENCE: gangue mineral in medium and low-temperature hydrothermal veins; product of chemical deposition from hot springs.

ASSOCIATED MINERALS AND LOCALITIES: in metalliferous veins, associated with various sulfides (galena, sphalerite, stibnite, siderite) and with fluorite, quartz, dolomite. Beautiful specimens come from the United States, Great Britain, Romania, and Morocco.

USES: principal mineral for the extraction of barium. It is also used as an additive in heavy muds for oil drilling, in the paper industry, in radiographic analysis, and for the preparation of a highly valued white pigment.

NOTES: The name is derived from the Greek for "heavy," in reference to its high specific gravity. It fuses with difficulty, decrepitates during heating and colors the flame green. It is insoluble in acids, and some varieties are fluorescent when exposed to ultraviolet light, emitting blue or green light; it is commonly phosphorescent. Its high specific gravity distinguishes it from celestite, aragonite, albite, gypsum, and calcite; because of this property, it was used in the past as an adulterating substance for flours and sugar.

Barite (ca. x 1), ▶
Felsobanja,
Romania.

ANGLESITE

COMPOSITION: $Pb(SO_4)$ lead sulfate.

CRYSTALLINE FORM: orthorhombic system, bipyramidal class. Appears as prismatic, tabular, or rarely as bipyramidal crystals; in granular or compact masses.

COLOR: colorless, white, emerald-green from inclusions of iron, gray and black from inclusions of galena.

HARDNESS: 2.5-3.

SPECIFIC GRAVITY: 6.38.

LUSTER: adamantine.

GEOLOGIC OCCURRENCE: a secondary mineral formed in the oxidation zone of lead mining deposits; very rarely as a sublimate of volcanic vapors.

ASSOCIATED MINERALS AND LOCALITIES: with galena, cerussite, sphalerite, smithsonite, hemimorphite, phosgenite. Beautiful specimens come from Namibia, Tunisia, the United States, New Caledonia, and Great Britain.

USES: of limited industrial interest as a lead mineral; the rare green crystals are very sought-after by collectors.

NOTES: The name is derived from the island of Anglesey (Scotland) where it was found for the first time. It fuses easily, decrepitating during heating, and is soluble with difficulty in nitric acid. It is fluorescent when exposed to ultraviolet light, emitting yellow light.

▼ *Anglesite (ca. x 0.7), Monteponi, Cagliari, Italy.*

GYPSUM

COMPOSITION: $Ca(SO_4)\cdot2H_2O$ hydrated calcium sulfate.

CRYSTALLINE FORM: monoclinic system, prismatic class. It appears as transparent, tabular crystals of notable size (selenite or donkey's mirror); in translucent fibrous aggregates (satin spar); in rosette-shaped lamellar aggregates (desert rose); in compact granular masses with a cerulean look, sometimes banded (alabaster). Twinned swallow-tail and lance-head crystals are typical.

COLOR: colorless if pure, yellow, brown, green, gray, pink.

HARDNESS: 2.

SPECIFIC GRAVITY: 2.32.

LUSTER: vitreous to silky.

GEOLOGIC OCCURRENCE: typical mineral of evaporative deposits where it can be formed through precipitation or through hydration of anhydrite; as a sublimation product in a fumarolic environment, and through precipitation from hot springs of volcanic origin; product of oxidation in metalliferous deposits.

ASSOCIATED MINERALS AND LOCALITIES: with rock salt, celestite, calcite, aragonite, dolomite, sulfur, quartz, pyrite. It is found in Mexico, Chile, the United States, Tunisia, Morocco, Canada, Italy, etc.

USES: used by sculptors, in the building materials industry, as a retarder in portland cement, as a fertilizer. The saccharoidal variety (alabaster) is used as an ornamental stone.

NOTES: The name is derived from the Greek. It fuses easily and becomes opaque in the flame through loss of water. It is soluble in warm water and in hydrochloric acid, and sometimes it is fluorescent.

▼ *Gypsum (ca. x 1), Sicily, Italy.*

POLYHALITE

COMPOSITION: $K_2MgCa_2(SO_4)_4\cdot2H_2O$ hydrated potassium magnesium calcium sulfate.

CRYSTALLINE FORM: triclinic system, pinacoidal class. Appears rarely as crystals, usually small and highly modified; in lamellar, fibrous or compact masses.

COLOR: colorless, white, gray, commonly salmon-pink or brick-red from inclusions of hematite.

HARDNESS: 3.5.

SPECIFIC GRAVITY: 2.78.

LUSTER: vitreous to resinous.

GEOLOGIC OCCURRENCE: found in saline deposits of marine origin, and as a sublimation product in fumarolic environments.

ASSOCIATED MINERALS AND LOCALITIES: with carnallite, glauberite, anhydrite, rock salt, sylvite. The German deposits and those in Kazakhstan are famous.

USES: an important mineral for the extraction of potassium.

NOTES: Polyhalite fuses easily. It is partially soluble in water, with dissociation and separation of gypsum. It is not fluorescent.

▼ *Polyhalite (ca. x 0.5), Calascibetta, Casazze Mine, Enna, Italy.*

CROCOITE

COMPOSITION: $PbCrO_4$ lead chromate.

CRYSTALLINE FORM: monoclinic system, prismatic class. It appears as very elongated prismatic crystals with striations parallel to the length; more rarely as pseudooctahedral and rhombohedral crystals and in acicular aggregates.

COLOR: orange-red, vermilion-red.

HARDNESS: 2.5-3.

SPECIFIC GRAVITY: 5.99.

LUSTER: very bright adamantine.

GEOLOGIC OCCURRENCE: secondary mineral in oxidation zones of lead deposits, especially if situated near ultrabasic eruptive rocks rich in chromite and chromium silicates.

ASSOCIATED MINERALS AND LOCALITIES: found with alteration minerals of lead deposits. Very beautiful specimens come from Tasmania, Brazil, the Philippines, France, and the Urals.

USES: rare mineral; it is especially sought-after by collectors because of the beauty of its crystals.

NOTES: The name is derived from the Greek for "saffron," in reference to the mineral's color. It fuses easily and is soluble in strong acids. When exposed to the light, it loses luster, and in ultraviolet light shows a weak, brown fluorescence. From this mineral elementary chromium was extracted for the first time.

▼ *Crocoite (ca. x 1), Tasmania.*

HUEBNERITE

COMPOSITION: $MnWO_4$ manganese tungstate.

CRYSTALLINE FORM: monoclinic system, prismatic class. The crystals are prismatic or tabular.

COLOR: black, yellowish brown, reddish.

HARDNESS: 4.5-5.5.

SPECIFIC GRAVITY: 7.1-7.3.

LUSTER: metallic to resinous, to adamantine.

GEOLOGIC OCCURRENCE: in pegmatites and high-temperature hydrothermal veins; in marine sands (placers).

ASSOCIATED MINERALS AND LOCALITIES: with wolframite. Beautiful specimens come from the United States.

USES: interest limited to researchers and collectors; in adequate concentrations it can be used for the extraction of tungsten.

NOTES: Huebnerite owes its name to the metallurgist Adolph Huebner. With ferberite ($FeWO_4$), it constitutes the isomorphous wolframite series with complete miscibility of Mn and Fe. It is semi-hard, heavy, cleavable, fuses with difficulty, producing a magnetic globule. It decomposes in hot sulfuric acid and in aqua regia with the separation of tungsten oxide. It is not fluorescent.

▲ *Huebnerite (ca. x 1.4), Trujillo, Peru.*

WOLFRAMITE

COMPOSITION: $(Fe,Mn)WO_4$ iron manganese tungstate.

CRYSTALLINE FORM: monoclinic system, prismatic class. Appears as tabular crystals, with the vertical faces commonly striated; sometimes in lamellar aggregates and also in granular masses.

COLOR: grayish black to brown-black, to reddish-chestnut based on the relative content of iron to manganese.

HARDNESS: 4.5-5.5.

SPECIFIC GRAVITY: 7.2-7.5.

LUSTER: metallic to submetallic, to adamantine and resinous depending on the relative content of iron to manganese.

GEOLOGIC OCCURRENCE: in pegmatites and in medium-high temperature hydrothermal veins; in alteration zones (greisen), in pneumatolytic veins; given its resistance to alteration, it can be concentrated in placer deposits

ASSOCIATED MINERALS AND LOCALITIES: with cassiterite, scheelite, stannite, fluorite, lepidolite, etc. Beautiful specimens come from China, Malaysia, Burma, Bolivia, Canada, and Australia.

USES: principal mineral for the extraction of tungsten (wolfram); used in the metallurgic industry, as a filament for electric lights, and for tools for punching and working metals.

NOTES: The name is derived from the German, *wolfram*, relating to the mineral's composition. It fuses with notable difficulty producing a magnetic globule. It is soluble in hot sulfuric acid and in aqua regia with separation of tungsten oxide, and it is not fluorescent. Constitutes an intermediate member of the solid solution between ferberite (iron tungstate) and huebnerite (manganese tungstate).

Wolframite (ca. x 1), ▶
Zinnwald, Germany.

SCHEELITE

COMPOSITION: $Ca(WO_4)$ calcium tungstate.

CRYSTALLINE FORM: tetragonal system, bipyramidal class. Appears as pseudooctahedral bipyramidal crystals, sometimes tabular and of noteworthy size; in granular or compact masses.

COLOR: yellow from the presence of molybdenum, greenish, gray, reddish, and white.

HARDNESS: 4.5-5.

SPECIFIC GRAVITY: 5.8-6.2.

LUSTER: vitreous to resinous.

GEOLOGIC OCCURRENCE: in high-temperature pneumatolytic veins, pegmatites; and in metasomatic skarn deposits which result when intrusive rocks come in contact with carbonate rocks.

ASSOCIATED MINERALS AND LOCALITIES: with wolframite, cassiterite, arsenopyrite, molybdenite, stibnite, calcite. Beautiful specimens come from Brazil, Bolivia, Burma, Malaysia, Japan, China, and the United States.

USES: useful mineral for the extraction of tungsten.

NOTES: The name honors the Swedish chemist K. W. Scheele (1742-1786), who ascertained the presence of tungsten oxide in scheelite. It fuses with notable difficulty and is soluble in acids. When exposed to ultraviolet light, it is fluorescent, emitting light blue light which becomes yellow in the molybdenum-rich variety (powellite). It alters easily, forming a hydrated oxide of tungsten called meymacite.

▼ *Scheelite (ca. x 1), Traversella, Turin, Italy.*

WULFENITE

COMPOSITION: $Pb(MoO_4)$ lead molybdate.

CRYSTALLINE FORM: tetragonal system, pyramidal class. Appears as tabular crystals, rarely as octahedral or pseudocubic crystals; in granular or earthy masses.

COLOR: orange, yellow, red-orange.

HARDNESS: 2.5-3.

SPECIFIC GRAVITY: 6.5-7.

LUSTER: resinous to adamantine.

GEOLOGIC OCCURRENCE: an alteration mineral formed in the oxidation zones of lead and molybdenum mine deposits.

ASSOCIATED MINERALS AND LOCALITIES: with cerussite, vanadinite, pyromorphite, mimetite. Very beautiful specimens come from the United States, Morocco, Mexico, the Congo, Algeria, Namibia, and Australia.

USES: mineral of secondary economic importance for the extraction of molybdenum; because of its beautiful crystals, it is very much appreciated by collectors.

NOTES: Wulfenite was named in honor of the Austrian mineralogist F. X. Wulfen (1728-1805). It fuses easily forming a globule of metallic lead. It dissolves slowly in acids, producing a solution that turns blue with the addition of alcohol, and it is not fluorescent.

▼ *Wulfenite (ca. x 1), Yugoslavia.*

MONAZITE

COMPOSITION: $(Ce,La,Y,Th)PO_4$ cerium, lanthanum, yttrium and thorium phosphate.

CRYSTALLINE FORM: monoclinic system, prismatic class. Appears as isolated, tabular, or prismatic crystals.

COLOR: yellow, brown, reddish.

HARDNESS: 5-5.5.

SPECIFIC GRAVITY: 4.6-5.4.

LUSTER: vitreous to resinous, to silky based on the degree of alteration to metamict as a result of the radioactivity of thorium.

GEOLOGIC OCCURRENCE: fairly rare mineral; found as an accessory in granitic, syenitic rocks, and in alkaline veins; the elevated hardness and high specific gravity allow the mineral to concentrate in coastal and fluvial placer deposits.

ASSOCIATED MINERALS AND LOCALITIES: with fergusonite, samarskite, columbite. Well-known deposits are found in Madagascar, Brazil, India, the United States, Sweden, and Norway.

USES: principal mineral for the extraction of thorium and cerium.

NOTES: The name is derived from the Greek for "solitary" and refers to the mineral's rarity. Infusible, it slowly dissolves in hot sulfuric acid and during heating turns gray. It is not fluorescent. The variety with a high thorium content is called cheralite. The indices of refraction and the specific gravity can vary as a function of its composition.

▼ *Monazite (ca. x 1.5), Arendal, Norestoe, Norway.*

APATITES

"Apatites" include a number of similar minerals which are differentiated on the basis of the anion that prevails at the end of the formula, or on the basis of the replacement of the PO_4 group by the CO_3F group or with CO_3OH.

TYPES OF APATITES	CHEMICAL FORMULA
Hydroxylapatite	$Ca_5(PO_4)_3OH$
Fluorapatite	$Ca_5(PO_4)_3F$
Chlorapatite	$Ca_5(PO_4)_3Cl$
Carbonate-fluorapatite	$Ca_5(PO_4,CO_3,F)_3F$
Carbonate-hydroxylapatite	$Ca_5(PO_4,CO_3,OH)_3OH$

Apatite crystals (Mexico).

WHERE THEY ARE FORMED

Only fluorapatites are present in igneous rocks; in other crystallizing conditions, the apatites (see page 54) may have various compositions. Hydroxycarbonate-apatite, in addition to being a mineral of the inanimate world, is also a fundamental component of the bones of many living things, including humans. This has allowed the formation of sedimentary deposits of phosphates, formed by an accumulation of organic remains. More important, however, are the sedimentary deposits of chemical origin in which fluorohydroxylapatite prevails. The great deposits of West Africa (the most important are those in the Western Sahara) are of this type.

THE NAME AND FORMS

The Greek name "to deceive" was assigned to this mineral because it can easily be confused with others such as tourmaline, olivine, aquamarine, amethyst, and fluorite, among others. This possibility for confusion is due to the form of the crystals; to the vitreous luster and to the chromatic variety in which they appear. The crystals can assume fairly varied forms; from stubby, hexagonal prisms with bipyramidal apices or with ends very rich in small faces referable to numerous other geometric shapes, to flattened prisms terminated by plane and parallel faces.

THE COLORS

Apatite can be colorless, but it is also found in many other colors such as yellow, honey yellow, brown, green-brown,

Brown apatite crystal.

intense green, pale green, violet, azure blue, gray, and pink. Because of its relatively low hardness, it is not valued as a gem except by collectors; it can be cut into a cabochon when it presents some chatoyancy. The various brown tones are the least valued; more attention is addressed, especially in the United States, to the dark green, greenish blue, light blue, and sapphire-blue colors. In England, a yellow green variety (asparagus

Chlorapatite (Norway).

Apatite (Germany).

Carbonate-apatite (the United States).

stone), which comes from Spain, is particularly valued.

APATITE

COMPOSITION: $Ca_5(PO_4)_3(OH,F,Cl)$ calcium hydroxyphosphate, chlorophosphate or fluorophosphate.

CRYSTALLINE FORM: hexagonal system, bipyramidal class. Appears as beautiful, fairly stubby, hexagonal, prismatic crystals in combination with bipyramids; in granular masses, earthy nodules, botryoidal incrustations, and cryptocrystalline aggregates (collophane).

COLOR: colorless, yellow, brown, green, violet; red and blue, rare.

HARDNESS: 5.

SPECIFIC GRAVITY: 3-3.2.

LUSTER: vitreous to subresinous.

GEOLOGIC OCCURRENCE: accessory in many acidic as well as basic igneous rocks; in pegmatite dikes and hydrothermal veins; as a gangue mineral in iron and in tungsten mineral deposits; in both contact and regional metamorphic rocks; in sedimentary rocks; in both organic deposits (phosphorites) and marine deposits of chemical origin.

ASSOCIATED MINERALS AND LOCALITIES: in pegmatite dikes, associated with lepidolite, spodumene, beryl, quartz; with adularite, chlorite, titanite, biotite, scapolite in metamorphic rocks. Very beautiful crystals come from the United States, Mexico, Canada, Morocco, Algeria, Tunisia, and Togo.

USES: principal mineral for the extraction of phosphorus and its salts.

NOTES: The name is derived from the Greek for "to deceive." It fuses with difficulty and some varieties lose color during heating. It is soluble in acids, and it is fluorescent, emitting yellow or pale violet light. Amorphous phosphate is the constituent of guano, a deposit resulting from the accumulation of bird excrement, present on some islands and along the South Pacific coasts. (See box, page 53.)

▼ *Apatite (ca. x 0.5), Mexico.*

PYROMORPHITE

COMPOSITION: $Pb_5(PO_4)_3Cl$ lead chlorophosphate.

CRYSTALLINE FORM: hexagonal system, bipyramidal class. Appears as prismatic crystals with a hexagonal base (stubby, sometimes with curved faces, swollen in the middle, in the shape of a keg); sometimes forms microcrystalline aggregates, small reniform masses, and incrustations.

COLOR: colorless, green, brown.

HARDNESS: 3.5-4.

SPECIFIC GRAVITY: 7.04.

LUSTER: resinous to adamantine.

GEOLOGIC OCCURRENCE: secondary mineral produced by alteration in the oxidation of lead deposits.

ASSOCIATED MINERALS AND LOCALITIES: with cerussite, smithsonite, hemimorphite, anglesite, malachite, leadhillite, wulfenite, mimetite. Very beautiful specimens come from Mexico, Australia, the United States, Germany, Great Britain.

USES: secondary mineral for the extraction of lead; of some interest to researchers and collectors.

NOTES: The name is derived from the Greek for "fire and form," relating to the appearance it assumes when fused (produces a globule of metallic lead). Together with mimetite and vanadinite, it constitutes the pyromorphite series of the apatite group. These minerals are isomorphous, with the possibility of complete anionic substitution. It fuses easily, is soluble in acids, and it is not fluorescent.

▼ *Pyromorphite (ca. x 2), Ural Mountains, Russia.*

MIMETITE

COMPOSITION: $Pb_5(AsO_4)_3Cl$ lead chlorarsenate.

CRYSTALLINE FORM: hexagonal system, bipyramidal class. Appears as prismatic crystals with a hexagonal base, commonly with curved faces like a keg; in globular aggregates (campylite variety) and in reniform crusts (ediphane variety).

COLOR: colorless, yellow, brown, orange.

HARDNESS: 3.5-4.

SPECIFIC GRAVITY: 7.24.

LUSTER: resinous to adamantine.

GEOLOGIC OCCURRENCE: secondary mineral produced by alteration in the oxidation zones of lead and zinc deposits.

ASSOCIATED MINERALS AND LOCALITIES: with cerussite, smithsonite, hemimorphite, anglesite, malachite, leadhillite, wulfenite, pyromorphite, arsenopyrite and other arsenic minerals. Beautiful crystals come from Australia, Namibia, Mexico, the United States, and Great Britain.

USES: secondary mineral for the extraction of lead; of interest to researchers and collectors.

NOTES: The name is derived from the Greek for "imitator" because of this mineral's remarkable resemblance to pyromorphite. It fuses easily, forming a globule of metallic lead; during heating in the reducing flame, emanates arsenical vapors with a garlicky odor. It is soluble in acids, it is not fluorescent, and it is isomorphous in the pyromorphite series.

▲ *Mimetite; campylite variety (ca. x 0.5), Idaho, the United States.*

VANADINITE

COMPOSITION: $Pb_5(VO_4)_3Cl$ lead chlorovanadate. (lead chloride and vanadate?)

CRYSTALLINE FORM: hexagonal system, bipyramidal class. It appears as prismatic crystals with a hexagonal base, generally small, commonly with central cavities and rounded edges; sometimes forms very beautiful druses with minute crystals.

COLOR: orange-red, scarlet-red, brown-yellow.

HARDNESS: 2.5-3.

SPECIFIC GRAVITY: 6.88.

LUSTER: resinous to adamantine.

GEOLOGIC OCCURRENCE: secondary mineral produced by alteration in the oxidation zones of lead deposits.

ASSOCIATED MINERALS AND LOCALITIES: with cerussite, descloizite, anglesite, mottramite, wulfenite, pyromorphite, mimetite. Beautiful specimens come from Morocco, Mexico, the United States, and South Africa.

USES: useful mineral for the extraction of vanadium; used in the metallugical and in the mordant industries.

NOTES: isomorphous in the pyromorphite series. It fuses easily forming a globule of metallic lead which, once oxidized and treated in the reducing flame, produces, in addition to phosphorus salt, an emerald-green pearl. It is soluble in acids; through evaporation from a nitric solution, produces a red residue, unlike the other minerals in the series which form a white residue. It is not fluorescent, and the variety rich in arsenic takes the name endlicheite.

Vanadinite (ca. x 2), ▶
Morocco.

ADAMITE

COMPOSITION: $Zn_2(AsO_4)(OH)$ hydrated zinc arsenate.

CRYSTALLINE FORM: orthorhombic system, bipyramidal class. Appears as rather small elongated crystals and in radial fibrous or fan-shaped aggregates.

COLOR: colorless, yellow, violet, green from the presence of copper, pink if cobalt is present.

HARDNESS: 3.5.

SPECIFIC GRAVITY: 4.32-4.48.

LUSTER: vitreous.

GEOLOGIC OCCURRENCE: secondary mineral produced by alteration in the oxidation zones of zinc deposits.

ASSOCIATED MINERALS AND LOCALITIES: with olivenite, smithsonite, hemimorphite, malachite, azurite. Beautiful specimens come from Algeria, Mexico, Namibia, Chile, Turkey, Greece, and France.

USES: interest limited to researchers and collectors.

NOTES: The name honors the French mineralogist G. I. Adam (1795-1881). Isomorphous with olivenite with which it makes up a continuous series due to the miscibility of zinc and copper. It fuses easily and decrepitates during heating, losing color. It is very soluble in dilute acids, and is sometimes fluorescent when exposed to ultraviolet light, emitting lemon-yellow light.

▼ Adamite (ca. x 0.4), Mexico.

OLIVENITE

COMPOSITION: $Cu_2(AsO_4)OH$ hydrous copper arsenate.

CRYSTALLINE FORM: orthorhombic system, bipyramidal class. Appears as elongated and tabular crystals; in granular or earthy masses, rarely compact; in fibrous structured nodules and in reniform aggregates.

COLOR: greenish blue, olive-green, or blackish green.

HARDNESS: 3.

SPECIFIC GRAVITY: 3.39-4.4.

LUSTER: adamantine to vitreous in the fibrous varieties.

GEOLOGIC OCCURRENCE: secondary mineral produced by alteration in the oxidation zones of copper and arsenic deposits.

ASSOCIATED MINERALS AND LOCALITIES: with adamite, smithsonite, malachite, azurite, scorodite. Very beautiful specimens come from Namibia, the United States, France, etc.

USES: interest limited to researchers and collectors.

NOTES: Olivenite is isomorphous with adamite with which it constitutes a continuous series of minerals with varying amounts of zinc and copper which are miscible and substitute for each other. It fuses easily, emanating arsenical vapors with a garlicky odor. It is soluble in acids; is not fluorescent.

▼ Olivenite (ca. x 1.5), Cornwall, Great Britain.

DESCLOIZITE

COMPOSITION: $Pb(Zn,Cu)(VO_4)(OH)$ hydrous lead vanadate.

CRYSTALLINE FORM: orthorhombic system, bisphenoidal class. Rare as crystals, it normally appears in subparallel aggregates, in mosaic-like or fan-shaped forms; in fibrous or botryoidal masses.

COLOR: from brown-red to brown-black, green due to the high mode of the copper content.

HARDNESS: 3-3.5.

SPECIFIC GRAVITY: 5.9-6.2.

LUSTER: resinous.

GEOLOGIC OCCURRENCE: secondary mineral produced by alteration in the oxidation zones of lead, zinc, and copper deposits.

ASSOCIATED MINERALS AND LOCALITIES: with vanadinite, cerussite, wulfenite, pyromorphite, mimetite. Beautiful crystals come from Namibia, the Congo, Zambia, Mexico, and the United States.

USES: useful vanadium mineral, rarely found in economically exploitable concentrations.

NOTES: The name honors the French mineralogist Alfred Des Cloizeau (1817-1897). Isomorphous with mottranite (lead and copper vanadate) with which it constitutes a continuous series, resulting from the complete miscibility of zinc and copper. It fuses easily, decrepitating during heating, is soluble in acids, and is not fluorescent.

▲ Descloizite (ca. x 0.3), Republic of South Africa.

LAZULITE

COMPOSITION: $(Mg,Fe)Al_2(PO_4)_2(OH)_2$ hydrous magnesium and aluminum phosphate.

CRYSTALLINE FORM: monoclinic system, prismatic class. Rare as pseudobipyramidal crystals; normally it appears in microgranular masses.

COLOR: blue.

HARDNESS: 5-6.

SPECIFIC GRAVITY: 3.06-3.22.

LUSTER: vitreous.

GEOLOGIC OCCURRENCE: found as an accessory mineral in silica-rich metamorphic rocks (quartzites); in pegmatites and quartz veins.

ASSOCIATED MINERALS AND LOCALITIES: with quartz, corundum, kyanite, sillimanite, garnet, sapphirine and dumortierite in quartzites; with andalusite and rutite in pegmatites. Beautiful specimens come from the United States, Bolivia, Brazil, Madagascar, Sweden, and Austria.

USES: of limited use primarily for ornamental objects.

NOTES: The name is derived from the Latin *lapis* (stone) and the Persian-Arabic *zuward* (blue) in relation to the mineral's color. It is infusible and loses its color during heating, breaking into pieces. It is moderately soluble in hot strong acids, and is not fluorescent.

▼ Lazulite (ca. x 2), Georgia, the United States.

VIVIANITE

COMPOSITION: Fe$_3$(PO$_4$)$_2$·8H$_2$O hydrated iron phosphate.

CRYSTALLINE FORM: monoclinic system, prismatic class. Appears as prismatic crystals sometimes of remarkable size (over a meter long); in radial aggregates, earthy nodules, patinas, and incrustations.

COLOR: colorless or white immediately after extraction; exposed to the air, it oxidizes taking on a greenish coloration with blue reflexions.

HARDNESS: 2.

SPECIFIC GRAVITY: 2.6-2.7.

LUSTER: vitreous to metallic from high iron content.

GEOLOGIC OCCURRENCE: secondary mineral in the oxidation zones of many sulfide deposits; alteration product of iron and magnesium phosphates in pegmatites; present in lacustrine sedimentary deposits as a result of the action of iron-rich waters on phosphatic organic remains.

ASSOCIATED MINERALS AND LOCALITIES: found with other alteration minerals. The large crystals of Cameroon are very famous; other specimens come from Bolivia, the United States, and Germany.

USES: exploitable only if in abundant concentration; used as a dye.

NOTES: Vivianite owes its name to the English mineralogist, J. G. Vivian, who first discovered it. With erythrite, annabergite, koettigite, and parasymplesite, it constitutes the isomorphous vivianite series, with replacement of the cation and/or the anion. It fuses easily, is soluble in strong acids, and it is not fluorescent.

▼ *Vivianite (ca. x 0.4), Kerc, Crimea, Ukraine.*

THE UNMISTAKABLE COLOR

TURQUOISE

Its color, the typical full and impressive azure which is certainly the most-known distinguishing mark of this stone, does not save turquoise from innumerable imitations.

THE NAME

The etymology of the name is uncertain and is still debated. According to some, it is derived from the Persian word *piruzeh* or from the Arabic *firuzeh*; according to others the name is derived from its deep blue color known in Italian as *turchino*. The most accredited interpretation takes the word back to the Turks; either because they circulated it in the West or because of their predilection for this stone (see page 57).

According to some scholars, turquoise could in fact be the eighth

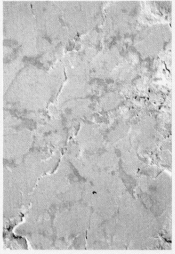

Turquoise (the United States).

stone of the Breastplate of the High Priest. Regardless, that turquoise was known to and widely used by ancient peoples is demonstrated by a great deal of archeological evidence; including the splendid pectoral of Pharaoh Sesostris II, made up almost entirely of turquoise, exhibited at the Metropolitan Museum of Art in New York.

THE LEGENDS

As with all ornamental stones, a variety of powers were attributed to the

Chinese turquoise sculpture (height, 25 cm).

turquoise in the past; including protection against the pain resulting from sudden falls and from spiritual disturbances. The stone was also considered to be a symbol of generosity and wisdom.

The Aztecs considered it so precious that a man's wealth was calculated on the basis of the number of these stones he possessed. To the Navajos in Arizona, the turquoise represents a piece of the sky fallen to earth. At the court of the eastern Roman Empire (the Byzantine Empire), supernatural powers were attributed to turquoise.

THE FAMOUS STONES

Turquoise is rarely found in large pieces. Among the largest is the one owned by the Grand Duke of Tuscany, on which the head of Julius Caesar is engraved. The piece of turquoise owned by the Duc d'Orleans, also engraved, and those pieces in the treasury of the Shah of Persia are also famous. In the East, pieces engraved with short verses from the Koran are common. The famous Ming-era "Kwanyin" statuette dated 1450, weighing 29 oz (820 grams) and completely made of turquoise adorned with a coral twig, is especially precious.

USE AND IMITATIONS

Turquoise is a much sought-after stone, and because of its high cost it has often been counterfeited. Two types of substitutes are on the market today: one, a synthetic and one an adulteration of the natural product.

The synthetic stone has a beautiful appearance, very similar to the natural one. However, it is identifiable by its honeycomb structure, which is visible with the aid of a lens. The second type of imitation is produced with natural stones rejected because of their high porosity. They are filled in with colored resins to improve their aesthetic properties. Even in this case, with a little experience and with the aid of a lens, it is possible to recognize them as fakes.

Prized turquoise sculptures.

ERYTHRITE

COMPOSITION: $Co_3(AsO_4)_2 \cdot 8H_2O$ hydrated cobalt arsenate.

CRYSTALLINE FORM: monoclinic system, prismatic class. Appears as small prismatic crystals, commonly joined in radial-fibrous aggregates; in earthy masses and in patinas known as "cobalt bloom."

COLOR: from carmine-red to violet-red; when altered, it takes on a pearl-gray coloration.

HARDNESS: 1.5-2.5.

SPECIFIC GRAVITY: 3.07.

LUSTER: vitreous to pearly.

GEOLOGIC OCCURRENCE: secondary mineral produced by alteration in the oxidation zones of cobaltic sulfides deposits.

ASSOCIATED MINERALS AND LOCALITIES: with adamite, azurite, malachite, corodite, symplesite. The localities in Morocco, Iran, Canada, the United States, Germany, and France are famous.

USES: its bright colors make it an excellent guide mineral in the hunt for cobalt minerals.

NOTES: The name is derived from the Greek for "red," in relation to its color; known also as "cobalt bloom." It fuses easily, changing its color to blue during heating. It is soluble in acids, producing a red-colored solution, and it is not fluorescent.

▼ *Erythrite (ca. x 0.4), Morocco.*

AUTUNITE

COMPOSITION: $Ca(UO_2)_2(PO_4)_2 \cdot 10H_2O$ calcium uranate phosphate.

CRYSTALLINE FORM: tetragonal system, ditetragonal bipyramidal class. It appears as tabular and lamellar crystals, with very easy basal cleavage, almost micaceous, and in fan-shaped aggregates.

COLOR: vivid yellow, greenish yellow.

HARDNESS: 2-2.5.

SPECIFIC GRAVITY: 3.05-3.19.

LUSTER: vitreous to pearly.

GEOLOGIC OCCURRENCE: secondary mineral produced by alteration of primary uranium minerals, present in pegmatites and hydrothermal veins.

ASSOCIATED MINERALS AND LOCALITIES: with torbernite and zeunerite. Very beautiful specimens come from the United States, Australia, Zaire, France, and Portugal.

USES: important uranium mineral, but rarely concentrated in economically exploitable deposits.

NOTES: The name is derived from the French locality Autun, where it was described for the first time. It is a member of the uranium mica group, so called because of their perfect cleavage and because they contain the ion uranyl. Through simple exposure to dry air, it changes to meta-autunite with a lattice containing molecules of water varying in number from zero to six. It is fusible and it is soluble in acids. When exposed to ultraviolet light, it is fluorescent, emitting yellowish green light, and it is highly radioactive.

▼ *Autunite (ca. x 2), Lurisia, Cuneo, Italy.*

TURQUOISE

COMPOSITION: $CuAl_6(PO_4)_4(OH)_8 \cdot 5H_2O$ hydrated copper aluminum phosphate.

CRYSTALLINE FORM: triclinic system, pinacoidal class. Very rare as crystals, it normally appears in nodules and microcrystalline veins with reniform surfaces; in compact, sometimes earthy, masses.

COLOR: soft azure, light green.

HARDNESS: 5-6.

SPECIFIC GRAVITY: 2.6-2.83.

LUSTER: silky.

GEOLOGIC OCCURRENCE: secondary mineral produced by alteration of aluminum-rich igneous rocks.

ASSOCIATED MINERALS AND LOCALITIES: with limonite, chalcedony, kaolin. The deposits in Iran, Egypt, Mexico, and the United States are famous.

USES: intense azure-colored stones, without shadings, are highly valued as gems.

NOTES: Turquoise owes its name–originally, Turkish stone–to the Turks who circulated it in the West. It is also known as "callais." It is infusible, dissolves in hydrochloric acid when hot, and the solution obtained in this fashion treated with ammonia yields a blue color; heated in the closed tube, becomes brown or black. When exposed to ultraviolet light, it is fluorescent, emitting green light of varying intensity. Imitations are common, among which are the "Western turquoise" or "odontolite" made of fossilized ivory and colored azure by vivianite, and turquoise paste. (See box, page 56).

▲ *Turquoise (ca. x 3), Iran.*

WAVELLITE

COMPOSITION: $Al_3(PO_4)_2(OH,F)_3 \cdot 5H_2O$ hydrated aluminum phosphate.

CRYSTALLINE FORM: orthorhombic system, bipyramidal class. Rare as rhombic, acicular crystals; it normally appears in microcrystalline aggregates, either in discoidal form or nodular with a radial fibrous structure.

COLOR: white, yellow, brown, blue, pale gray.

HARDNESS: 3.5-4.

SPECIFIC GRAVITY: 2.35.

LUSTER: vitreous to silky.

GEOLOGIC OCCURRENCE: low-temperature secondary mineral, present in low-grade metamorphic rocks rich in apatite and in cracks in aluminum-rich rocks; in limonitic or phosphatic deposits.

ASSOCIATED MINERALS AND LOCALITIES: with turquoise, limonite, quartz. Beautiful specimens come from Brazil, Bolivia, the United States, and Great Britain.

USES: in economically exploitable deposits, it is used as a primary material for phosphatic fertilizers.

NOTES: Wavellite is named in honor of its discoverer, the English doctor William Wavell. The mineral is infusible. It swells and crumbles when heated and colors the flame green. It is soluble in strong acids, and it is not fluorescent.

▼ *Wavellite (ca. x 1.5), Vogtland, Germany.*

OLIVINE

COMPOSITION: $(Mg,Fe)_2SiO_4$ iron magnesium nesosilicate.

CRYSTALLINE FORM: Orthorhombic system, bipyramidal class. The crystals have a prismatic, tabular form; sometimes appears in microcrystalline masses.

COLOR: olive-green, green-yellow, dark green, brownish yellow, black in iron-rich end members .

HARDNESS: 6.5-7.

SPECIFIC GRAVITY: 3.16-4.39.

LUSTER: vitreous.

GEOLOGIC OCCURRENCE: the magnesian end members are accessories of many igneous rocks, fundamental constituents of peridotites and other ultrabasic rocks; in high-grade metamorphic rocks (dolomitic marbles). The iron-rich end members are rarer and are occasionally present in granites and in pegmatites.

ASSOCIATED MINERALS AND LOCALITIES: covers the walls of geodes, commonly without associated minerals, in ultrabasic rocks, or crystals, even large ones, are scattered in the rocks. Beautiful specimens come from Germany, the Urals, the United States, Egypt, Burma, etc.

USES: the golden-yellow-colored, transparent variety (chrysolite) is a valuable gem; rocks rich in magnesiferous olivine are used for the extraction of magnesium and for the making of refractories.

NOTES: The name alludes to the color of the mineral, also called peridot. It is a solid solution from forsterite, $Mg_2(SiO_4)$ and fayalite $Fe_2(SiO_4)$, miscible in all proportions; it easily alters to serpentine, chlorite, iddingsite. Its properties vary according to the iron and magnesium content; as the iron content increases, the indexes of refraction and the solubility in acids rise, but the melting point decreases. It is not fluorescent.

▼ *Olivine (ca. x 1.4), Zabargad, Red Sea, Egypt.*

ZIRCON

COMPOSITION: $Zr(SiO_4)$ zirconium nesosilicate.

CRYSTALLINE FORM: tetragonal system, ditetragonal bipyramidal class. Appears as prismatic, stubby, sometimes bipyramidal crystals.

COLOR: colorless, yellow, green, brown, gray, reddish from the presence of rare earths (alvite variety).

HARDNESS: 7.5.

SPECIFIC GRAVITY: 3.9-4.86.

LUSTER: adamantine.

GEOLOGIC OCCURRENCE: accessory mineral of acidic plutonic rocks and metamorphic rocks; in pneumatolytic veins and pegmatites; because of its elevated hardness, it is concentrated in alluvial deposits.

ASSOCIATED MINERALS AND LOCALITIES: with quartz, beryl, tourmaline, mica, etc. Beautiful specimens come from Brazil, Sri Lanka, Japan, the United States, and the Urals.

USES: used for the extraction of zirconium, and the preparation of refractory materials (thorium, hafnium).

NOTES: The name is of uncertain etymology; it may be derived from the name "jargon" for the colorless or smoky variety from Sri Lanka or possibly from the Persian *zargun* meaning "gold-colored." Zircon was first described by the German naturalist A. G. Werner (1750-1817). It is infusible, soluble in concentrated sulfuric acid, and in ultraviolet light some types of zircon display an orange-red to pale-blue fluorescence. The zirconium in zircon may be replaced by hafnium, uranium, and other rare earth elements which may lead to structural changes (metamict stage) with a consequent change in physical properties (e.g., the lessening of density and birefringence). (See box, page 59).

▲ *Zircon (ca. x 1), Jusel Seiland, Norway.*

SPHENE

COMPOSITION: $CaTiSiO_5$ calcium titanium nesosilicate.

CRYSTALLINE FORM: monoclinic system, prismatic class. Occurs as crystals of very varied appearance: prismatic, lamellar, wedge-shaped, stubby, and elongated. Contact and penetration twins are common as are microcrystalline aggregates.

COLOR: rarely colorless, white, yellow, green, dark brown, black, pink from the presence of manganese (greenovite).

HARDNESS: 5-5.5.

SPECIFIC GRAVITY: 3.4-3.6.

LUSTER: adamantine to resinous.

GEOLOGIC OCCURRENCE: accessory mineral in many igneous rocks, both intrusive (syenites, diorites) and extrusive (porphyrs, quartz rich rocks and rhyolites); also present in metamorphic rocks such as gneiss, amphibolites, and mica schists. It can accumulate in alluvial sands where the surfaces of the granules show chloritic alteration patinas.

ASSOCIATED MINERALS AND LOCALITIES: with garnet, diopside, apatite, etc. Beautiful specimens come from the United States, Canada, Norway, Madagascar, and Italy.

USES: in abundant deposits, it is exploited for the extraction of titanium; high cost gems are obtained from cabochon-cut, clear and colored varieties.

NOTES: Sphene, also known as titanite, was discovered at the end of the eighteenth century by the German mineralogist M. H. Klaproth. It was called "sphene" by the abbott Hauy because of the wedge-shaped form of its crystals (from the Greek for "wedge"). It fuses easily, changing into a dark green, vitreous mass. Scarcely affected by hydrochloric acid, it decomposes in concentrated sulfuric acid and when warm; with oxygenated water, the solution turns yellow from the presence of titanium; shows a slight bluish-gray fluorescence.

▼ *Sphene (ca. x 0.5), Hofenhorn, Binn Valley, Switzerland.*

ALMANDINE

COMPOSITION: $Fe_3^{+2}Al_2(SiO_4)_3$ iron aluminum nesosilicate.

CRYSTALLINE FORM: cubic system, hexoctahedral class. Appears as rhombic dodecahedral crystals, sometimes in combination with the hexoctahedron and the trapezohedron.

COLOR: red, dark red with violet tones.

HARDNESS: 7.

SPECIFIC GRAVITY: 4.1-4.3.

LUSTER: vitreous to resinous, to adamantine.

GEOLOGIC OCCURRENCE: typical mineral of regional metamorphic schists, also present in contact metamorphic rocks; accessory constituent of eruptive rocks (granites, andesites, diorites); because of its resistance to alteration, it is concentrated in sedimentary deposits.

ASSOCIATED MINERALS AND LOCALITIES: with quartz, muscovite, biotite, zircon, andalusite, cordierite, epidotes. It is a fairly common mineral, very beautiful specimens of which come from Sri Lanka, India, Japan, Brazil, Mexico, the United States, Greenland, Madagascar, Australia, Norway, Sweden, Switzerland, Italy, and Russia.

USES: as a medium-hard abrasive for cloth and paper; when used as a low-cost gem, the red and wine-red varieties are generally emerald-cut.

NOTES: The name is derived from the ancient city of Alabanda in Asia Minor, where it was widely used. It is also known under the name noble garnet. It is a member of a family of garnet minerals made up of two series within which the miscibility between components is complete. It fuses easily, is insoluble in acids, and, it is not fluorescent. (See box, page 63.)

▼ *Rhombic dodecahedral almandine (ca. x 1), Tyrol, Austria.*

ANDRADITE

COMPOSITION: $Ca_3Fe_2^{+3}(SiO_4)_3$ calcium iron nesosilicate.

CRYSTALLINE FORM: cubic system, hex-octahedral class. Appears as rhombic dodecahedral crystals, sometimes in combination with the hexoctahedron and the trapezohedron.

COLOR: brown, reddish, black (melanite), yellow (topazolite), green (demantoid).

HARDNESS: 6-6.5.

SPECIFIC GRAVITY: 3.7-3.9.

LUSTER: vitreous to adamantine.

GEOLOGIC OCCURRENCE: in limestones affected by contact metamorphism and in skarns; melanite in volcanic rocks; demantoid and topazolite in serpentinous rocks.

ASSOCIATED MINERALS AND LOCALITIES: andradite is found with hedenbergite, magnetite, wollastonite, quartz, topazolite; demantoid with chlorite, serpentine, talc; melanite with nepheline and feldspars. Beautiful specimens come from Norway, Italy, Switzerland, the United States, and Russia.

USES: demantoid and topazolite are used as medium-priced, emerald-cut gems.

NOTES: the name commemorates the Portuguese mineralogist D'Andrada. Member of a family of minerals made up of two series within which the miscibility between components is complete. Demantoid takes its name from the brilliance and play of light that make it similar to the diamond. It fuses easily, yielding a magnetic black globule, is slowly corroded by acids, and sometimes emits orange-red light in ultraviolet light. (See box, page 63.)

▼ *Demantoid andradite (ca. x 1), Val Malenco, Sondrio, Italy.*

A FALSE DIAMOND

ZIRCON

Zircon appears in a wide variety of colors (blue, red, yellow, green), but the colorless is perhaps the best known, even though it is not the most widespread. Because it is so brilliant, zircon has become popular as the principal substitute for the diamond.

HISTORY AND LEGEND

In the past, zircon (see page 58) was called "hyacinth" and was differentiated into two varieties: one red and one blue. This name was probably used to designate minerals differing either in composition or in color. The blue variety of hyacinth, in fact, was probably the sapphire (corundum). Scrodero, however, describes the hyacinth as a very luminous gem that "faintly turns red in the blond and imitates the flames of the fire" – words well-suited to the red variety of zircon. This mineral was considered to be rare until the end of the eighteenth century when Klaproth proved its correspondence to the "Ceylon hyacinth." In the past, zircon was used as an amulet and was thought to safeguard against the plague. A different power was attributed to each color:

Small wedding ring with zircons and rubies.

the violet variety was a symbol of wisdom, the red symbolized fury and was used in propitiatory magic rites, while the colorless variety was used against the evil eye and spells. According to Catholic tradition, colorless zircon was also the symbol of modesty.

PROPERTIES

The zircon, because of its remarkable chromatic variations, has many similarities with corundum, but it is distinguished from corundum by its lower hardness (7.5 versus 9) and by its higher index of refraction (1.92 versus 1.77). Zircon may contain some radioactive elements, but the stones commonly offered on the market, beautiful both in color and in transparency, have not been damaged by radioactivity and are called "high zircons."

Single colored zircons of different carats.

Because of its high index of refraction, the colorless variety of zircon has often been used as a substitute for the diamond. However, it can be distinguished from the diamond by its birifringence (the diamond is singly refractive). This can be detected with the aid of a lens; observing the stone through the table (the upper facet) and focusing on the lower edges which will appear doubled.

COLORS

Without a doubt, the most common and well-known variety of zircon is the transparent one, which can also be obtained through heat treatment of some brown and reddish varieties. Imitation zircons can also be found on the market. Among these are synthetic spinel, YAG (yttrium aluminum garnet), and the "cubic zirconia" which, however similar it may be in name, is not related to the natural stone.

The blue or azure variety is called starlite (from the English star) and

can also be obtained by heat treatment of other colored zircons. The preferred shade is electric-blue, fairly typical in the gem field, with a slight greenish pleochroism.

The red variety was once called hyacinth (from the Greek *hyacinthos* or the Arabic *zarkun*) and can be mistaken for some spinels and red garnets, which are, however, singly refractive. Its high bifringence also distinguishes it from the ruby (corundum).

The name jargon (from the Persian *zargùn*) denotes the yellow variety, and the green variety, called malacon, is relatively common and has variable optical properties due to its different metamict stages. It is distinguishable

Blue zircon.

from the green sapphire, from olivine, and from the green tourmalines as clear zircons have high bifringence, while the muddy ones (high metamitic stage) are milky and have close parallel striations.

Green zircon (above) and reddish zircon (below).

UVAROVITE

COMPOSITION: $Ca_3Cr_2(SiO_4)_3$ calcium chromium nesosilicate.

CRYSTALLINE FORM: cubic system, hexoctahedral class. Appears as small crystals of complex form.

COLOR: emerald-green, dark green.

HARDNESS: 7.5.

SPECIFIC GRAVITY: 3.41-3.52.

LUSTER: vitreous to resinous.

GEOLOGIC OCCURRENCE: in serpentinous rocks and in skarns.

ASSOCIATED MINERALS AND LOCALITIES: with talc, chlorite, serpentine, chromite, calcite, diopside, quartz. Beautiful crystals come from Turkey, Russia, Italy, and Canada.

USES: as an emerald-cut gem.

NOTES: Uvarovite is named in honor of Count Uvarov (1785-1855), a minister at the czar's court and a mineral collector. It belongs to a family of garnet minerals made up of two series within which miscibility between components is complete. It fuses easily, is unaffected by acids, and it is not fluorescent. (See box, page 63.)

▲ *Uvarovite (ca. x 0.5), Finland.*

ANDALUSITE

COMPOSITION: Al_2SiO_5 aluminum nesosilicate.

CRYSTALLINE FORM: orthorhombic system, bipyramidal class. Crystals are prismatic, stubby, in some cases elongated with square sections; acicular and bacillary aggregates are common.

COLOR: colorless, white, gray, yellow, brown, rosy; green-brown from carbonaceous inclusions in the chiastolite variety.

HARDNESS: 7.5.

SPECIFIC GRAVITY: 3.16-3.20.

LUSTER: vitreous.

GEOLOGIC OCCURRENCE: in argillaceous rocks affected by low-pressure regional metamorphism or by contact metamorphism.

ASSOCIATED MINERALS AND LOCALITIES: with cordierite, quartz, calcite. Beautiful crystals come from Spain, Austria, Brazil, the United States, and Australia.

USES: employed in the preparation of refractories and in acid resistant ceramic products; transparent varieties are used as low cost gems.

NOTES: The name is derived from Andalusia, a region of Spain. The compound Al_2SiO_5 is trimorphous with kyanite and sillimanite; andalusite is the polymorph that is stable at low pressures. It is infusible, unaffected by acids, and it is fluorescent only in cathode rays, emitting a greenish yellow light.

▼ *Andalusite (ca. x 1), Vipiteno, Bolzano, Italy.*

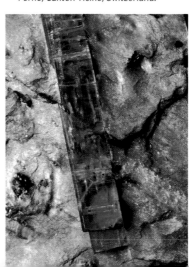

KYANITE

COMPOSITION: Al_2SiO_5 aluminum nesosilicate.

CRYSTALLINE FORM: triclinic system, pinacoidal class. Generally appears as elongated and flattened crystals, sometimes twinned; in fibrous aggregates, slightly curved and with a radial structure (reticite variety).

COLOR: generally blue with whitish or yellowish brown spots.

HARDNESS: 5-7.5.

SPECIFIC GRAVITY: 3.59.

LUSTER: vitreous.

GEOLOGIC OCCURRENCE: mineral of medium-temperature, high-pressure metamorphic rocks (mica schists, gneiss, amphibolites, and sometimes eclogites) derived from sedimentary pelitic rocks rich in aluminum; less commonly, as a secondary mineral in corundum-rich alluvial sands.

ASSOCIATED MINERALS AND LOCALITIES: commonly found with staurolite, garnets, corundum, paragonite. Beautiful specimens come from India, Burma, Brazil, the United States, Madagascar, Italy, etc.

USES: basic mineral in the porcelain and refractory industries; for electrical insulators and acid-resistant products. The transparent blue varieties are sometimes used as low-cost gems, called "sapparé."

NOTES: The name is derived from the Greek for "blue," in relation to the mineral's color. Also known as disthene, kyanite was first described by the German mineralogist A. G. Werner (1750-1817). It is the high pressure polymorph of Al_2SiO_5. It is infusible, insoluble in acids, and in ultraviolet light it exhibits a faint fluorescence. Its hardness is very high in the direction perpendicular to the length of the crystals, but is notably less in the other orthogonal directions.

▼ *Kyanite in mica schist (ca. x 1), Pizzo Forno, Canton Ticino, Switzerland.*

TOPAZ

COMPOSITION: $Al_2(SiO_4)$ (F,OH)$_2$ hydrous fluorine aluminum nesosilicate.

CRYSTALLINE FORM: orthorhombic system, bipyramidal class. Appears as stubby, prismatic crystals, sometimes striated; bacillary aggregates with a radial fibrous structure are rare.

COLOR: colorless, yellow, ivory-white, pink, blue, reddish.

HARDNESS: 8.

SPECIFIC GRAVITY: 3.49-3.57.

LUSTER: vitreous to adamantine.

GEOLOGIC OCCURRENCE: typical mineral of pneumatolytic veins and pegmatites; because of its hardness and resistance to alteration, it is concentrated in placer deposits.

ASSOCIATED MINERALS AND LOCALITIES: with quartz, tourmaline, beryl, cassiterite. Beautiful specimens come from Brazil, Burma, Sri Lanka, Mexico, the United States, and the Urals.

USES: of no use in the industrial field; much appreciated and very high-priced gems are obtained from the transparent varieties with uniform and intense coloration (dark yellow and blue).

NOTES: The name is probably is derived from Topazos, an island in the Red Sea from which the ancient Romans acquired various colored gems. It is infusible and slowly soluble in sulfuric acid. Some varieties exhibit a faint fluorescence emitting green, reddish light. Yellow topazes, when heated to 572-842° F (300-450° C) take on a rosy-reddish color (burned topaz); pyroelectric. (See box, p. 61.)

▼ *Topaz (ca. x 1.5), Saxony, Germany.*

STAUROLITE

COMPOSITION: $(Fe,Mg,Zn)_2Al_9(Si,Al)_4O_{22}$ $(OH)_2$ iron aluminum nesosilicate.

CRYSTALLINE FORM: orthorhombic system, bipyramidal class. Appears as prismatic, sometimes stubby crystals; twins, with two interpenetrated crystals in the shape of an upright cross (Greek cross) or an oblique cross (Saint Andrew's cross), are common.

COLOR: red, red-brown, black.

HARDNESS: 7-7.5.

SPECIFIC GRAVITY: 3.65-3.80.

LUSTER: subvitreous.

GEOLOGIC OCCURRENCE: common accessory mineral in medium-temperature regional metamorphic rocks (e.g., mica schists) because of its high resistance to alteration, it is also found scattered in sediments.

ASSOCIATED MINERALS AND LOCALITIES: with almandine, garnet and kyanite; at low pressure with cordierite, andalusite and sillimanite. Beautiful crystals come from the United States, Brazil, Switzerland, Scotland, Bavaria, France, etc.

USES: interest limited to researchers and collectors.

NOTES: The name is derived from the Greek for cross and stone, relating to the typical shape exhibited by the twins; so typical that in the past it was called *lapis crucifer*. In some areas of America, the cross-shaped twins are used as amulets. Infusible in the flame and slightly corroded by sulfuric acid. It is not fluorescent.

▼ *Staurolite: twin crystal, above, and single, below, (ca. x 1), Brazil.*

TOPAZ

Contrary to what is commonly believed, topaz is not only found in nature in the yellow color which its name might indicate. It was not until the eighteenth century that itopaz was precisely defined and its modern usage was initiated.

HISTORY AND LEGEND

The topaz (see page 60) has certainly been known since ancient times, although it is likely that confusion arose among different yellow-colored

Colorless topaz, 1.95 c. (Brazil).

stones. According to the Greek philosopher, Archelaus of Miletus, the mineral was found by shipwrecked troglodyte pirates on the island of Cytis (Arabia). This legend is connected to one of several possible etymologies of the name which ties it to the island Topazos in the Red Sea, now called Zabargad. According to the Jewish cabala, the second name of God (Gomel, the one who rewards) could only be written on the topaz. This mineral also has a historical role in the Catholic tradition as a symbol of knowledge and, during the Middle Ages, it was one ot the seven stones that symbolized the gifts of the Holy Spirit. Topaz was thought to guard against and cure the plague. It is said that Pope Clement VI wore a large topaz on his finger with which he touched the boils of plague victims, who seemed to recover in a short time. As F. Hermann related, another power attributed to the topaz was that it inspired a horror of blood and, therefore, all men who had the power to declare wars should possess one.

FAMOUS GEMS

Topaz is often found in crystals of remarkable size. The largest one in the world is held in the American Museum of Natural History in New York (blue colored, it weighs about 1457 lbs. or 300 kg). Two topazes are preserved in the British Museum in London, one of 614 carats, blue and square step-cut, and the other 1300 carats, colorless, in an oval brilliant-cut. The Natural History Museum in Chicago holds a 5,890 carat stone obtained from a raw topaz weighing over 10,000 carats. Other large stones – some hundreds of carats – are found in various museums around the world. The topaz is cut into a variety of shapes, but the most common are squares, rectangles, or ovals, all generally very multi-faceted.

COLORS

The typical topaz color is honey-yellow, of which there are many more or less intense shades. According to the customary scale of values, the most sought-after is the imperial topaz (golden-colored with soft pink shadings). Other yellow shades are called: golden topaz (yellow), sherry topaz (intense yellow, almost brown), and

"Madera" (caramel colored). The blue-colored stones, in nature commonly soft-toned, are also very beautiful. An intense blue can also be

Yellow topaz, 150 c. (Brazil).

Pink topaz, 12.50 c.

given to colorless or soft-toned stones through radiation treatment. These can be confused with the aquamarine (beryl). Because of its rarity, the natural pink variety is the most highly prized and can be confused with kunzite. However, for some time, it has been a widespread practice to obtain pink stones from the heat treatment of yellow stones with pink shadings. In gemology this practice is not considered to be fraudulent and is often accepted. Colorless topaz crystals are the most common and the value of these gems is very low. They can be confused with many stones of similar appearance, and it is necessary to measure the index of refraction in order to verify their identity.

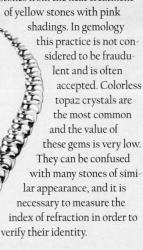

Blue topaz.

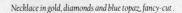

Necklace in gold, diamonds and blue topaz, fancy-cut.

PHENAKITE

COMPOSITION: Be$_2$(SiO$_4$) beryllium nesosilicate.

CRYSTALLINE FORM: trigonal system, rhombohedral class. The crystals are prismatic in form, stubby, commonly rhombohedral, usually with many very shiny faces. Twins and intergrowths are common.

COLOR: colorless, white, yellow, pink.

HARDNESS: 7.5-8.

SPECIFIC GRAVITY: 2.97-3.

LUSTER: vitreous to adamantine.

GEOLOGIC OCCURRENCE: found in high-temperature pegmatites and in mica schists.

ASSOCIATED MINERALS AND LOCALITIES: with quartz, chrysoberyl, beryl, apatite, topaz. Beautiful specimens come from Brazil, Mexico, Sri Lanka, Tanzania, the United States, the Urals, and Norway.

USES: clear and transparent specimens are used as low-cost gems.

NOTES: The name is derived from the Greek for "deceiver," since it is easily confused with other minerals. It is infusible, insoluble, and colored specimens exposed to ultraviolet light exhibit a faint green fluorescence. Sometimes exhibits needle-shaped inclusions which produce chatoyancy.

▼ Phenakite (ca. x 1.5), Minas Gerais, Brazil.

DATOLITE

COMPOSITION: CaB(SiO$_4$)(OH) hydrous calcium boron nesosilicate.

CRYSTALLINE FORM: monoclinic system, prismatic class. Appears as highly modified stubby crystals; in microcrystalline aggregates with a porcelain-like appearance.

COLOR: colorless, white, yellowish, green, reddish.

HARDNESS: 5-5.5.

SPECIFIC GRAVITY: 2.99.

LUSTER: vitreous.

GEOLOGIC OCCURRENCE: secondary mineral of hydrothermal origin, found in veins and cavities within rocks of low silica content: diabase, basalt, serpentinite.

ASSOCIATED MINERALS AND LOCALITIES: with zeolite, prehnite, calcite; beautiful specimens come from Germany, Norway, the United States, and Italy.

USES: in economically exploitable deposits, it is extracted and used to obtain boron; cut as a cabochon or faceted; it is a gem of little value, of interest only to the collector.

NOTES: the name is derived from the Greek for "to divide," referring to the mineral's granular variety. It fuses easily and colors the flame green (boron). It is soluble in strong acids with the formation of silica gel, and in ultraviolet light some varieties exhibit a blue fluorescence. Together with homilite and gadolinite, it makes up the datolite group; all have similar lattice structures, but distinct chemical make-up with no intermediate compositions.

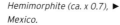

▼ Datolite (ca. x 1), Ciano Val d'Enza, Reggio Emilia, Italy.

HEMIMORPHITE

COMPOSITION: Zn$_4$(Si$_2$O$_7$)(OH)$_2$.H$_2$O hydrous zinc sorosilicate.

CRYSTALLINE FORM: orthorhombic system, pyramidal class. Rarely occurs as small crystals, which generally assume a prismatic or tabular form, and are hemimorphous, that is, they terminate in different ways at either extremity of the vertical axis; more commonly in granular, compact masses or as stalactites.

COLOR: colorless, yellow, green, blue, brown.

HARDNESS: 4.5-5.

SPECIFIC GRAVITY: 3.4-3.5.

LUSTER: vitreous.

GEOLOGIC OCCURRENCE: secondary mineral in the oxidized zones of lead and zinc sulfide deposits.

ASSOCIATED MINERALS AND LOCALITIES: with smithsonite, cerussite, anglesite, sphalerite, galena. Beautiful specimens come from Mexico, the United States, Romania, Austria, Great Britain etc..

USES: as a source of zinc.

NOTES: the former name, calamine, is probably derived from the Greek for "cadmium stone," which then became calamine from the Latin calamus; the name hemimorphite refers to the different terminations at the prism ends. It fuses with difficulty, is soluble in strong acids with the formation of silica gel, and is not fluorescent. When heated it becomes pyroelectric, and when subjected to pressure, piezoelectric, because it develops opposite electrical charges at the two extremities of the crystal's vertical axis.

Hemimorphite (ca. x 0.7), ▶ Mexico.

ILVAITE

COMPOSITION: CaFe^{2+}$_2$Fe^{3+}(Si$_2$O$_7$)O(OH) hydrous calcium divalent and trivalent iron sorosilicate.

CRYSTALLINE FORM: orthorhombic system, dipyramidal class. Appears as prismatic, elongated, and striated crystals; in compact masses, and in bacillary aggregates with a radial fibrous structure.

COLOR: dark gray to black.

HARDNESS: 5.5-6.

SPECIFIC GRAVITY: 4.03.

LUSTER: submetallic to vitreous.

GEOLOGIC OCCURRENCE: found in contact metamorphic rocks and in metasomatic skarn deposits; rarely, as an accessory in syenites.

ASSOCIATED MINERALS AND LOCALITIES: with hedenbergite, magnetite, andradite, pyrite. Beautiful specimens come from Greenland, the United States, the Urals, Greece, Italy etc.

USES: interest limited to researchers and collectors.

NOTES: the name is derived from ilva, the ancient Latin name for the island of Elba, the place where it was found by Fleurian de Bellevue in 1796. Also called lievrite. It fuses easily, changing into a magnetic black mass with a vitreous appearance, is soluble in hydrochloric acid with the formation of silica gel, and is not fluorescent.

▼ Ilvaite (ca, x 1.2), Campiglia, Livorno, Italy.

EPIDOTE

COMPOSITION: $Ca_2Fe^{+3}Al_2O(SiO_4)(Si_2O_7)$ OH hydrous calcium iron aluminum sorosilicate.

CRYSTALLINE FORM: monoclinic system, prismatic class; in elongated prismatic crystals with striated and shiny crystalline faces; in granular masses, bacillary and radial fibrous aggregates.

COLOR: pistachio-green, green-yellow, almost black, green when it has a high iron content, rosy red.

HARDNESS: 6.

SPECIFIC GRAVITY: 3.38-3.49.

LUSTER: vitreous to resinous.

GEOLOGIC OCCURRENCE: accessory mineral in regional metamorphic rocks (amphibolite, gneiss, prasinite) and contact metamorphic rocks; in igneous rocks through alteration of plagioclase (saussuritization).

ASSOCIATED MINERALS AND LOCALITIES: with amphibole, apatite, asbestos. Beautiful specimens come from Madagascar, Greenland, Brazil, the United States, and France.

USES: transparent specimens with beautiful colors are often emerald-cut but are of little value; of interest to scientists and collectors.

NOTES: the name is derived from the Greek for "increase" and was assigned by the mineralogist Haüy; also known as pistacite. It fuses with a certain degree of difficulty, is insoluble in acids, and is not fluorescent.

▼ *Epidote (ca. x 0.7), Bramberg, Knappenwand, Salzburg, Austria.*

POMEGRANATE SEEDS

GARNETS

The word "garnet" does not denote a single mineral, but rather an entire family whose members show compositional and structural similarities.

The word was coined by the philosopher Albertus Magnus in the 12th century because of the similarity between red garnets and the seeds of the pomegranate. The chief minerals are pyrope, almandine (see page 58), spessartine, uvarovite (see page 60), grossular, and andradite (see page 59).

HISTORY AND LEGEND

In ancient times red garnets, which are the most common, were called "carbuncles." Since the time of the ancient Egyptians, they were valued as an ornamental stone and were considered to be the bearers of well-being and family harmony. The Greeks and Romans believed that by wearing a jewel with these stones for a long time, an individual would be

the time of the Moorish occupation. In nineteenth century Europe, its use was remarkably widespread because of its low cost and because of the fashion for light complexions against which these stones would stand out.

THE FAMILIES

The garnets are subdivided into two principal groups: pyralspite and ugrandite, which are acronyms of the names of the principal members in each group. The first name is derived from *pyrope*, *almondine*, *spessartite*, the second from *uvarovite*, *grossular*, *andradite*. To pyralspite belongs a variety called rhodolite, characterized by an intense pink color, whose composition is intermediate between those of pyrope and almandine. Numerous varieties with characteristic colors belong to the ugrandite group.

Tsavorite, 0.72c. (Tanzania).

Green grossular, 1.70c. (Tanzania).

Top: Essonite.
Below: ring with classic dark red almandine garnet in a trapezoidal cut.

favored with inheritances. Pliny the Elder said that they inspired joy. In the Jewish cabala, carbuncle was the stone on which one of the names of the Eternal could be written (*Eloha*, which means "God is strong"). It was the symbol of the city of Granada at

---

VESUVIANITE

COMPOSITION: $Ca_{10}(Mg,Fe)_2Al_4(SiO_4)_5(Si_2O_7)_2(OH)_4$ hydrous calcium magnesium iron aluminum sorosilicate.

CRYSTALLINE FORM: tetragonal system, ditetragonal dipyramidal class. Crystals are prismatic, stubby, sometimes with dipyramidal terminations; lengthwise striations on the faces are common; in microcrystalline masses.

COLOR: brown, yellow, light green from the presence of boron (wiluite variety), rarely blue from the presence of copper (cyprine variety).

HARDNESS: 6-7.

SPECIFIC GRAVITY: 3.27-3.45.

LUSTER: vitreous to resinous.

GEOLOGIC OCCURRENCE: typical of contact metamorphism of calcareous rocks and in skarns; in nepheline syenites and in metasomatically-altered basic igneous rocks (rodingites).

ASSOCIATED MINERALS AND LOCALITIES: with grossular, andradite, wollastonite, diopside. Beautiful specimens come from Mexico, Canada, the United States, the Urals, Italy, and Switzerland.

USES: ornamental objects and medium-priced gems are obtained from the transparent, bright-colored varieties; in particular californite (compact variety with a green color similar to jade).

NOTES: owes its name to Mount Vesuvius where it was found for the first time; also called idocrase. It fuses easily and changes into a brown or greenish bubbly glass. It is insoluble in acids, and is not fluorescent.

▼ *Vesuvianite (ca. x 1.5), Bellecombe, Aosta, Italy.*

AXINITE

COMPOSITION: $Ca_2(Fe,Mn)Al_2BO_3(Si_4O_{12})$ (OH) hydrous calcium iron manganese aluminum boron cyclosilicate.

CRYSTALLINE FORM: triclinic system, pinacoidal class. Appears as wedge-shaped crystals with very sharp edges; in microcrystalline aggregates, granular and lamellar masses.

COLOR: reddish brown, light violet, chestnut-brown, yellow (from high manganese content), greenish when covered with very minute chloritic lamellae.

HARDNESS: 6.5-7.

SPECIFIC GRAVITY: 3.27-3.29.

LUSTER: vitreous.

GEOLOGIC OCCURRENCE: found in marbles metamorphosed by contact with acidic magmas; rarely in metamorphic rocks in green schist facies.

ASSOCIATED MINERALS AND LOCALITIES: with prehnite, zoisite, datolite, pumpellyite, tourmaline, actinolite, calcite. Beautiful specimens come from Japan, Tasmania, the United States, France, Switzerland etc.

USES: interest limited to researchers and collectors.

NOTES: the name is derived from the Greek for "axe," refers to the sharpness of the crystal edges. It fuses easily and colors the flame green from the presence of boron. It is insoluble in acids and some specimens, when exposed to ultraviolet light, show a faint red fluorescence. It acquires electrical properties through rubbing and is pyroelectric.

▼ Axinite (ca. x 0.4), Bourg d'Oisans, Balme d'Auris, Dauphiné, France.

BENITOITE

COMPOSITION: $BaTi(Si_3O_9)$ barium titanium cyclosilicate.

CRYSTALLINE FORM: trigonal system, ditrigonal dipyramidal class. Occurs as stubby crystals with a dipyramidal habit.

COLOR: blue, violet-blue.

HARDNESS: 6-6.5.

SPECIFIC GRAVITY: 3.64-3.65.

LUSTER: vitreous.

GEOLOGIC OCCURRENCE: a very rare mineral; found in small veins within blue schist rocks intercalated with serpentinite.

ASSOCIATED MINERALS AND LOCALITIES: it has only been discovered in the Diablo Range in California, where it is found with natrolite and neptunite.

USES: because of its rarity and beauty, it is used in the United States in jewelry making as (emerald-cut) highly valued gems.

NOTES: the name is derived from San Benito County, California where it was found for the first time in 1907. It fuses easily, forming a glossy, colorless mass, and is soluble in hydrochloric and hydrofluoric acids. In ultraviolet light, it is fluorescent, emitting very bright blue light. It possesses remarkable pleochroism, visible even to the naked eye: a crystal observed through the prism faces appears blue, but through the base it seems colorless.

▼ Benitoite (ca. x 1), San Benito, California, the United States.

BERYL

COMPOSITION: $Be_3Al_2(Si_6O_{18})$ beryllium aluminum cyclosilicate.

CRYSTALLINE FORM: hexagonal system, dihexagonal dipyramidal class. Crystals present a hexagonal prismatic habit, truncated at the end (basal pinacoid) or in combination with hexagonal dipyramids; striations parallel to the length of the prism are common.

COLOR: colorless, white, yellowish, pink, green (from the presence of chromium and vanadium), blue (from the presence of ferrous iron).

HARDNESS: 7.5-8.

SPECIFIC GRAVITY: 2.63-2.8.

LUSTER: vitreous to resinous.

GEOLOGIC OCCURRENCE: found in druses in granitic rocks and in pegmatites where crystals can reach metric proportions; in metamorphic rocks (mica schists, marbles); because of its resistance to alteration, it is concentrated in alluvial deposits.

ASSOCIATED MINERALS AND LOCALITIES: with quartz, feldspar, muscovite, lepidolite, topaz, tourmaline, spodumene, cassiterite, tantalite. Very beautiful specimens come from Colombia, Brazil, Madagascar, India, and the United States.

USES: beryllium is extracted from muddy and opaque specimens and is used in the aeronautical and nuclear industries; precious, sought-after gems are obtained from the clear and transparent green (emerald) and blue (aquamarine) varieties.

NOTES: the name is derived from the Greek. In addition to emerald and aquamarine, other prized, colored varieties are morganite, heliodor, and golden beryl. It fuses with difficulty and is insoluble in acids. In ultraviolet light, some specimens exhibit blue or green fluorescence. (See box, page 65.)

▼ Beryl (ca. x 1), Val Codera, Sondrio, Italy.

CORDIERITE

COMPOSITION: $(Mg, Fe)_2Al_2Si(Si_4Al_2O_{18})$_nH_2O$ hydrated magnesium iron aluminum cyclosilicate.

CRYSTALLINE FORM: orthorhombic system, dipyramidal class. Appears as prismatic, stubby crystals; penetration twins which simulate hexagonal symmetry are common; in microcrystalline aggregates.

COLOR: whitish, blue, blue-violet, yellow.

HARDNESS: 7-7.5.

SPECIFIC GRAVITY: 2.53-2.78.

LUSTER: vitreous.

GEOLOGIC OCCURRENCE: common accessory mineral in low-pressure contact metamorphic rocks (hornfels) and in rocks derived from regional metamorphism (mica schists, gneiss); less common in igneous rocks.

ASSOCIATED MINERALS AND LOCALITIES: with garnet, sillimanite, spinel, feldspar, sapphirine. Beautiful specimens come from Brazil, Madagascar, Sri Lanka, Norway, and Finland.

USES: clear and transparent crystals are emerald-cut to obtain low-cost gems valued for their bright color and pleochroism.

NOTES: also known as iolite or dichroite. Characterized by remarkable pleochroism: a prismatic crystal observed through the base seems blue or violet, while in the perpendicular direction it appears colorless. According to legend, the Vikings used the "sun stone" (as they called cordierite) for navigation: it was said to appear clear when viewed in the direction of the sun even on cloudy days, thus providing a constant reference point. It fuses with difficulty and is partially soluble in acids. In ultraviolet light it exhibits a greenish blue fluorescence.

▼ Cordierite (ca. x 1.5), Kragerö, Norway.

FROM KING SOLOMON'S MINES

BERYL

This mineral (see page 64) has two varieties that are highly esteemed as gems: the emerald and the less valuable aquamarine. With the diamond, the ruby, and the sapphire, the emerald is one of the four most precious stones. The most beautiful ones are from Colombia (Muzo mines).

HISTORY AND LEGEND

The emerald has been known since antiquity: the Egyptians extracted it from the famous mines of King Solomon near the Red Sea, and held it in very high esteem because of its pleasing and captivating color. According to the Jewish cabala, the third name of God (*Adar*, the magnificent) had to be engraved on an emerald. Aristotle attributed numerous therapeutic powers to this stone and to the powder derived from it: it was used as an antidote to poison, to heal bites, and to cure hair loss and epilepsy. In India, it was prized by the powerful because of the belief that it could unmask traitors. It is thought that the emerald was the stone set in the Holy Grail (the cup used in the last supper of Jesus Christ).

Even though emerald mines were known in Europe since antiquity, the great prominence of this gem in the West coincides with the first Spanish conquests of Mexico by Cortés, and of Central and South American territory by other explorers. In the Mantu Valley in Peru, the Indians worshipped an emerald as big as an ostrich egg which they called the "Emerald Goddess" and the "Goddess of Chastity." The priests, taking advantage of the naiveté of the natives, made them hand over the emeralds found in the territory, declaring that the stones were beloved by the Goddess and that these gifts would safeguard the chastity of young girls and the fidelity of wives. During the Spanish invasion, the priests hid "the Emerald Goddess" and it was never found again.

Aquamarine was known as long ago as the time of Pliny the Elder, who simply called it "beryl." According to him, the most beautiful ones were those which became invisible when submerged in the sea. According to the philosopher Albertus Magnus, the stone guaranteed victory over enemies and drove troubles away. Pope Julius II had an aquamarine 2 inches (5 centimeters) in diameter mounted on his tiara.

FAMOUS STONES

Apart from Colombia, beautiful emeralds come from Brazil, South Africa, Zimbabwe, Zambia, Pakistan, and Russia. The best cut for both the emerald and the aquamarine is one with a rectangular table with the corners smoothed by a series of rectangular facets (also known as the emerald-cut), but other shapes are also common for aquamarines. One of the most beautiful and one of the largest (2,205 carats) emeralds is in the Crown Treasury of Vienna. Displayed in the Kunsthistorisches Museum of the same city is a small pomade jar (2,680 carats) made from an emerald cut by Dionisio Misseroni in 1642 for the Hapsburgs. A large raw crystal from Colombia, in a hexagonal prism shape and weighing 632 carats, is exhibited in the American Museum of Natural History in New York City.

A 1,400 carat emerald adorning one of the Grand Sultan's garments is exhibited in the Museum of the Grand Seraglio in Constantinople.

Some remarkable specimens of aquamarine, among which is one weighing 8,000 lbs. (3,629 kg), are preserved in the Institute of Mining in Leningrad. A raw aquamarine, weighing 216lbs. (98 kg) and of an azure-light green coloration, is on

Above, golden beryl, 1.37 c. (Brazil); Below, bixbite 0.85 c. (the United States).

display in the Museum of Natural History in Florence. An oval-cut stone weighing 875 carats is exhibited in the British Museum in London.

THE LESSER SISTERS

The emerald, green from the presence of chromium, and the aquamarine, blue from the presence of iron, are so famous that they overshadow the other colored varieties of beryl. The chromatic variations of beryl are still fine quality gems despite their lower cost and similarity to other gemstones.

Top: aquamarine gem.
Left: earrings with two Colombian drop-cut emeralds, seven carats each.

TOURMALINE

COMPOSITION: (Na, Ca)(Mg, Fe^{+2}, Fe^{+3}, Al, Mn, Li)$_3$Al$_6$(BO$_3$)$_3$(Si$_6$O$_{18}$)(OH,F)$_4$ hydrous alkaline and alkaline-earth aluminum boron cyclosilicate.

CRYSTALLINE FORM: trigonal system, ditrigonal pyramidal class. The crystals are elongated and striated with a prismatic form sometimes terminated differently at the ends of the three-fold or vertical axis; in parallel or radiating bacillary aggregates.

COLOR: black and blue (iron), various shades of yellow-brown (magnesium), pink (lithium); in addition, colorless, red, and polychrome green in either concentric bands or along the length of the crystal.

HARDNESS: 7-7.5.

SPECIFIC GRAVITY: 3-3.2.

LUSTER: vitreous to resinous.

GEOLOGIC OCCURRENCE: very widespread accessory mineral. The iron-rich members are common both in granites and in pegmatite and pneumatolytic veins associated with the granite; magnesium-rich members in metamorphic rocks and metasomatic deposits; because of its resistance to alteration, tourmaline is concentrated in clastic sediments.

ASSOCIATED MINERALS AND LOCALITIES: with quartz, topaz, lepidolite, spodumene, fluorite, apatite. Beautiful specimens come from Brazil, Sri Lanka, Namibia, Madagascar, and the United States.

USES: some clear and transparent varieties are cut for use as semi-precious gems. The most sought-after are rubellite (red) and indicolite (blue).

NOTES: Tourmaline refers to a group of minerals made up of isomorphous mixtures of very varied composition. The principal members are elbaite (pink, with lithium and aluminum), dravite (brown, with magnesium) and schorl (black, with iron and manganese). It fuses with difficulty, is insoluble in acids, and is generally not fluorescent. It is strongly pyroelectric and piezoelectric. (See box, page 67.)

▼ *Polychrome tourmaline (ca. x 1.5), Island of Elba, Livorno, Italy.*

DIOPTASE

COMPOSITION: $Cu_6(Si_6O_{18})\_6H_2O$ hydrous copper cyclosilicate.

CRYSTALLINE FORM: trigonal system, rhombohedral class. The stubby crystals have a prismatic form, sometimes terminated at the ends by rhombohedrons.

COLOR: emerald-green.

HARDNESS: 5.

SPECIFIC GRAVITY: 3.28-3.35.

LUSTER: vitreous.

GEOLOGIC OCCURRENCE: a secondary mineral, it forms in the oxidation zones of copper deposits.

ASSOCIATED MINERALS AND LOCALITIES: Found with limonite and chrysocolla. Beautiful specimens come from Namibia, the Congo, Chile, Zaire, and the United States.

USES: rare and beautiful, similar to the emerald but with low hardness and easily cleaved; it is worked to produce sought-after gems and is valued by collectors.

NOTES: the name is derived from the Greek for "through" and "optics" and means "to see through," which refers to the fractures that are present in the crystals. It is infusible and when heated turns black, swells up, and dehydrates. It is soluble in hydrochloric acid with the formation of silica gel, soluble in ammonia, and it is not fluorescent.

▼ *Dioptase (ca. x 0.7), Republic of South Africa.*

SILLIMANITE

COMPOSITION: $Al_2O(SiO_4)$ aluminum nesosilicate.

CRYSTALLINE FORM: orthorhombic system, dipyramidal class. Appears as elongated, acicular prismatic crystals and in silky fibrous aggregates (fibrolite variety); needle-shaped crystals are often included in other minerals such as quartz and feldspars.

COLOR: colorless, white, greenish, brown, gray.

HARDNESS: 6.5-7.5.

SPECIFIC GRAVITY: 3.23-3.24.

LUSTER: vitreous.

GEOLOGIC OCCURRENCE: accessory mineral in metamorphosed pelitic rocks (gneiss and hornfels) subjected to high-grade regional metamorphism or to contact metamorphism.

ASSOCIATED MINERALS AND LOCALITIES: with cordierite, quartz, biotite, corundum. Beautiful specimens come from Burma, India, Sri Lanka, the United States, Brazil, Germany etc.

USES: used in industry for the manufacture of refractory materials; use as a gem is rare, limited to blue and green varieties that show chatoyancy.

NOTES: the mineral was identified in 1824 by Bowen and named in honor of the American chemist Benjamin Silliman; the high-temperature polymorph of Al_2SiO_5. It is infusible, insoluble, and is not fluorescent. When heated to 2813°F (1545° C), it decomposes, forming mullite (aluminum silicate).

▼ *Sillimanite (ca. x 1), provenance unknown.*

ENSTATITE

COMPOSITION: $Mg_2Si_2O_6$ magnesium inosilicate (pyroxene).

CRYSTALLINE FORM: orthorhombic system, dipyramidal class. Stubby crystals with a prismatic form are rare; lamellar and radial fibrous aggregates, more common.

COLOR: colorless, yellowish gray, green, olive-brown.

HARDNESS: 5-6.

SPECIFIC GRAVITY: 3.2-3.9.

LUSTER: vitreous to pearly.

GEOLOGIC OCCURRENCE: common mineral in plutonic rocks (gabbros, peridotites) and volcanic rocks (basalts, andesites) with a basic or ultrabasic composition; in high-grade metamorphic rocks (granulites).

ASSOCIATED MINERALS AND LOCALITIES: with olivine, diopside, spinel. Beautiful specimens come from Burma, India, South Africa, the United States, Greenland, and Switzerland.

USES: some inclusion-rich specimens are cabochon-cut; the gems are chatoyant and exhibit a beautiful four-rayed asterism; of interest to researchers and collectors.

NOTES: the name is derived from the Greek for "opponent" or "contrary." With ortho-ferrosilite (iron-rich end member), constitutes a pyroxene solid solution series with bronzite (70-88% enstatite) and hypersthene (50-70% enstatite) as intermediate members. Orthorhombic pyroxenes, especially those rich in magnesium, alter to serpentine and talc. It fuses with difficulty, slowly decomposes in cold hydrofluoric acid, and is not fluorescent.

▼ *Enstatite (ca. x 1.5), Greenland.*

DIOPSIDE

COMPOSITION: $CaMg(Si_2O_6)$ calcium magnesium inosilicate (pyroxene).

CRYSTALLINE FORM: monoclinic system, prismatic class. Appears as prismatic crystals, sometimes striated, stubby; in bacillary or radial fibrous aggregates.

COLOR: green, whitish, yellowish, violet in the manganese-rich variety (violane), dark green from the presence of chromium (chrome diopside) or vanadium (lavrovite).

HARDNESS: 5-6.

SPECIFIC GRAVITY: 3.2-3.38.

LUSTER: vitreous.

GEOLOGIC OCCURRENCE: typical mineral of contact metamorphosed limestones and dolomites (marbles) and of basic rocks; also present in kimberlites and in skarns.

ASSOCIATED MINERALS AND LOCALITIES: with albite, epidote, wollastonite, serpentine, talc. Beautiful specimens come from Austria, Switzerland, Sweden, the Urals, and the United States.

USES: interest limited to researchers and collectors.

NOTES: the name is derived from the Greek for "double" and "vision" because the prism can be viewed in two directions. With hedenbergite (iron-rich end member), it forms a complete solid solution series and has intermediate compositions called salite (rich in magnesium) and ferrosalite (rich in iron). It fuses with difficulty, is soluble in hydrofluoric acid, and in ultraviolet light exhibits fluorescence, emitting dark violet light.

Diopside (ca. x 0.4), ▶ Madagascar.

HEDENBERGITE

COMPOSITION: $CaFe(Si_2O_6)$ calcium iron inosilicate (pyroxene).

CRYSTALLINE FORM: monoclinic system, prismatic class; rarely found as crystals with a prismatic and stubby habit; common in bacillary and radial aggregates.

COLOR: dark green to brown, to almost black.

HARDNESS: 6.

SPECIFIC GRAVITY: 3.5-3.6.

LUSTER: vitreous to submetallic.

GEOLOGIC OCCURRENCE: in contact metamorphic rocks that have other iron-bearing minerals and in skarns.

ASSOCIATED MINERALS AND LOCALITIES: with ilvaite, garnets, sulfides, epidote, and calcite. Beautiful specimens come from Japan, Nigeria, Australia, Norway, and Italy.

USES: interest limited to researchers and collectors.

NOTES: the name honors the Swedish chemist Ludwig Hedenberg. With diopside (magnesium end member), it forms a complete solid solution series (see notes under diopside). Fuses with relative ease, producing a black magnetic globule. It is soluble in hydrofluoric acid and is not fluorescent.

▼ *Hedenbergite (ca. x 2), Zillertal, Greiner, Tyrol, Austria.*

TOURMALINE

This is one of the showiest of stones, because of its variety of colors and tones and its striking pleochroism which makes the mineral appear a different color when from different directions.

HISTORY AND LEGEND

Known since antiquity, but probably often confused with other stones, tourmaline *(see page 65)* was described by Pliny in his monumental *Natural History* under the name "lynx stone" because in India it was a common belief that the mineral was formed from the urine of these animals. It was introduced in Europe at the

Watermelon tourmaline crystals.

beginning of the eighteenth century by the Dutch. Because of its ability to become electrically charged, tourmaline was used to clean ashes from pipes. It enjoyed a period of great popularity as a gem at the end of the nineteenth century. Today, it is undergoing a period of rediscovery and reevaluation.

PROPERTIES

Before the advent of synthetic materials tourmaline was used to make the famous tourmaline tongs, an instrument which polarized light. Polarization results from the strong absorption of light in one direction by some tourmalines, as manifested in their striking pleochroism.

COLORS AND VARIETIES

Because of its remarkable chromatic

variation, generally produced by the substitution of one element for another, tourmaline has a number of named varieties. Achroite (from the Greek for "without color") is colorless and rare. Dravite (from the Austrian river, the Drave) is yellow to dark yellow, to chestnut. The very rare indicolite is greenish blue to dark blue and its market name is the "Brazilian sapphire." Rubellite, with an etymology similar to that of the ruby (from the Latin *rubinus* from *rudere*, to turn red) is pink to red and marketed under the name "Siberian ruby." Schorl (from the Schorl region in Saxony) is black. The green, pistachio-green, and dark

Yellow tourmaline, 1c. (Mozambique).

Polychrome tourmaline, 4c. (Brazil).

green varieties are the most abundant and the most beautiful. They do not have specific names, but in the past were known as Brazilian emerald, chrysolite, or Ceylon (Sri Lanka) peridot.

The colors listed only hint at the remarkable variety of hues of this mineral. This chromatic variation reaches its height in polychromatic crystals such as Moor's head tourmalines, green at the base and black at the top, and watermelon tourmalines, pink or red on the inside and green on the outside.

AUGITE

COMPOSITION: $(Ca,Na)(Mg,Fe,Al,Ti)$ $(Si,Al)_2O_6$ calcium magnesium iron aluminum inosilicate (pyroxene).

CRYSTALLINE FORM: monoclinic system, prismatic class. Appears as crystals with a prismatic form, generally stubby; contact twins fairly common.

COLOR: green, yellow-green, brown, violet, black.

HARDNESS: 5-6.

SPECIFIC GRAVITY: 3.2-3.6.

LUSTER: vitreous to resinous.

GEOLOGIC OCCURRENCE: most widespread pyroxene in nature. Essential constituent of basic igneous rocks (gabbros, basalts), also common in ultrabasic rocks (peridotites, pyroxenites); found in high-grade metamorphic rocks (granulites), in contact metamorphic rocks, and in pyroclastic rocks.

ASSOCIATED MINERALS AND LOCALITIES: found with plagioclase, olivine, and others. Beautiful specimens come from South Africa, Greenland, the United States, and Italy.

USES: interest limited to researchers and collectors.

NOTES: fuses with relative difficulty, forming a black magnetic globule. It is insoluble in acids, with the exception of the titaniferous variety (titanaugite) which slowly dissolves in hot hydrochloric acid. It is not fluorescent and often occurs altered to uralite, a blue hornblende amphibole, or to chlorite; other alteration products are epidote and carbonates. Member of the diopside-hedenbergite and the aegirine-augite solid solution series.

▼ *Augite (ca. x 1.5), Ariccia, Rome, Italy.*

AEGIRINE

COMPOSITION: $NaFe^{3+}(Si_2O_6)$ sodium trivalent iron inosilicate (pyroxene).

CRYSTALLINE FORM: monoclinic system, prismatic class. Crystals have a prismatic form, elongated and sometimes striated, irregularly terminated or pointed; in radial fibrous aggregates.

COLOR: green, black; crystals generally have hourglass-shaped zoning (visible only in a thin section under a polarizing microscope).

HARDNESS: 6-6.5.

SPECIFIC GRAVITY: 3.5-3.6.

LUSTER: vitreous to resinous.

GEOLOGIC OCCURRENCE: common mineral in alkaline igneous rocks (syenites, nepheline syenites, phonolites); also found in high-pressure, regionally metamorphosed rocks (glaucophane schists and granulites).

ASSOCIATED MINERALS AND LOCALITIES: with sodium-rich amphibole, alkalifeldspar, nepheline, sodalite. Beautiful specimens come from Canada, Greenland, and the United States.

USES: interest limited to researchers and collectors.

NOTES: fuses easily, forming a weakly magnetic globule; during heating, colors the flame yellow (sodium). It is slightly corroded by acids, is not fluorescent, and forms a solid solution series with augite (aegirine-augite).

▼ *Aegirine (ca. x 1), Egersund, Norway.*

JADEITE

COMPOSITION: $NaAl(Si_2O_6)$ sodium aluminum inosilicate.

CRYSTALLINE FORM: monoclinic system, prismatic class. Crystals are extremely rare; generally appears in irregular compact masses with a waxy look in which the crystals are in fibrous or lamellar aggregates.

COLOR: colorless, white with emerald-green spots, green, greenish blue, dark blue to black in the iron-rich variety (chloromelanite).

HARDNESS: 6.5.

SPECIFIC GRAVITY: 3.3-3.5.

LUSTER: subvitreous to pearly.

GEOLOGIC OCCURRENCE: found in low-temperature, high-pressure metamorphic rocks (blue schist facies).

ASSOCIATED MINERALS AND LOCALITIES: with albite, glaucophane, lawsonite, pumpellyite. Beautiful specimens come from Burma, China, Japan, Mexico, Venezuela, Costa Rica, the United States, and New Zealand.

USES: also known as jade, a commercial term for a prized material used, especially in the Orient, for the production of ornamental objects of great value.

NOTES: the name is derived from the Spanish *piedra de ijada* or *hijada* (stone of the flank or loins) because of its supposed efficacy in the treatment of kidney ailments. Not all objects that are called jade are actually jadeite. Some are made from other less valuable minerals such as nephrite (variety of actinolite). Fuses with relative ease, forming an almost transparent globule. It is insoluble in acids, and in ultraviolet light exhibits a faint bluish-gray fluorescence. *(See box, page 70.)*

▼ *Jadeite (ca. x 2), oriental craftsmanship, Mineralogy Museum, Bologna, Italy.*

SPODUMENE

COMPOSITION: $LiAl(Si_2O_6)$ lithium aluminum inosilicate (pyroxene).

CRYSTALLINE FORM: monoclinic system, prismatic class. Crystals are prismatic, striated along their length, and can reach metric proportions; also appears in microcrystalline masses and bacillary aggregates.

COLOR: colorless, white, grayish, violet-pink from the presence of manganese (kunzite variety), emerald-green from the presence of chromium (hiddenite variety).

HARDNESS: 6.5-7.

SPECIFIC GRAVITY: 3.2.

LUSTER: vitreous.

GEOLOGIC OCCURRENCE: typical mineral of lithium-rich granitic pegmatites.

ASSOCIATED MINERALS AND LOCALITIES: found with quartz, albite, lepidolite, beryl, tourmaline, petalite. Beautiful specimens come from Brazil, Madagascar, the United States, Canada, Mexico, Sweden etc.

USES: important mineral for the extraction of lithium which is used in the metallurgical and pharmaceutical industries and in the manufacture of ceramics and lubricants. Gems are obtained from the clear, colored varieties (kunzite, hiddenite).

NOTES: the name is derived from the Greek for "ashes" and refers to the mineral's generally pale coloration. Fuses easily, coloring the flame a lovely carmine-red (lithium). It is insoluble in acids, is not fluorescent, and when exposed to ultraviolet light the kunzite variety exhibits an intense reddish-yellow fluorescence.

▼ *Spodumene: kunzite variety (ca. x 1.5), Brazil.*

TREMOLITE

COMPOSITION: $Ca_2Mg_5(Si_8O_{22})(OH)_2$ hydrous calcium magnesium inosilicate (amphibole).

CRYSTALLINE FORM: monoclinic system, prismatic class. Appears as elongated, prismatic crystals, commonly in fibrous aggregates with a radial structure (amphibole asbestos).

COLOR: colorless, white, gray-green.

HARDNESS: 5-6.

SPECIFIC GRAVITY: 3.

LUSTER: vitreous to silky in fibrous aggregates.

GEOLOGIC OCCURRENCE: in contact or low-grade, regionally metamorphosed limestones (marbles, tremolite schists).

ASSOCIATED MINERALS AND LOCALITIES: with calcite, magnesite, talc, serpentine. Beautiful specimens come from Switzerland, the Urals, Italy, and the United States.

USES: interest limited to researchers and collectors.

NOTES: the name was assigned by the physicist Pini, after the Tremola Valley in Switzerland where the mineral is particularly abundant. Fuses with relative difficulty, turning red during heating. It is soluble only in hydrofluoric acid, and it is not fluorescent.

▼ *Tremolite (ca. x 0.6), Switzerland.*

ACTINOLITE

COMPOSITION: $Ca_2(Mg,Fe)_5(Si_8O_{22})(OH)_2$ hydrous calcium magnesium divalent iron inosilicate (amphibole).

CRYSTALLINE FORM: monoclinic system, prismatic class. Appears as prismatic, elongated crystals, sometimes striated; in fibrous aggregates.

COLOR: brilliant green, gray-green.

HARDNESS: 5-6.

SPECIFIC GRAVITY: 3-3.4.

LUSTER: vitreous.

GEOLOGIC OCCURRENCE: very widespread metamorphic mineral; present in contact metamorphosed calcareous rocks (lime schists, marbles), and in basic rocks metamorphosed under low and medium-grade conditions (prasinites, amphibolites); in gabbros and diabases by the alteration of pyroxenes.

ASSOCIATED MINERALS AND LOCALITIES: with epidote, chlorite, hornblende. Beautiful specimens come from Austria, Poland, Italy, Germany, the United States etc.

USES: interest limited to researchers and collectors. Nephrite, a very tough, greenish, microcrystalline variety is used in ornamental objects which are less valuable than those made of jade (jadeite, a pyroxene).

NOTES: the name is derived from the Greek for "ray" and "stone." Actinolite is the intermediate member in the solid solution series with the endmembers tremolite and ferroactinolite ($Ca_2Fe_5(Si_8O_{22})$ $(OH)_2$).

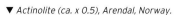

▼ *Actinolite (ca. x 0.5), Arendal, Norway.*

HORNEBLENDE

COMPOSITION: $(Ca, Na)_2(Mg,Fe,Al)_5(AlSi)_8$ $O_{22}(OH,F)_2$ hydrous calcium magnesium divalent iron aluminum inosilicate (amphibole).

CRYSTALLINE FORM: monoclinic system, prismatic class. Appears as prismatic crystals with a needle-shaped or stubby habit; in fibrous or parallel aggregates.

COLOR: green, green-brown, black.

HARDNESS: 5-6.

SPECIFIC GRAVITY: 2.9-3.4.

LUSTER: vitreous to resinous.

GEOLOGIC OCCURRENCE: very widespread mineral: in medium-grade (amphibolites) and high-grade (granulites) metamorphic rocks; in intermediate composition plutonic rocks (gabbros, diorites), but also in ultrabasic (hornblendites) and acid plutonic rocks; in volcanic rocks (basalts).

ASSOCIATED MINERALS AND LOCALITIES: with anthophyllite, cummingtonite, plagioclase, olivine, pyroxene. Beautiful specimens come from Japan, Greenland, the United States, Canada, Finland, and Germany.

USES: interest limited to researchers and collectors.

NOTES: the name is derived from the German *horn* and *blenden*, and refers both to its color and to the fact that it is not an economically important mineral. Horneblende's composition is variable since in many cases sodium, potassium, trivalent iron and aluminum partially replace other elements in the formula. Varieties of hornblende include: edenite, which is light green in color and iron-poor, and basaltic hornblende, which is brown in color, rich in ferric iron, and occurs in volcanic rocks (andesites).

▼ *Horneblende (ca. x 1), Predazzo, Trento, Italy.*

GLAUCOPHANE

COMPOSITION: $Na_2Mg_3Al_2(Si_8O_{22})(OH)_2$ hydrous sodium magnesium aluminum inosilicate (amphibole).

CRYSTALLINE FORM: monoclinic system, prismatic class. Crystals are prismatic, striated along the length; in fibrous and bacillary aggregates.

COLOR: colorless, lavender-blue, dark blue.

HARDNESS: 6.

SPECIFIC GRAVITY: 3.08-3.30.

LUSTER: vitreous to pearly.

GEOLOGIC OCCURRENCE: mineral present in basaltic rocks subjected to low-temperature, high-pressure regional metamorphism; essential component of glaucophane schists; also present in eclogites.

ASSOCIATED MINERALS AND LOCALITIES: with lawsonite, jadeite, epidote, pumpellyite, omphacite. Beautiful specimens come from Italy, Norway, France, the United States etc.

USES: interest limited to researchers and collectors.

NOTES: the name is derived from the Greek for "blue" and "to appear" and refers to the mineral's color. Constitutes a complete solid-solution series with riebeckite by the substitution of divalent iron for magnesium and trivalent iron for aluminum; intermediate members are called crossite. It fuses fairly easily, is insoluble in acids, and is not fluorescent.

▼ *Glaucophane (ca. x 1.5), Piedmont, Italy.*

RIEBECKITE

COMPOSITION: $Na_2Fe_3^{2+}Fe_2^{3+}(Si_8O_{22})$ $(OH)_2$ hydrous sodium di- and trivalent iron inosilicate (amphibole).

CRYSTALLINE FORM: monoclinic system, prismatic class. Crystals are prismatic, elongated; in lamellar or bacillary aggregates; in finely fibrous aggregates (crocidolite variety).

COLOR: from dark blue to black.

HARDNESS: 5.5-6.

SPECIFIC GRAVITY: 3.38.

LUSTER: vitreous to silky.

GEOLOGIC OCCURRENCE: mineral present in low-temperature and medium-high-pressure regional metamorphic rocks; as an accessory mineral in sodium-rich acid plutonic rocks (granites and alkaline syenites); crocidolite is found in veins associated with banded iron formations (BIF).

ASSOCIATED MINERALS AND LOCALITIES: found with quartz, alkali feldspar. Beautiful specimens come from the United States and Spain.

USES: interest limited to researchers and collectors. In some cases the crocidolite variety is replaced by quartz (tiger's eye quartz); the fibrous texture gives tiger's eye its beautiful chatoyancy.

NOTES: the name honors the mineralogist Emil Riebeck. Constitutes a complete solid-solution series with glaucophane (see notes for glaucophane). It fuses easily and colors the flame green (sodium). It is unaffected by acids and is not fluorescent.

▼ *Riebeckite: crocidolite variety (ca. x 0.4), Republic of South Africa.*

WOLLASTONITE

COMPOSITION: $Ca(SiO_3)$ calcium inosilicate.

CRYSTALLINE FORM: triclinic system, pinacoidal class. Crystals are elongated, tabular in form; sometimes appears in fibrous or acicular masses.

COLOR: generally white, colorless, pale green, gray.

HARDNESS: 4.5–5.

SPECIFIC GRAVITY: 2.87–3.09.

LUSTER: vitreous to pearly, to silky when fibrous.

GEOLOGIC OCCURRENCE: typical mineral of contact metamorphosed limestones; rarely, in low to medium-temperature, low-pressure regional metamorphic rocks (amphibolites).

ASSOCIATED MINERALS AND LOCALITIES: with tremolite, diopside, anorthite, scapolite, calcite. Beautiful specimens come from Mexico, the United States, Germany, France, and Romania.

USES: used in the ceramics industry.

NOTES: the name honors the English chemist and mineralogist W. H. Wollaston (1766–1828). Two other polymorphs of $Ca(SiO_3)$ exist: parawollastonite (monoclinic), like wollastonite, stable at temperatures lower than 2059°F (1126° C); pseudowollastonite (triclinic), stable above 2059°F (1126° C). It fuses easily, with the formation of silica gel. It is soluble in strong acids and is not fluorescent.

▼ *Wollastonite (ca. x 1.5), Monte Somma, Naples, Italy.*

THE STONE OF IMMORTALITY

JADEITE

Jadeite is a sodium pyroxene which, in microcrystalline masses, is the principal component of the material commonly known as jade. Today, fine quality jade is fairly rare.

MATERIALS

Other materials similar in appearance and color to jade are found on the market under this name. However, these materials have a different composition and are more common. Generally, nephrite (an amphibole) and, more rarely, serpentine (a phyllosilicate) are used in place of jade. Even in antiquity these substitutes were used as long as they came from particularly beautiful fragments. This may imply an unwitting confusion resulting from the remarkable similarities between nephrite and jadeite (see page 68). Both minerals are characterized by a high degree of toughness, similar to that of metals. It was not until 1863 that a French mineralogist recognized that jade objects do not all consist of the same mineral. Today, these ornamental minerals are known as jadeite-jade and nephrite-jade.

HISTORY AND LEGEND

Jade has been known since antiquity, as is demonstrated by paleolithic evidence from Europe and New Zealand. In China, a multitude of virtues (charity, modesty, courage, justice, wisdom) were attributed to jade. It was also thought to give the gift of immortality. The philosopher Khi Van Gungh asserted that jade's qualities reflected those of man: its softness corresponded to benevolence; its smoothness, to knowledge; its stability, to justice; its harmlessness, to virtue; its lack of stains, to purity; its toughness, to constancy; the visibility of its defects, to frankness; the ability to work it without staining it, to rectitude; and finally, its ability to emit sounds, to internal harmony.

The ancient Chinese believed that jade could restore life and used it for burial garments. At the famous archaeological site of Manch'eng two bodies from about two thousand years ago were found wrapped in garments made of small pieces of jade sewn together with gold threads.

In Central America, jade was considered more precious than gold and a man's worth was judged on the amount and quality of the stones he possessed. Jade was commonly used to adorn the statues of divinities and funeral vestments.

Top: small imperial jade amulet: this is the rarest type, with an intense emerald-green color that may be translucent at the edges. Left: leaves and little birds in an oriental sculpture that takes advantage of the many shades of jade.

RHODONITE

COMPOSITION: $(Mn,Fe,Mg)(SiO_3)$ manganese divalent iron magnesium inosilicate.

CRYSTALLINE FORM: triclinic system, pinacoidal class. Crystals are rare and have a tabular shape; generally appears in compact masses with black veins from the oxidation of manganese.

COLOR: pink, brown-red.

HARDNESS: 5.5–6.5.

SPECIFIC GRAVITY: 3.40–3.68.

LUSTER: vitreous.

GEOLOGIC OCCURRENCE: found as a metamorphic and metasomatic product in manganese deposits.

ASSOCIATED MINERALS AND LOCALITIES: with manganite, haussmanite, braunite, pyrolusite, quartz, calcite. Beautiful specimens come from Brazil, Australia, India, Japan, the Urals, Sweden, and Italy.

USES: used as a facing stone or worked to form very beautiful objects; colored and transparent varieties are cabochon-cut and used for necklaces.

NOTES: the name is derived from the Greek for "rose" and refers to the mineral's color. It fuses easily and darkens during heating as it also does when exposed to air as a result of the oxidation of manganese. It is insoluble in acids, slowly decomposes in hydrochloric acid, and is not fluorescent.

▼ *Rhodonite (ca. x 1), Sussex Co., Franklin Furnace, New Jersey, the United States.*

APOPHYLLITE

COMPOSITION: $KCa_4(Si_4O_{10})_2F \cdot 8H_2O$ hydrated fluorine potassium calcium phyllosilicate.

CRYSTALLINE FORM: tetragonal system, ditetragonal class. Crystals combine the prism with the dipyramid and the pinacoid; thus the crystal's appearance depends on which of these forms is best developed.

COLOR: colorless, white, pink, yellow, green.

HARDNESS: 4.5-5.

SPECIFIC GRAVITY: 2.33-2.37.

LUSTER: vitreous to pearly on pinacoid faces.

GEOLOGIC OCCURRENCE: secondary mineral lining the vesicles of basaltic rocks; product of hydrothermal alteration in skarns.

ASSOCIATED MINERALS AND LOCALITIES: with stilbite, scolecite, zeolites, prehnite, datolite, calcite. Beautiful specimens come from Greenland, Iceland, Germany, Canada, Brazil etc.

USES: interest limited to researchers and collectors.

NOTES: the name was assigned by the French mineralogist Haüy and is derived from the Greek for "from" and "leaf," because of the mineral's tendency to flake when heated. Fuses easily and colors the flame violet (potassium). It is soluble in hydrochloric acid with the production of silica gel, and it is not fluorescent.

▼ *Apophyllite (ca. x 1.5), Poona, India.*

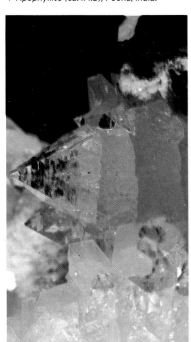

PREHNITE

COMPOSITION: $Ca_2Al(AlSi_3O_{10})(OH)_2$ hydrous calcium aluminum phyllosilicate.

CRYSTALLINE FORM: orthorhombic system, pyramidal class. Crystals are rare, commonly tabular, sometimes forming fan-shaped or rosette-shaped aggregates. Generally appears in mammilary, botryoidal, or stalactitic microcrystalline masses.

COLOR: colorless, white, gray, yellowish pale green.

HARDNESS: 6-6.5.

SPECIFIC GRAVITY: 2.80-2.95.

LUSTER: vitreous.

GEOLOGIC OCCURRENCE: secondary mineral lining cavities in basic volcanic rocks; in veins in plutonic rocks; common in burial metamorphism of graywacke and argillites, and in low-grade regional metamorphic settings.

ASSOCIATED MINERALS AND LOCALITIES: with zeolites and calcite in volcanic cavities; associated with quartz, pumpellyite, lawsonite, laumontite, in metamorphic rocks. Beautiful specimens come from the United States, South Africa, France, and Italy.

USES: interest limited to researchers and collectors.

NOTES: the name honors the Dutch army Colonel Prehn (1790), who found the mineral on the Cape of Good Hope. It fuses easily, forming a glassy mass with a bubbly appearance and a white-yellowish color. It slowly dissolves in hydrochloric acid, and is not fluorescent.

▼ *Prehnite (ca. x 1),*
Siusi Alps, Bolzano,
Italy.

PYROPHYLLITE

COMPOSITION: $Al_2(Si_4O_{10})(OH)_2$ hydrous aluminum phyllosilicate.

CRYSTALLINE FORM: monoclinic system, prismatic class. Never appears as isolated crystals; generally in lamellar, foliated, radial aggregates, or in compact masses (agalmatolite).

COLOR: white, yellowish, light green, brown-green.

HARDNESS: 1.5.

SPECIFIC GRAVITY: 2.79.

LUSTER: pearly to greasy.

GEOLOGIC OCCURRENCE: found in aluminum-rich rocks regionally metamorphosed at low-grade (metapelites, metabauxites); hydrothermal alteration product of muscovite and feldspars.

ASSOCIATED MINERALS AND LOCALITIES: with quartz, micas, kyanite. Beautiful specimens come from South Africa, China, and the United States.

USES: utilized in the paper, textile, and ceramics industries; as a thermal and electrical insulator; the compact variety (agalmatolite) is worked to create ornamental objects.

NOTES: the name is derived from the Greek for "fire" and "leaf," from the mineral's tendency to flake when heated. Infusible, it is slowly soluble in sulfuric or hydrofluoric acids; greasy to the touch, and it is not fluorescent. Distinguishable from talc by X-ray diffraction or by chemical analysis; a specimen of pyrophillite, soaked in a solution of cobalt nitrate and heated red-hot, takes on a dark blue coloration, while talc appears light violet.

▲ *Pyrophyllite (ca. x 0.7),*
Belgium.

TALC

COMPOSITION: $Mg_3(Si_4O_{10})(OH)_2$ hydrous magnesium phyllosilicate.

CRYSTALLINE FORM: monoclinic system, prismatic class. Never appears as isolated crystals; generally in pseudohexagonal leaves, in scaly, foliated, stellate aggregates or in felted compact masses (steatite or soapstone).

COLOR: white, greenish, gray, brownish.

HARDNESS: 1.

SPECIFIC GRAVITY: 2.78.

LUSTER: pearly to greasy.

GEOLOGIC OCCURRENCE: product of medium-temperature hydrothermal alteration of magnesium silicates (olivine, serpentine) present in ultrabasic rocks, or from metasomatism of dolomitic marbles; essential constituent of talc schists.

ASSOCIATED MINERALS AND LOCALITIES: with tremolite, anthophyllite, dolomite, quartz. Large masses are found in Canada, the United States, India, Korea, Australia, Austria, and Spain.

USES: as an electrical and thermal insulator; in the paper, textile, cosmetic and dye industries; ornamental objects are obtained from the compact variety.

NOTES: infusible, it is insoluble in acids, and is not fluorescent. It is a poor conductor of heat and greasy to the touch. (cf. pyrophyllite).

▼ *Talc (ca. x 1), Russia.*

MUSCOVITE

COMPOSITION: $KAl_2(AlSi_3O_{10})(OH)_2$ hydrous potassium aluminum phyllosilicate (white mica).

CRYSTALLINE FORM: monoclinic system, prismatic class. Appears as tabular crystals with a pseudohexagonal or triangular outline, sometimes of large size, with striations on the faces of the prism; in scaly, lamellar, foliated aggregates.

COLOR: silver-white, white-yellow, brown from inclusions of needles of rutile.

HARDNESS: 2–2.5.

SPECIFIC GRAVITY: 2.83.

LUSTER: vitreous to silky.

GEOLOGIC OCCURRENCE: very common; chiefly in low- or medium- to high-grade metamorphic rocks (schists, phyllites, amphibolites); in silica- and aluminum-rich plutonic rocks (peraluminous granites) and in pegmatites; less common in acid extrusive rocks (rhyolites); as an accessory in sedimentary rocks (sandstones, marls).

ASSOCIATED MINERALS AND LOCALITIES: found with a large variety of other minerals. Beautiful specimens come from Brazil, India, Canada, and the United States.

USES: as an electrical and thermal insulator; in the paper and rubber industries; as a lubricant; used in ceramic products.

NOTES: muscovite and biotite are the most common micas and are called, respectively, "white mica" and "black mica" (the name is derived from the Latin *micare*). Muscovite displays the phenomenon of polytypism; that is, it crystallizes in different lattice types while maintaining the same basic atomic arrangement and chemical composition. Several varieties are produced by the partial replacement of aluminum by other elements: fuchsite (chromium), roscoelite (vanadium), and phengite (iron, magnesium, and silicon).

▼ *Muscovite (ca. x 0.5), Valtellina, Alpe Sommafiume M.ra, Sondrio, Italy.*

PARAGONITE

COMPOSITION: $NaAl_2(AlSi_3O_{10})(OH)_2$ hydrous sodium aluminum phyllosilicate (white mica).

CRYSTALLINE FORM: monoclinic system, prismatic class. Appears in small laminae with a pseudohexagonal outline; in lamellar, scaley aggregates.

COLOR: colorless, light yellow.

HARDNESS: 2.5.

SPECIFIC GRAVITY: 2.89.

LUSTER: vitreous to pearly.

GEOLOGIC OCCURRENCE: found in low to medium-temperature, medium-pressure regional metamorphic rocks (schists, phyllites, gneisses).

ASSOCIATED MINERALS AND LOCALITIES: with quartz, muscovite, pyrophyllite, staurolite, kyanite. Beautiful specimens come from Italy and Switzerland.

USES: interest limited to researchers and collectors.

NOTES: belongs to the muscovite family and is a fairly common constituent of rocks; displays polytypism (see notes under muscovite). Together with muscovite, it constitutes a complete solid solution series at high temperatures, but not at low to medium temperatures where substitution of potassium and sodium is more difficult. It fuses with difficulty, is insoluble in acids, and is not fluorescent.

▲ *Paragonite (ca. x 0.8), Pizzo Forno, Canton Ticino, Switzerland.*

ZINNWALDITE

COMPOSITION: $K(Li,Al,Fe)_3(AlSi_3O_{10})(OH,F)_2$ hydrous potassium lithium trivalent iron aluminum phyllosilicate (white mica).

CRYSTALLINE FORM: monoclinic system, domatic class. Appears as pseudohexagonal lamellae or in parallel lamellar or radially oriented aggregates.

COLOR: silver-gray to brown.

HARDNESS: 2.5–4.

SPECIFIC GRAVITY: 2.97.

LUSTER: vitreous to pearly.

GEOLOGIC OCCURRENCE: rare mineral found in granitic pegmatites and in tin-rich veins.

ASSOCIATED MINERALS AND LOCALITIES: with spodumene, topaz, beryl, tourmaline, monazite, fluorite, cassiterite. Beautiful specimens come from Canada, Greenland, the United States, Madagascar, Germany etc.

USES: useful mineral for the extraction of lithium.

NOTES: exhibits the phenomenon of polytypism (see notes under muscovite). It fuses easily, forming (when iron-rich) a lightly magnetic, blackish globule; during heating, colors the flame red from the presence of lithium. It is corroded by strong acids, and it is not fluorescent.

▼ *Zinnwaldite (ca. x 1.5), Zinnwald, Germany.*

BIOTITE

COMPOSITION: $K(Mg,Fe)_3(AlSi_3O_{10})(OH,F)_2$ hydrous potassium magnesium divalent iron phyllosilicate (mica).

CRYSTALLINE FORM: monoclinic system, prismatic class. Appears as tabular crystals, sometimes of remarkable size, with a pseudohexagonal outline; in irregular laminae and in scaly aggregates.

COLOR: black, brown, dark green, yellow.

HARDNESS: 2.5–3.

SPECIFIC GRAVITY: 2.9–3.3.

LUSTER: vitreous.

GEOLOGIC OCCURRENCE: very widespread mineral; common in many igneous rocks and regional metamorphic rocks, accessory in detrital sedimentary ones.

ASSOCIATED MINERALS AND LOCALITIES: given its occurrence in many different rock types, it is found with a variety of minerals. Beautiful specimens come from Brazil, the United States, Greenland, Madagascar, Scandinavia etc.

USES: interest limited to researchers and collectors.

NOTES: the name honors the French physicist and astronomer J.-B. Biot. Its composition is more complex than described above since it may contain small amounts of other elements such as titanium, manganese, and trivalent iron. Exposed to air and water, it oxidizes easily; it alters to other minerals, most commonly chlorite. Biotite has three polytypes, and constitutes a solid solution with phlogopite in which iron is replaced by magnesium. It fuses with difficulty; is soluble only in hot concentrated sulfuric acid, and is not fluorescent.

▼ *Biotite (ca. x 0.8), Mount Vesuvius, Naples, Italy.*

KAOLINITE

COMPOSITION: $Al_2(Si_2O_5)(OH)_4$ hydrous aluminum phyllosilicate (clay mineral).

CRYSTALLINE FORM: triclinic system, pinacoidal class. Crystals are microscopic and have a pseudohexagonal lamellar form; generally appears in earthy compact masses, greasy to the touch.

COLOR: white, gray, yellowish, brown.

HARDNESS: 2-2.5.

SPECIFIC GRAVITY: 2.63.

LUSTER: pearly when crystalline, otherwise dull earthy.

GEOLOGIC OCCURRENCE: very common mineral: found in clay-rich rocks, in igneous rocks by low-temperature hydrothermal alteration of feldspars and other aluminum-rich silicates; the largest deposits are in clays of lagoons or deltaic environments.

ASSOCIATED MINERALS AND LOCALITIES: found with other clay minerals. Large deposits are found in China, France, Germany, England etc.

USES: basic material for the manufacture of porcelains and fine ceramics; used in the paper industry.

NOTES: principal constituent of the clay minerals group to which it lends its name, and which includes the polymorphs dickite and nacrite, both rare; halloysite is the hydrated form of kaolinite. It is infusible, dehyrates at 734°-842° F (390-450° C), and mixed with water it becomes plastic and can be modeled. It is soluble only in hot concentrated sulfuric acid and is not fluorescent.

▼ *Kaolinite (ca. x 1), France.*

SERPENTINE

COMPOSITION: $Mg_3(Si_2O_5)(OH)_4$ hydrous magnesium phyllosilicate.

CRYSTALLINE FORM: monoclinic or orthorhombic system. Occurs as three polymorphs: antigorite (monoclinic and orthorhombic), with a lamellar appearance or in scaly aggregates; chrysotile (monoclinic and orthorhombic), in silky fibers; lizardite (orthorhombic), finely fibrous in compact aggregates.

COLOR: green, blackish green, yellowish white, yellowish brown, gray.

HARDNESS: 2.5-3.5.

SPECIFIC GRAVITY: 2.55-2.6.

LUSTER: silky to resinous.

GEOLOGIC OCCURRENCE: a product of metamorphism by hydrothermal alteration of magnesium silicates (olivine, pyroxene); lining cavities or in incrustations in ultrabasic rocks; fundamental constituent of sepentinite.

ASSOCIATED MINERALS AND LOCALITIES: with talc, brucite, chlorite, magnetite. Large serpentine masses are found in orogenic mountain chains; Canada, the United States, the Urals and Italy are among the principal localities.

USES: the fibrous variety (serpentine asbestos) was used in the past as a thermal and acoustic insulator. At present, because it is highly noxious to humans, its use has been prohibited.

NOTES: the name alludes to the surface appearance of serpentine-bearing rocks: they are similar to snakeskin. The prized variety, called "noble serpentine," is made up of a compact, translucent, microcrystalline mass with a pale olive-green color used in the past for necklaces and small ornamental objects. It is generally infusible; only antigorite fuses with difficulty at its edges. It is soluble in acids and is not fluorescent.

▼ *Noble serpentine (ca. x 1), Val Malenco, Sondrio, Italy.*

CHRYSOTILE

COMPOSITION: $Mg_3(Si_2O_5)(OH)_4$ hydrous magnesium phyllosilicate.

CRYSTALLINE FORM: two polymorphs of the mineral exist, one monoclinic and the other orthorhombic; in flexible fibers sometimes elongated in thin, very delicate threads (asbestos or amianthus).

COLOR: white, green-gray, yellowish.

HARDNESS: 3-4.

SPECIFIC GRAVITY: 2.51.

LUSTER: silky.

GEOLOGIC OCCURRENCE: secondary mineral (see serpentine); lines fractures in ultrabasic rocks with densely intertwined fibers either parallel or perpendicular to the fracture.

ASSOCIATED MINERALS AND LOCALITIES: with antigorite, talc, brucite, chlorite, demantoid, magnetite. Chrysotile deposits are found in serpentine rocks in Canada, Zimbabwe, Cyprus, the United States, and the Urals.

USES: interest limited to researchers and collectors.

NOTES: the name is derived from the Greek for "golden" and "fibrous" and refers to the mineral's appearance. Because of its insulating properties, it was used in the past in many industrial sectors. However, because inhalation of asbestos dust is very dangerous for humans (causing asbestosis and pleural mesothelioma), its use in Western countries is now prohibited; synthetic asbestos is used as a substitute. It is infusible, soluble in strong acids with the formation of silica gel, and is not fluorescent.

▼ *Chrysotile: amianthus variety (ca. x 0.4), Val Malenco, Sondrio, Italy.*

CHLORITE

COMPOSITION: $(Mg,Fe,Al)_6(Si,Al)_4O_{10}(OH)_8$ hydrous magnesium iron aluminum phyllosilicate.

CRYSTALLINE FORM: monoclinic system, prismatic class. Crystals appear in laminae with a pseudohexagonal outline, in compact, scaly, or parallel lamellar aggregates.

COLOR: colorless, green-gray, bluish green, brown (oxidized varieties).

HARDNESS: 2-2.5.

SPECIFIC GRAVITY: 2.64-2.74.

LUSTER: vitreous to pearly.

GEOLOGIC OCCURRENCE: very common mineral in low to medium-grade metamorphic rocks (chlorite schists, phyllites, prasinites); in igneous rocks from hydrothermal alteration of ferromagnesian silicates (biotite, amphiboles); in argillaceous rocks and in iron-rich sediments.

ASSOCIATED MINERALS AND LOCALITIES: with muscovite, stilbite, actinolite, albite, epidote, quartz. Beautiful specimens come from Italy, Switzerland, Austria, and the United States.

USES: interest limited to researchers and collectors.

NOTES: the name is derived from the Greek for "green" and refers to the mineral's color. The chlorites are a group of minerals of diverse composition: the most common varieties are clinochlore and penninite (magnesium), pennantite (manganese), and chamosite (iron). Infusible, it generally tends to flake during heating. It is soluble in hot sulfuric acid (iron-rich varieties also in hydrochloric acid), and it is not fluorescent.

▼ *Chlorite: clinochlore variety (ca. x 1.5), Brewster, New York, the United States.*

QUARTZ

COMPOSITION: SiO_2 silica group mineral, tectosilicate.

CRYSTALLINE FORM: trigonal system, trapezohedral class. Crystals are generally prismatic, hexagonal, terminated with a double rhombohedron so as to simulate a hexagonal dipyramid; faces of the prisms are sometimes striated; commonly forms microcrystalline compact masses or veins.

COLOR: colorless, white, pink, violet, yellow, brown, red, green.

HARDNESS: 7.

SPECIFIC GRAVITY: 2.65.

LUSTER: vitreous.

GEOLOGIC OCCURRENCE: ubiquitous mineral, constitutes about 12% by volume of the earth's crust. Found in acidic igneous, metamorphic, and sedimentary rocks; gangue mineral in mining deposits.

ASSOCIATED MINERALS AND LOCALITIES: found in association with numerous minerals. Very beautiful specimens come from the Alps and from Brazil, Madagascar, the United States, Russia, China, and Japan.

USES: constitutes the primary material of the glass, enamel, abrasive (silicon carbide), refractory, machine (special supports made of agate), and advanced technology industries that exploit its numerous properties: piezoelectricity, rotary polarization, transparent to X-rays, etc. Also used in jewelry making for semiprecious objects.

NOTES: The name is probably of Germanic origin. It is infusible, soluble only in hydrofluoric acid, and is not fluorescent. (*See box, page 75.*)

◄ *Smoky quartz (ca. x 1), Saint Gotthard, Switzerland.*

TRIDYMITE

COMPOSITION: SiO_2 silica group mineral, tectosilicate.

CRYSTALLINE FORM: orthorhombic system (alpha-tridymite) or hexagonal system (beta-tridymite). Appears in laminae with a hexagonal outline; commonly joined in radial threelings with a spherulitic appearance.

COLOR: colorless, white.

HARDNESS: 6.5-7.

SPECIFIC GRAVITY: 2.28.

LUSTER: vitreous to pearly.

GEOLOGIC OCCURRENCE: as a sublimate in cavities and vesicles in acidic volcanic rocks (rhyolites, obsidian, trachytes); common in impure limestones or arkose metamorphosed by intrusions of basic magmas.

ASSOCIATED MINERALS AND LOCALITIES: with sanidine, sometimes with augite and fayalite. Beautiful specimens come from Italy, France, Germany, Mexico etc.

USES: interest primarily limited to researchers and collectors. Artificially produced tridymite is employed in the manufacture of refractories and in ceramics resistant to high temperatures.

NOTES: Tridymite is a polymorph of the SiO_2 compound. There are two forms of tridymite: alpha-tridymite which is stable at low temperature, and beta-tridymite which is stable at high temperature. Alpha-tridymite originally forms as beta-tridymite and inverts to the lower temperature form after cooling. It is infusible, soluble in hydrofluoric acid, and is not fluorescent.

▼ *Tridymite in trachyte (ca. x 2), Zovon, Padua, Italy.*

OPAL

COMPOSITION: $SiO_2 \cdot nH_2O$ hydrated silica.

CRYSTALLINE FORM: non-crystalline; found in shapeless masses.

COLOR: colorless, milk-white, yellow, red, green, blue, black.

HARDNESS: 5.5-6.5.

SPECIFIC GRAVITY: 2.01-2.16.

LUSTER: vitreous to pearly.

GEOLOGIC OCCURRENCE: in volcanic rocks from hydrothermal alteration of silicates; as a precipitate from hot springs; as a constituent of living organisms, such as the shells of diatoms (a type of plankton), and spicules of poriferans whose accumulated remains form important sedimentary deposits (fossil flour); during the process of silicification of petrified forests (wood opal).

ASSOCIATED MINERALS AND LOCALITIES: beautiful specimens come from Australia, the United States, Mexico, Egypt, Turkey, Italy etc.

USES: some varieties used as precious stones of great value: precious opal, from dark gray to black in color, with iridescent effects; fire opal, transparent with red reflections; harlequin opal, gray or blueblack.

NOTES: The name is derived from the Sanskrit *upala* (precious stone). Because it has no defined crystalline structure, opal is not actually a mineral in the strict sense. Opal tends to transform into chalcedony or microcrystalline quartz. (*See box, p. 76.*)

▲ *Opal: fire opal variety (ca. x 1), Queretaro, Mexico.*

SANIDINE

COMPOSITION: $KAlSi_3O_8$ potassium tectosilicate (alkaline feldspar).

CRYSTALLINE FORM: monoclinic system, prismatic class. Crystals assume a prismatic, tabular form; commonly twinned, generally through penetration (Carlsbad law).

COLOR: colorless, whitish, gray, light yellow.

HARDNESS: 6.

SPECIFIC GRAVITY: 2.53-2.56.

LUSTER: vitreous.

GEOLOGIC OCCURRENCE: typical mineral of extrusive rocks of acidic or intermediate composition (trachyte, syenite); found as microcrystalline groundmass in basic extrusive rocks such as basanite, shonkinites, and in some lamprophyres.

ASSOCIATED MINERALS AND LOCALITIES: with plagioclases, biotite, diopside, leucite. Beautiful specimens come from Italy, Germany, the Caucasus etc.

USES: interest limited to researchers and collectors.

NOTES: The name is derived from the Greek for "table" and "appearance," in relation to the characteristic tabular form presented by the crystals. It is the high-temperature polymorph of potash feldspar and has a disordered crystalline structure. At high temperatures, sanidine forms a complete isomorphous mixture with albite (sodium feldspar). It is infusible, completely decomposes in hydrofluoric acid, and is not fluorescent.

▼ *Sanidine (ca. x 0.7), Val della Madonna, Trento, Italy.*

ORTHOCLASE

COMPOSITION: KAlSi₃ O₈ potassium tectosilicate (alkaline feldspar).

CRYSTALLINE FORM: monoclinic system, prismatic class. Prismatic, elongated, tabular crystals; commonly twinned either through contact (Manebach law, Baveno law) or penetration (Carlsbad law); in compact microcrystalline masses.

COLOR: colorless, whitish, gray, reddish.

HARDNESS: 6-6.5.

SPECIFIC GRAVITY: 2.56.

LUSTER: vitreous to pearly.

GEOLOGIC OCCURRENCE: very widespread mineral in nature; essential constituent of acid and alkaline intrusive rocks (granites, syenites, monzonites) and in pegmatites; common in medium high-grade metamorphic rocks (gneiss, migmatites) and in detrital sedimentary rocks.

ASSOCIATED MINERALS AND LOCALITIES: with quartz, albite, biotite, muscovite and accessory minerals such as beryl, apatite, topaz, zircon. Beautiful specimens come from Brazil, Madagascar, Germany, Italy etc.

USES: in the porcelain industry: a wet mixture of orthoclase, quartz, and powdered kaolin is easily modeled; baked at 2012°-2372° F (1100-1300° C), it vitrifies forming a light, not very porous material called porcelain. The golden-yellow and lemon-yellow varieties, if clear and transparent, can be emerald-cut.

NOTES: The name is derived from the Greek for "straight" and "fracture," relating to the two cleavage planes at right angles to one another. Orthoclase is the medium-temperature polymorph of potash feldspar and has a partially ordered crystalline structure. It easily alters to kaolin, sericite, epidote, and zeolites. It fuses with difficulty and colors the flame violet (potassium). It is soluble only in hydrofluoric acid and is not fluorescent.

▼ *Orthoclase (ca. x 0.7), Brazil.*

QUARTZ

Along with calcite, quartz is the most common mineral on the earth's crust. Its ubiquity is a result of the fact that it can crystallize in any environment – igneous, metamorphic, and sedimentary.

HISTORY AND LEGEND

There have been numerous finds of paleolithic flint (chert, or microcrystalline quartz), a material our ancestors used to make arms, to get food, or as a defense against the hostility of nature. Because of their numerous crystalline forms and vast range of colors, some varieties of quartz (see page 74) were also used as gems. In Hebrew tradition, the twelve attributes of God were represented by twelve stones mounted on the Breastplate, an ornament of the Jewish Chief Priest. Of these, no less than six are varieties of quartz: *Melek* (King) was represented by sard, *Tolchai* (living God) by jasper, *Elohim* (omnipotent God) by cornelian, *El* (strong) by agate, *Iaho* (God) (Jehovah) by amethyst, and *Adonai* (Lord) by onyx.

Quartz is a mineral that has been used as an ornament in almost all eras, despite the fact that in the past its varieties were muddled because of its chromatic similarity to other minerals. It is said that the Emperor Nero possessed beautiful rock crystal goblets and slabs on which the events of the *Iliad* were engraved. Ideas concerning the powers of rock crystals flourished in the Middle Ages.

Brooch with a cascade of Brazilian amethysts.

According to one of the most widespread beliefs, when pulverized and mixed with honey, this mineral had the property of being an excellent stimulus for wet nurses' milk.

Amethyst is probably the most well-known variety of quartz. In the

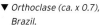

Rock crystal brooch, surrounded by coral, onyx, and diamonds.

past, it may have been confused with other stones of similar color. Various properties were attributed to it, such as guarding against drunkenness, protecting fields from plagues of grasshoppers and from storms and, when worn on the middle finger of the right hand, dispersing anger and passion. In the Catholic tradition, it is thought that the stone Saint Joseph gave to the Virgin Mary as a pledge of his faith was an amethyst. For this reason, it is the only gem used by the Princes of the Church. In the Middle Ages and the Renaissance, it was valued both as a gem and as a material for marvelous sculptures, vases, and goblets.

A VERY SPECIAL MINERAL

Quartz has two important properties: piroelectricity and piezoelectricity. Both are linked to the production of electrical energy: the first through heating, the second through pressure on two opposite faces of a crystal. The latter property, by far the more often applied, was made famous by the quartz gas lighters which, through simple pressure on an appropriately positioned crystal, created a difference in potential which produced sparks.

But if the quartz gas lighter is now a museum object, this mineral has found its use in many advanced technology fields. Its ability to vibrate when excited by alternating current is the basis for the operation of oscillators, radios, clocks, etc. However, high technology requires very pure crystals. To obviate the difficulty of finding them in nature, processes for synthesizing quartz crystals have been developed. Practically perfect crystals can be formed in the laboratory using these processes.

COLORS

When pure, quartz crystals are limpid and transparent and are called rock crystals. Large crystals have always captured the imagination of artists, who often wrought fabulous pieces from them. Goblets of various styles, vases, glasses, and other objects were produced by patiently cutting one of these crystals.

This production system should not be confused with that of today. The famous Bohemian crystals, though beautiful, are only glass, produced by melting very pure siliceous sands which may be supplemented by additives to yield pleasing color shades. When impure, quartz can take on very varied colorations, and transparent crystals or masses can be cut as gems.

Small amethyst statue.

MICROCLINE

COMPOSITION: $KAlSi_3O_8$ potassium tectosilicate (alkaline feldspar).

CRYSTALLINE FORM: triclinic system, pinacoidal class. Crystals are prismatic, stubby; polysynthetic twinning following the albite-pericline law is common.

COLOR: white, gray, faded yellow, pink, blue-green (amazonite variety).

HARDNESS: 6-6.5.

SPECIFIC GRAVITY: 2.56.

LUSTER: vitreous to pearly.

GEOLOGIC OCCURRENCE: found in acid intrusive rocks (granites) and in pegmatites; present in high-grade metamorphic rocks (gneiss); in detrital sedimentary rocks (sands and sandstones).

ASSOCIATED MINERALS AND LOCALITIES: with quartz, orthoclase, biotite, zircon, pyrite, tourmaline. Beautiful specimens come from Brazil, India, Madagascar, Tanzania, Canada, the United States, France etc.

USES: interest limited to researchers and collectors; the amazonite variety, also called Amazon stone, has a beautiful blue-green color, and is used as a prized ornamental stone or is cabochon-cut for use as a gem.

NOTES: The amazonite variety of microcline was used by the ancient Egyptians to make goblets and small statues. The largest crystal 1543 lbs. (700 kg) is in the Museum of Natural History, in Paris. Microcline is the low-temperature polymorph of potash feldspar; like orthoclase, it is found intergrown with quartz or albite. It is infusible, soluble only in hydrofluoric acid, and is not fluorescent.

▼ *Microcline: amazonite variety (ca. x 6), Teller Co., Cripple Creek, Colorado, the United States.*

ALBITE

COMPOSITION: $NaAlSi_3O_8$ sodium tectosilicate (alkaline feldspar).

CRYSTALLINE FORM: triclinic system, pinacoidal class. Appears as crystals with a prismatic, tabular habit, sometimes elongated; common in twins made up of two or more lamellar individuals, joined following both simple and complex laws (albite, pericline, Manebach, albite-Carlsbad).

COLOR: colorless, white, light gray.

HARDNESS: 6-6.5.

SPECIFIC GRAVITY: 2.62-2.65.

LUSTER: vitreous to pearly.

GEOLOGIC OCCURRENCE: found either through primary crystallization or by alteration (albitization) of calcium-rich plagioclases; essential constituent of many igneous, intrusive, and extrusive rocks; in low-grade regional metamorphic rocks; in sedimentary rocks as a detrital or autogenetic mineral.

ASSOCIATED MINERALS AND LOCALITIES: with quartz, other feldspars, biotite, chlorite, calcite, epidote. Beautiful specimens come from Brazil, the United States, Italy, Switzerland, Austria etc.

USES: employed in the industrial field for the manufacture of refractory materials.

NOTES: The name is derived from the Latin *albus* (white). Together with anorthite, constitutes the complete isomorphic plagioclase series; intermediate mixtures are conventionally described in terms of the percentage of anorthite (the Ca-rich end-member): albite (An 0-10%), oligoclase (An 10-30%), labradorite (An 50-70%), bytownite (An 70-90%), anorthite (An 90-100%). It is soluble in hydrofluoric acid, fuses with difficulty, and colors the flame yellow. It is not fluorescent.

▼ *Albite (ca. x 0.7), Switzerland.*

THE STONE OF THE GODS

OPAL AND CHALCEDONY

Opal is one of the most charming and imitated of stones and the value of the precious and black varieties is very high. The name is derived from the Sanskrit upala, (precious stone) and in India it was the sacred stone of the gods.

OPAL

The opal (see page 74) is also called amorphous silica or silica gel and is closely related to quartz. Opal tends to dehydrate, forming the cryptocrystalline silica form chalcedony. Chalcedony and quartz have very similar physical properties.

Because of its compositional variability and possible impurities, opal can assume various tones and plays of color: precious opal of a milky-white color (rarely yellow, red, green) with a play

Black precious opal, 1.30 c. (Australia).

opal of a white or soft gray color. It has been used as an ornamental stone since antiquity and in the past the value attributed to it was certainly higher than that assigned to it today.

CHALCEDONY

Chalcedony is generally semitransparent in bulk and its color is rather homogeneous and remarkably varied: green (chrysoprase), yellow-brown (sard), red (cornelian) and even brown, azure, blue, white, or gray alternated with black (onyx). The variegrated kinds (agate) are due to variations of impurities during the deposition of small needle-shaped crystals, arranged perpendicular to the surface on which they crystallize. Chalcedony has always been used as an ornamental stone, often for making valuable furnishings. It is currently most commonly used as a semiprecious stone for jewelry.

Another trick of nature is silicified fossil wood. Over time, by a process of chemical substitution, entire fossil plants have been replaced by opal (wood opal variety) which subsequently changes into chalcedony. The color is variable and the plants, even though silicified, retain many of the physical peculiarities of the wood.

Agate

Chrypoprase

Onyx

of color in large, shaded spots; black opal with an onyx-black or grayish color and a soft play of color; harlequin opal, gray, blue-black in color with small spots of iridescence; girasol opal with blue reflections; fire opal, yellow-orange, reddish in color with minimal play of color; common

Agate cameo.

LABRADORITE

COMPOSITION: $CaAl_2Si_2O_8$–$NaAlSi_3O_8$ calcium sodium tectosilicate (plagioclase).

CRYSTALLINE FORM: triclinic system, pinacoidal class. Crystals are prismatic, tabular, commonly twinned; in masses, even of remarkable size.

COLOR: gray with bluish reflections; iridescent effects (labradorescence), resulting from the mineral's structure and from minute inclusions of sphene, magnetite, and ilmenite, are common.

HARDNESS: 6–6.5.

SPECIFIC GRAVITY: 2.70–2.72.

LUSTER: pearly to vitreous.

GEOLOGIC OCCURRENCE: generally found in intrusive and extrusive rocks with a basic composition (gabbros, norites, anorthosites, basalts); in medium-grade metamorphic rocks (gneiss).

ASSOCIATED MINERALS AND LOCALITIES: with olivine, pyroxenes, chromite, magnetite, ilmenite, apatite. Beautiful specimens come from Canada, Madagascar, Finland, and Norway.

USES: utilized in the ceramics and refractories industries; in the building trade as a covering stone; specimens which exhibit iridescence are worked to obtain ornamental objects.

NOTES: The name is derived from the Canadian region of Labrador in Newfoundland, where it was found for the first time in 1770. It is soluble in hydrofluoric acid and, slowly, in hydrochloric acid. It fuses with relative difficulty and is not fluorescent. It constitutes an intermediate member of the plagioclase series and has a medium anorthite content, between 50 and 70%.

▼ *Labradorite (ca. x 0.6), Labrador, Canada.*

ANORTHITE

COMPOSITION: $CaAl_2Si_2O_8$ calcium tectosilicate (plagioclase).

CRYSTALLINE FORM: triclinic system, pinacoidal class. The rare crystals appear with a prismatic, tabular habit, sometimes elongated and twinned according to the albite law and the albite-Carlsbad law.

COLOR: white, gray, greenish, pink.

HARDNESS: 6–6.5.

SPECIFIC GRAVITY: 2.74–2.76.

LUSTER: vitreous.

GEOLOGIC OCCURRENCE: found in regionally metamorphosed rocks and in some basalts.

ASSOCIATED MINERALS AND LOCALITIES: with feldspathoids, pyroxenes, biotite. Beautiful specimens come from Japan and Italy.

USES: interest generally limited to researchers and collectors.

NOTES: The name is derived from the Greek for "slanting," in relation to its triclinic structure. Clear, light-colored varieties may be cut; in Val di Fassa, Italy, a pink variety that can be cut as a gem has been found sporadically. Anorthite, together with albite, is an end member of the plagioclase series. It fuses with difficulty, is soluble in hydrofluoric acid and hydrochloric acid with the formation of silica gel, and is not fluorescent.

▲ *Anorthite (ca. x 1), Japan.*

HYALOPHANE

COMPOSITION: $(K,Ba)(Si,Al)_3O_8$ potassium barium tectosilicate (alkaline feldspar).

CRYSTALLINE FORM: monoclinic system, prismatic class. Appears as crystals of prismatic form.

COLOR: colorless, white, light red.

HARDNESS: 6–6.5.

SPECIFIC GRAVITY: 2.58–2.82.

LUSTER: vitreous.

GEOLOGIC OCCURRENCE: rare mineral; found in manganese deposits; may also be present as an accessory in some gneisses.

ASSOCIATED MINERALS AND LOCALITIES: with dolomite, pyrolusite, manganite. Beautiful specimens come from Japan, Australia, Canada, the United States, and Switzerland.

USES: interest limited to researchers and collectors.

NOTES: The name is derived from the Greek for "transparent" and "shining." Similar to adularite, it is an intermediate member between orthoclase (potash feldspar) and celsian (barium feldspar). It fuses with difficulty, is soluble in hydrofluoric acid, and is not fluorescent.

▼ *Hyalophane (ca. x 2), Courmayeur, Tour des Romains, Aosta, Italy.*

DANBURITE

COMPOSITION: $CaB_2(SiO_4)_2$ calcium boron tectosilicate.

CRYSTALLINE FORM: orthorhombic system, dipyramidal class. Appears as prismatic crystals, commonly with wedge-shaped terminations, striated in the length-wise direction of the prism.

COLOR: colorless, yellow, whitish, brown.

HARDNESS: 7–7.5.

SPECIFIC GRAVITY: 2.97–3.02.

LUSTER: vitreous.

GEOLOGIC OCCURRENCE: rare gangue mineral, in metalliferous deposits; in some dolomites with hydrothermal quartz; in some evaporative sediments.

ASSOCIATED MINERALS AND LOCALITIES: with quartz, cassiterite, orthoclase, fluorite, wollastonite, andradite. Beautiful specimens come from the United States, Mexico, Madagascar, Burma, Japan, Switzerland etc.

USES: interest generally limited to researchers and collectors. Clear and transparent crystals of various shades of yellow, may be cut as gems.

NOTES: The name is derived from the United States locality Danbury, Connecticut, where it was discovered in the last century. Color and crystalline form are similar to topaz and citrine quartz. It fuses easily, forming a colorless glass, and during heating colors the flame green. It is soluble only in hydrochloric acid and, if previously calcinated, forms silica gel. It is not fluorescent, but becomes phosphorescent with a reddish coloration when heated.

▼ *Danburite (ca. x 1.5), Russia.*

SCAPOLITE

COMPOSITION: solid solution between *marialite* $Na_4 (Al_3Si_9O_{24})Cl$ and *meionite* $Ca_4(Al_6Si_6O_{24})CO_3$.

CRYSTALLINE FORM: tetragonal system, dipyramidal class. Crystals are prismatic, elongated, sometimes striated; in microcrystalline masses and in fibrous aggregates.

COLOR: colorless, whitish, yellowish, bluish gray, pink, violet.

HARDNESS: 5.5–6.

SPECIFIC GRAVITY: 2.5–2.8.

LUSTER: vitreous.

GEOLOGIC OCCURRENCE: generally found in metamorphic rocks; in skarns; secondary mineral, through hydrothermal alteration of basic igneous rocks. Rare in pegmatites.

ASSOCIATED MINERALS AND LOCALITIES: with almandine, wollastonite, calcite, diopside, andradite, and actinolite. Beautiful specimens come from Canada, the United States, Madagascar, Namibia, Brazil etc.

USES: interest limited to researchers and collectors. Pink and yellow transparent crystals are rare and in some cases are cut for gems. A pleochroic violet variety was recently discovered.

NOTES: the name is derived from the Greek for "shaft," referring to the elongated form of the crystals. Sodium-rich scapolite is practically insoluble in hydrochloric acid, while calcium-rich scapolite decomposes in it, with the formation of silica gel. It fuses to a bubbly mass, and is fluorescent in ultraviolet light, emitting pink, bluish light; in some cases with aventurescence (a spangled appearance produced by inclusions of hematite).

◄ *Scapolite (ca. x 0.6), Arendal, Norway.*

NEPHELINE

COMPOSITION: $(Na,K)AlSiO_4$ sodium tectosilicate (feldspathoid).

CRYSTALLINE FORM: hexagonal system, dipyramidal class. Hexagonal, prismatic crystals; in compact aggregates.

COLOR: colorless, white, gray, green, brick red (eleolite, microcrystalline variety).

HARDNESS: 5.5–6.

SPECIFIC GRAVITY: 2.55–2.65.

LUSTER: vitreous.

GEOLOGIC OCCURRENCE: typical mineral in silica-poor igneous, extrusive, and intrusive rocks; as a product of metasomatism in limestones, gneiss.

ASSOCIATED MINERALS AND LOCALITIES: with analcime, zeolites, apatite, albite. Large masses of nepheline-bearing rocks are found in Canada, the United States, South Africa, and Norway; very beautiful crystals are found in the volcanic blocks ejected from Mount Vesuvius, Italy.

USES: in the ceramic, dye, textile, and glassworks industries.

NOTES: The name is derived from the Greek for "cloud," pertaining to the mineral's behavior in hydrochloric acid, where it is soluble with the separation of a cloud of silica gel. It fuses easily, forming a colorless, glassy, bubbly mass and during heating colors the flame yellow; It is not fluorescent. and easily alters to analcime, cancrinite, and sodalite.

▼ *Nepheline (ca. x 1), Monte Somma, Naples, Italy.*

LEUCITE

COMPOSITION: $KAlSi_2O_6$ potassium tectosilicate (feldspathoid).

CRYSTALLINE FORM: tetragonal system, dipyramidal class. Appears as pseudocubic, roundish crystals with a trapezohedral habit and with striations on the crystals' faces.

COLOR: white.

HARDNESS: 5.5–6.

SPECIFIC GRAVITY: 2.45–2.50.

LUSTER: vitreous.

GEOLOGIC OCCURRENCE: typical mineral of potassium-rich basic and ultrabasic extrusive rocks (basanites, leucite tephrytes, leucitites); not found in ancient lavas.

ASSOCIATED MINERALS AND LOCALITIES: with augite, haüynite, analcime, nepheline, olivine. Beautiful specimens come from Australia, the United States, Zaire, Uganda, Italy etc.

USES: interest limited to researchers and collectors; in the past, it was used as a fertilizer.

NOTES: The name is derived from the Greek for "white" or "bright," referring to the appearance of its crystals. Dimorphous, it occurs in two forms: a cubic form stable at temperatures over 1121° F (605° C), and a tetragonal form stable at lower temperatures. The cubic crystals formed at high temperatures in magmas transform as the temperature drops, taking on the tetrahedral symmetry while maintaining the cubic morphology. It is infusible, soluble in hot hydrochloric acid with the separation of silica gel and in cold sulfuric acid, and it is not fluorescent. It alters easily to analcime or to kaolin.

▼ *Leucite (ca. x 1), Ariccia, Rome, Italy.*

ANALCIME

COMPOSITION: $Na(AlSi_2O_6 \cdot H_2O)$ sodium tectosilicate (feldspathoid).

CRYSTALLINE FORM: cubic system, hexoctahedral class. Icositetrahedral crystals, sometimes cubic with vertices truncated by icositetrahedrons; in radial aggregates and in microcrystalline masses.

COLOR: colorless, white, pink, grayish.

HARDNESS: 5.5.

SPECIFIC GRAVITY: 2.25.

LUSTER: vitreous.

GEOLOGIC OCCURRENCE: primary mineral in silica-poor intrusive rocks; as a filling in cavities in volcanic rocks; autogenous mineral in clasitc rocks (sandstones).

ASSOCIATED MINERALS AND LOCALITIES: with zeolites, calcite, prehnite. Beautiful specimens come from Italy, the United States, Australia, Iceland etc.

USES: interest limited to researchers and collectors.

NOTES: The name is derived from the Greek for "weak" and pertains to its weak pyroelectricity, a property brought out when it is heated or rubbed. It fuses easily, forming a colorless glass and during heating colors the flame yellow (sodium). It is soluble in hydrochloric acid with the separation of silica gel, and it is not fluorescent.

▼ *Analcime (ca. x 1), Siusi Alps, Bolzano, Italy.*

SODALITE

COMPOSITION: $Na_4Al_3Si_3O_{12}Cl$ sodium tectosilicate (feldspathoid).

CRYSTALLINE FORM: cubic system, hextetrahedral class. Rare rhombic dodecahedral crystals; generally in compact aggregates.

COLOR: pale pink, gray, yellow, blue, green.

HARDNESS: 5-6.

SPECIFIC GRAVITY: 2.30.

LUSTER: vitreous.

GEOLOGIC OCCURRENCE: found in silica-poor plutonic rocks (nepheline syenites, phonolites); in calcareous rocks affected by metasomatic processes.

ASSOCIATED MINERALS AND LOCALITIES: with nepheline, leucite, cancrinite, melanite, fluorite. Large masses are found in Canada, the United States, Brazil, Bolivia, Greenland, and Burma.

USES: uniformly-colored compact masses are worked to obtain necklaces, bracelets, and other ornamental objects; pre-Columbian sodalite objects have been found in the Tiahuanaco Mountains in Bolivia.

NOTES: Sodalite also refers to the group of isomorphous minerals which includes sodalite, noselite, haüynite, and lazurite. In compact masses with an azure-blue coloration, it can be mistaken for lazurite (lapis lazuli). However, they can be distinguished from each other: when polished, sodalite masses appear violet. It fuses easily, forming a colorless glass and during heating colors the flame yellow from the presence of sodium. It decomposes in hydrochloric acid with the separation of silica gel, is not fluorescent, and alters easily to natrolite, kaolin, sericite, and calcite. The hackmanite variety of sodalite exhibits the phenomenon of tenebrescence: when freshly cut, it exhibits a pink color that fades when exposed to the light. However, if the mineral is kept in the dark for a few weeks or is bombarded with X-rays, the color reappears.

▼ *Sodalite (ca. x 0.4), Herculaneum, Naples, Italy.*

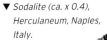

HAÜYNITE

COMPOSITION: $(Na,Ca)_{4-8}Al_6Si_6O_{24}$ $(SO_4,S)_{1-2}$ calcium sodium tectosilicate (feldspathoid).

CRYSTALLINE FORM: cubic system, hextetrahedral class. Rare crystals have an octahedral or rhombic dodecahedral habit; commonly in rounded granules.

COLOR: white, gray, green, blue, red, yellow.

HARDNESS: 5-6.

SPECIFIC GRAVITY: 2.4-2.5.

LUSTER: vitreous.

GEOLOGIC OCCURRENCE: found in silica-poor alkaline extrusive rocks (phonolites, syenite, nephelinic); constituent of haüynophyres (haüynite-rich volcanic rocks).

ASSOCIATED MINERALS AND LOCALITIES: with garnets, leucite, melilite. Beautiful specimens come from Italy, France, Morocco, Germany etc.

USES: interest limited to researchers and collectors.

NOTES: Haüynite, also called haüyne, is named in honor of the French mineralogist abbé René-Just Hay. Haüynite belongs to the sodalite series and is distinguishable from sodalite by the fact that on dissolving the minerals in nitric acid and evaporating the solutions, sodalite forms crystals of calcium chloride, while haüynite forms small needles of gypsum. It fuses easily, yielding a glassy mass of a greenish blue color, decomposes in hydrochloric acid with the separation of gelatinous silica, and it is not fluorescent.

▼ *Haüynite (ca. x 1), Ariccia, Rome, Italy.*

LAZURITE

COMPOSITION: $(Na,Ca)_8(Al, Si)_{12}O_{24}$ (S,SO_4) calcium sodium tectosilicate (feldspathoid).

CRYSTALLINE FORM: cubic system, hextetrahedral class. Octahedral crystals extremely rare; commonly appears in compact microcrystalline masses.

COLOR: intense bright blue, sometimes with greenish reflections (lapis lazuli), rarely violet. In the microcrystalline masses, inclusions of pyrite and calcite are very common and account for their typical heterogeneous coloration.

HARDNESS: 5.5-6.

SPECIFIC GRAVITY: 2.38-2.42.

LUSTER: vitreous.

GEOLOGIC OCCURRENCE: found in contact metamorphosed calcareous rocks and in high-temperature granulites.

ASSOCIATED MINERALS AND LOCALITIES: with calcite, pyrite, diopside. Deposits are found in Afghanistan, China, the Chilean Andes, Iran, Canada, the United States, Burma, and Angola.

USES: compact microcrystalline masses (lapis lazuli) are worked to create high-value ornamental objects. In the past, it was reduced to powder to obtain a dyestuff (ultramontane blue).

NOTES: The name is derived from the Arabic *lazaward* (sky), referring to the mineral's color. Lazurite is an essential component of lapis lazuli (azure stone), a rock made up of an aggregate of lazurite, calcite, pyroxenes, and pyrite. It belongs to the sodalite series and fuses with relative ease, forming a whitish glassy mass. It rapidly decomposes in acids, evolving sulfuric acid from inclusions of pyrite, and it is not fluorescent.

▼ *Lazurite (ca. x 1), Afghanistan.*

NATROLITE

COMPOSITION: $Na_2Al_2Si_3O_{10} \cdot 2H_2O$ sodium tectosilicate (zeolite).

CRYSTALLINE FORM: orthorhombic system, pyramidal class. Appears as slender prismatic crystals, striated; commonly in spherulitic aggregates with a radial fibrous structure or in compact aggregates with a felted look.

COLOR: colorless, white, pink, yellow.

HARDNESS: 5-5.5.

SPECIFIC GRAVITY: 2.2.

LUSTER: vitreous to pearly.

GEOLOGIC OCCURRENCE: filling up amygdules in basaltic lavas; produced by the alteration of plagioclases in syenites and aplites, and of sodalite to nepheline in basic extrusive rocks.

ASSOCIATED MINERALS AND LOCALITIES: with calcite, prehnite, zeolites. Beautiful specimens, some of remarkable size, come from Canada, the United States, Brazil, Greenland, and India.

USES: its use in the industrial field is linked to the capacity of its crystalline structure to filter molecular fluids.

NOTES: The word is derived from the Latin *natrium* (sodium), an element present in the mineral. Natrolite also refers to a mineral group which includes mesolite, thomsonite, scolecite, gonnardite, and edingtonite; a group belonging to the zeolite family. The zeolites are characterized by a relatively open crystalline structure with channels that allow the passage of organic molecules called "molecular sieves." Natrolite fuses easily, forming a colorless glassy mass, and during heating colors the flame yellow. It is soluble in hydrochloric acid with the separation of silica gel, and sometimes exhibits orange fluorescence. Heated to about 572° F (300°C), it loses water and changes into metanatrolite; when exposed to humidity in the air, it easily regains water.

▼ *Natrolite in basalt (ca. x 1), Gambellara, Vicenza, Italy.*

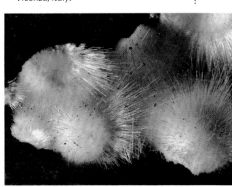

MESOLITE

COMPOSITION: $Na_2Ca_2Al_6Si_9O_{30} \cdot 8H_2O$ sodium calcium tectosilicate (zeolite).

CRYSTALLINE FORM: monoclinic system, sphenoidal class. Appears as prismatic, elongated, needle-shaped crystals; in aggregates with a radial fibrous structure.

COLOR: white, gray, yellowish.

HARDNESS: 5.

SPECIFIC GRAVITY: 2.2-2.4.

LUSTER: vitreous to silky.

GEOLOGIC OCCURRENCE: filling up cavities in basaltic volcanic rocks.

ASSOCIATED MINERALS AND LOCALITIES: with datolite, prehnite, calcite and other zeolites. Beautiful specimens come from the United States, Scotland, Siberia etc.

USES: use of these minerals in the industrial field is linked to the crystalline structure's ability to filter molecular fluids; they are used to absorb odors, as purifiers of petroleum products, for the removal of radioactive isotopes from nuclear waste, as polishes in fluoride toothpastes, as dietary food supplements for poultry and pigs, and to purify natural gases with a low heat content. However, they have now been replaced by synthetic products.

NOTES: Mesolite belongs to the natrolite group with a composition between natrolite and scolecite. Needle-shaped crystalline aggregates with a felted appearance are called "cotton stone" by the English. It fuses easily, is soluble in hydrochloric acid, and is not fluorescent.

SCOLECITE

COMPOSITION: $Ca(Al_2Si_3O_{10}) \cdot 3H_2O$ calcium tectosilicate (zeolite).

CRYSTALLINE FORM: monoclinic system, sphenoidal class. Appears as prismatic, elongated and striated crystals, sometimes twinned with a pseudohexagonal symmetry; in aggregates with a radial fibrous structure.

COLOR: colorless, white.

HARDNESS: 5-5.5.

SPECIFIC GRAVITY: 2.3.

LUSTER: vitreous to silky.

GEOLOGIC OCCURRENCE: generally, filling up cavities in basaltic volcanic rocks; sometimes in cracks in contact metamorphosed calcareous rocks or as an alteration product in druses of syenitic rocks.

ASSOCIATED MINERALS AND LOCALITIES: with calcite and other zeolites. Beautiful specimens come from Iceland, Scotland, the United States, India, and Brazil.

USES: use in the industrial field is linked to the ability of the crystalline structure to filter molecular fluids. (*see mesolite.*)

NOTES: The name is derived from the Greek for "worm," inspired by the fact that when the mineral is subjected to the appropriate degree of heat, before fusing to a bubbly glass, it curls up in contorted shapes with a worm-like appearance. It belongs to the natrolite group, is easily soluble in hydrochloric acid with the separation of silica gel, and it is not fluorescent. When heated, it changes into highly pyroelectric metascolecite.

THOMSONITE

COMPOSITION: $NaCa_2Al_5Si_5O_{20} \cdot 6H_2O$ sodium calcium tectosilicate (zeolite).

CRYSTALLINE FORM: orthorhombic system, dipyramidal class. Appears as prismatic, acicular crystals, sometimes lamellar, generally joined in aggregates with a radial fibrous structure.

HARDNESS: 5-5.5.

SPECIFIC GRAVITY: 2.3.

LUSTER: vitreous to pearly.

GEOLOGIC OCCURRENCE: generally filling up amygdules in volcanic rocks with a basaltic composition or in silica-poor alkaline extrusive rocks (phonolites).

ASSOCIATED MINERALS AND LOCALITIES: with calcite, analcime, and other minerals of the zeolite family. Beautiful specimens come from Scotland, the Faeroe Islands, Greenland, and the United States.

USES: usage in the industrial field is linked to the ability of the crystalline structure to filter molecular fluids (*see mesolite*).

NOTES: Thomsonite is also known as comptonite or faröelite; belongs to the natrolite group. It fuses easily, is soluble in hydrochloric acid with the separation of silica gel, and it is not fluorescent. The term metathomsonite is used for the dehydrated variety.

LAUMONTITE

COMPOSITION: $Ca(Al_2Si_4O_{12} \cdot 4H_2O$ calcium tectosilicate (zeolite).

CRYSTALLINE FORM: monoclinc system, prismatic class. Prismatic, elongated crystals, striated along the length of the prism; in fibrous aggregates.

COLOR: white, yellow, pink.

HARDNESS: 3-3.5.

SPECIFIC GRAVITY: 2.3.

LUSTER: vitreous.

GEOLOGIC OCCURRENCE: filling up druses and geodes in acid intrusive rocks (basalts, diabases); in cavities in basic volcanic rocks (basalts, diabases); in veins of metamorphic rocks and in metalliferous veins.

ASSOCIATED MINERALS AND LOCALITIES: with calcite, analcime, and other zeolites. Beautiful specimens come from the United States, New Zealand, Norway, etc.

USES: usage in the industrial field is linked to the ability of the crystalline structure to filter molecular fluids.

NOTES: Laumontite is named in honor of the French mineralogist Gillet de Laumont (1747-1834), who discovered the mineral at Huelgoat in Brittany (France) in 1785. Together with mordenite and dachiardite, it forms the laumontite group. It fuses easily, is soluble in hydrocholoric acid with the separation of gelatinous silica, and is not fluorescent. When heated or exposed to dry air and light, it easily gives up water, becoming dehydrated and changing into the opaque and friable variety called leonhardite.

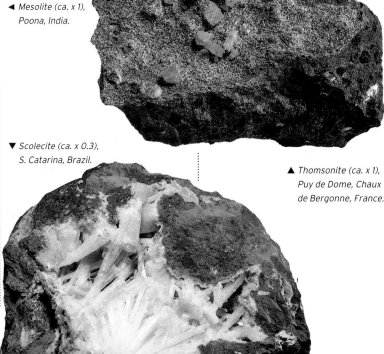

◄ *Mesolite (ca. x 1), Poona, India.*

▼ *Scolecite (ca. x 0.3), S. Catarina, Brazil.*

▲ *Thomsonite (ca. x 1), Puy de Dome, Chaux de Bergonne, France.*

▼ *White laumontite with chabazite (ca. x 1), New Jersey, United States.*

HEULANDITE

COMPOSITION: $(Ca,Na_2,K_2)Al_2Si_7O_{18} \cdot 6H_2O$ calcium sodium and potassium tectosilicate (zeolite).

CRYSTALLINE FORM: monoclinic system, prismatic class. Appears as tabular crystals and in parallel aggregates.

COLOR: colorless, white, yellow, green, red-orange.

HARDNESS: 3.5-4.

SPECIFIC GRAVITY: 2.20.

LUSTER: vitreous to pearly.

GEOLOGIC OCCURRENCE: filling up cavities in basic volcanic rocks (basalts, diabases, andesites); in veins in metamorphic rocks (schists, gneiss) and in metalliferous veins.

ASSOCIATED MINERALS AND LOCALITIES: with calcite and other minerals in the zeolite family. Beautiful specimens come from the United States, Canada, Brazil, India, Iceland etc.

USES: usage in the industrial field is linked to the crystalline structure's ability to filter molecular fluids. (*see mesolite, p. 80.*)

NOTES: Heulandite is named in honor of the Englishman Henry Heuland, a mineral collector. Gives its name to the so-called heulandite mineral group which includes stilbite, clinoptilolite, ferrierite, and brewsterite. It fuses easily, and during heating swells up from the loss of water; dissolves easily in hydrochloric acid with the formation of silica gel, and it is not fluorescent.

STILBITE

COMPOSITION: $NaCa_2Al_5Si_{13}O_{36} \cdot 14H_2O$ sodium calcium tectosilicate (zeolite).

CRYSTALLINE FORM: monoclinic system, prismatic class. Prismatic crystals, commonly joined in sheaf-shaped aggregates; in radial, spherulitic aggregates; in cruciform twins simulating a pseudorhombic symmetry.

COLOR: white, gray, brown, reddish.

HARDNESS: 3.5-4.

SPECIFIC GRAVITY: 2.15.

LUSTER: vitreous to pearly.

GEOLOGIC OCCURRENCE: filling up amygdules in basaltic rocks; more rarely, in fissures in metamorphic rocks.

ASSOCIATED MINERALS AND LOCALITIES: with calcite and other minerals of the zeolite family. Beautiful specimens come from the United States, Canada, Brazil, India, Iceland, Scotland etc.

USES: usage in the industrial field is linked to the crystalline structure's ability to filter molecular fluids. (*see mesolite, p. 80.*)

NOTES: The name is derived from the Greek for "to shine," pertaining to the mineral's luster. It belongs to the heulandite group, and fuses easily, forming an opaque glassy mass with a white color. It is easily soluble in hydrochloric acid with the separation of silica gel, and it is not fluorescent.

PHILLIPSITE

COMPOSITION: $(Ca_{0.5},Na,K)_3(Al_3Si_5O_{16}) \cdot 6H_2O$ potassium calcium and sodium tectosilicate (zeolite).

CRYSTALLINE FORM: monoclinic system, prismatic class. Crystals are prismatic, bacillary, stubby, commonly in cruciform twins with thin striations.

COLOR: colorless, white, yellowish, grayish.

HARDNESS: 4-4.5.

SPECIFIC GRAVITY: 2.15.

LUSTER: vitreous.

GEOLOGIC OCCURRENCE: filling up cavities in volcanic basic rocks (basalts); present in deep ocean sediments as an alteration product of feldspars or of volcanic ash.

ASSOCIATED MINERALS AND LOCALITIES: with chabazite, calcite, and other minerals of the zeolite family. Beautiful specimens come from Germany, Iceland, and the United States; in the Pacific Ocean as a nucleus of iron and manganese nodules.

USES: usage in the industrial field is linked to the crystalline structure's ability to filter molecular fluids. (*see mesolite, p. 80.*)

NOTES: Phillipsite is named in honor of the English mineralogist W. Phillips. Together with harmatome and gismondine, it composes the phillipsite group. When heated, it breaks up, swells, and finally fuses to a white glassy mass. It is soluble in hydrochloric acid with the separation of silica gel, and it is not fluorescent.

CHABAZITE

COMPOSITION: $CaAl_2Si_4O_{12} \cdot 6H_2O$ calcium tectosilicate (zeolite).

CRYSTALLINE FORM: trigonal system, ditrigonal scalenohedral class. Crystals are rhombohedral, pseudocubic, form penetration twins.

COLOR: white, greenish, reddish.

HARDNESS: 4-5.

SPECIFIC GRAVITY: 2.10.

LUSTER: vitreous.

GEOLOGIC OCCURRENCE: filling up cavities in basic extrusive rocks (basalts); as a precipitate from hot springs.

ASSOCIATED MINERALS AND LOCALITIES: with calcite, analcime and other minerals of the zeolite family. Beautiful specimens come from Germany, Scotland, the Faeroe Islands, Canada, the United States, and and Australia.

USES: usage in the industrial field is linked to the crystalline structure's ability to filter molecular fluids. (*see mesolite, p. 80.*)

NOTES: The name is derived from an archaic Greek word meaning "stone." It gives its name to the so-called chabazite group together with erionite and gmelinite; very rare zeolites in nature. It fuses easily, swelling up, and forming a glassy, bubbly mass of a whitish color. It dissolves in hydrochloric acid with the separation of silica gel, and it is not fluorescent.

▼ *Stilbite (ca. x 0.5), Rio das Antas, Rio Grande do Sul, Brazil.*

▼ *Chabazite (ca. x 0.4), Laneariz, Bohemia, Czech Republic*

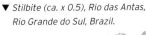

Phillipsite (ca. ▶ x 1.5), Aci Castello, Catania, Italy.

▲ *Heulandite (ca. x 1), Teigarhorn, Iceland.*

OPAQUE MINERALS

OPAQUE MINERALS

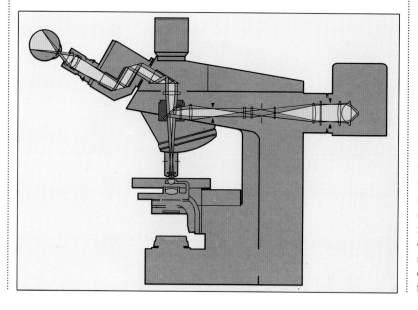

A specimen with beautiful stibnite crystals and, to the right, a few crystals embedded in a small cylinder filled with resin whose upper surface is polished to a mirror finish for observation under a reflected light microscope (a polished section).

People have long been drawn to minerals because of the possibility of extracting from them both useful metals and beautiful gems. Crushing minerals into powders for dyes is also a very ancient practice. The widespread use of objects made from certain materials has led scholars to designate eras which marked steps in human evolution according to the materials used. Thus the Stone Age is followed by the Copper Age, which is then succeeded by the Bronze Age and finally by the Iron Age (bronze is an alloy of copper and tin).

If we analyze the development of the principal ancient civilizations, we see that some of them became powerful just after the discovery of deposits of minerals that were useful and of great commercial value. For example, Egyptian civilization underwent a great period of expansion soon after the discovery of the gold mines in the Upper Nile. Under the Athenian supremacy, Hellenic civilization profited from the exploitation of the silver mines in Laurion (Sunion peninsula). The Etruscans had the rich iron mines of Tuscany, a territory which also held the only cassiterite (a tin-bearing mineral) mine on the Italian peninsula.

Moreover, the territorial conquests of the Romans were partially inspired by the existence of mines in far-off lands. As handed down by Pliny the Elder, the most important mineral for Roman civilization was gold, at that time known in the native form as visible gold grains or nuggets. Following gold, the minerals in order of decreasing importance were: silver-bearing galena; the copper minerals; the iron minerals, especially magnetite whose magnetic powers were astonishing; and finally, cassiterite.

Many of the most important mines were discovered prior to the Roman conquests. At that point in history, the silver-bearing galena mines of Sardinia and Iberia, where gold could also be found, were already very well-known. The island of Elba was the principal source for iron, just as Cornwall was for tin (from cassiterite) which was needed in the manufacture of bronze. Territories lacking these riches were destined for oblivion.

This politics of conquest has survived to the present day, although it is practiced now in a less direct fashion through multinational corporations. However, the present age is not without its wars, albeit rare, inspired by motives similar to those of the ancient Romans. As everyone knows, the most recent war in the Persian Gulf was closely linked to control of the oil fields in Kuwait. Even though hydrocarbons do not belong to the field of mineralogy, they are nevertheless an extremely precious natural resource.

Given the great, long-standing interest in economic minerals, scientists have attempted to understand the origin of the deposits in which these minerals occur. Economic geology is a branch of geology that specifically studies these deposits and their genesis. The opaque minerals constitute a large part of these rocks. This group of minerals includes the native elements and the sulfide and oxide minerals.

The name "opaque minerals" refers to the fact that they do not allow light to pass through them, not even when cut into extremely thin sections. This property distinguishes them from the transparent minerals. However, like the transparent minerals, opaque minerals absorb and reflect some light. The proportion of light that is reflected (the reflectance) is specific to each species. As in the study of transparent minerals, the polarizing microscope is a very widely used instrument for the study of opaque minerals. In the case of the latter group of minerals, the microscope is set up in such a way that light is directed onto the mineral's upper surface and any light reflected by the mineral travels through the eyepiece for observation. The mineral sample chosen for observation is prepared as a "polished section": a fragment of the mineral is put into a cylindrical mold that is then filled with resin and allowed to set. The cylinder is cut crosswise so that the mineral intersects a circular plane parallel to the base of the cylinder. The new surface is carefully polished to a mirror finish so that even the smallest scratches are eliminated. The polishing is carried out with diamond paste, an abrasive oil-based paste containing very fine diamond particles of uniform size.

Observations of the optical properties of opaque minerals by use of the reflected-light microscope are helpful in identifying the minerals. As with transparent minerals, the optical properties of the opaque minerals depend upon their chemical composition and crystalline structure. The determinations of other properties such as specific gravity, hardness, color, etc., are carried out just as they are for transparent minerals.

Some opaque minerals are very important sources or *ores* of useful metals. *Useful* in this context refers to those metals required by the technological structure of a society, whether they are of common or rare use. The common metals such as iron, copper, zinc, and lead, are extracted from minerals such as hematite, magnetite, chalcopyrite, cuprite, chalcocite, sphalerite, smithsonite, and galena; the abundance of these minerals in rocks of the earth's crust varies greatly. In order for exploitation to be economically profitable, the extractable minerals must constitute a greater than average proportion of the total rock mass, and the metal content must exceed a certain percentage which depends on the type of metal. For example, for copper and tin a 1% concentration (by weight) is sufficient; for lead and zinc, 4%; for iron, no less than 25%. These relatively high concentrations of metals are called *ore deposits*.

For rare minerals such as gold, silver, and platinum, deposits containing seemingly low concentrations are economically profitable. For example, a gold concentration lower than 0.001% of the total rock weight is sufficient. This quantity is terribly small, and the metal is so scattered in the deposits as to be nearly always invisible to the naked eye. Visible aggregates such as grains and small nuggets are very rare. Therefore, one can see how the famous gold rushes of the past are the stuff of legends and unlikely to be repeated.

Initially, gold prospectors only collected visible gold which they were able to separate from stream gravel with very simple tools like the *pan* or the *batea*, a conical pan made of wood or iron used chiefly in Latin America. Eventually, they learned to use mercury which is the only substance capable of alloying easily with gold at low temperatures. The gold was then separated by evaporating the mercury through simple heating. The extraction process was made more efficient by the use of small, inclined sluices that allowed the deposits to be *washed* as they passed along these channels. Today this type of activity is a hobby for the weekend prospector, and there are even clubs which organize gold-hunting competitions.

These gold deposits, known as "placer gold," are very distinctive: for millenia, streams and rivers washed over and eroded rocks containing very widely dispersed bits of gold and carried the rock fragments downstream. The

gold, because of its high specific gravity, became concentrated wherever the water could no longer transport it due to the decreased velocity of the current.

Deposits generally consist of an aggregate of many minerals that in some cases crystallized under similar environmental conditions. Economically useful minerals, the object of mining operations, are contained in a rock mass with other, generally less *useful* minerals, which last are referred to as "gangue." The expression "to mine" means the extraction from the subsoil of any type of material, whether accomplished in a tunnel or in an open-pit quarry.

Silver and copper, Mexico.

Meneghinite, Bottino, Lucca, Italy.

Octahedral magnetite, Binn Valley, Switzerland.

GOLD

COMPOSITION: Au *native gold*.
CRYSTALLINE FORM: cubic system, hexoctahedral class. Octahedral, rhombic dodecahedral, and cubic crystals are rare; lamellar aggregates common, sometimes dendritic; in threads, grains, and small rounded masses (nuggets).
COLOR: golden yellow.
HARDNESS: 2.5-3.
SPECIFIC GRAVITY: 15.2-19.3.
LUSTER: metallic.
GEOLOGIC OCCURRENCE: in hydrothermal veins associated with acid volcanic rocks; found in placer deposits, both recent (fluvial sands) and ancient (matrix of conglomerates).
ASSOCIATED MINERALS AND LOCALITIES: with quartz, pyrite, pyrrhotite, sphalerite, and heavy minerals. The largest deposits are found in conglomerates, among which the ones in South Africa, the United States, Canada, Mexico, India, and Siberia are famous; important hydrothermal deposits are found in Brazil, Japan, and Australia.
USES: native gold is the principal source of the metal; used in jewelry; for coins and as a monetary standard; in dental technology; for electronic and scientific equipment.
NOTES: gold gives its name to a group of minerals with similar structure and chemistry (gold, lead, silver, and copper). Rare in nature in its pure state, it is commonly associated with other elements in natural alloys: silver (electrum), palladium (porpezite), rhodium (rhodite). It fuses at 1942°F (1061°C), is insoluble in acids with the exception of aqua regia, and is not fluorescent. It is an excellent conductor of electricity and heat. (*see box, p. 87*.)

SILVER

COMPOSITION: Ag *native silver*.
CRYSTALLINE FORM: cubic system, hexoctahedral class. Octahedral, rhombic dodecahedral, and cubic crystals are rare, small in size, and usually malformed; common in filamentary masses with odd shapes and in arborescent and stellate aggregates.
COLOR: silvery white, surfaces often tarnished by patinas of sulfides.
HARDNESS: 2.5-3.
SPECIFIC GRAVITY: 10.1-11.1.
LUSTER: metallic.
GEOLOGIC OCCURRENCE: in cementation zones of deposits containing silver salts; in both low- and high-temperature hydrothermal veins.
ASSOCIATED MINERALS AND LOCALITIES: with various sulfides, native copper, calcite, uraninite. Beautiful specimens come from Norway, Mexico, Bolivia, the United States, Canada, and Australia.
USES: excellent mineral source of the metal, but fairly rare in nature; utilized in the chemical and electronics industries, in jewelry making, and in alloys for minting coins.
NOTES: the origin of the word *silver* is uncertain, perhaps of Anglo-Saxon derivation. Very ductile and malleable, it is the best natural conductor of electricity and heat and has one of the highest reflectances (95%). It fuses at 1760°F (960°C), is soluble in nitric acid, and the resultant solution, when treated with hydrochloric acid, precipitates silver chloride which initially is white in color but darkens with exposure to light. It is not fluorescent. (*see box, p. 87*.)

COPPER

COMPOSITION: Cu *native copper*.
CRYSTALLINE FORM: Cubic system, hexocahedral class. Cubic, octahedral, and tetrahexahedral crystals are rare, generally with rounded edges; common in compact or spongy masses, in arborescent aggregates; sometimes as a pseudomorph after calcite or cuprite.
COLOR: typical copper-red color, commonly with a greenish (malachite) or blackish alteration patina.
HARDNESS: 2.5-3.
SPECIFIC GRAVITY: 8.95.
LUSTER: metallic.
GEOLOGIC OCCURRENCE: in oxidation zones of deposits of copper sulfides; rarely, in cavities in basaltic rocks.
ASSOCIATED MINERALS AND LOCALITIES: with cuprite, bornite, calcite, malachite, azurite, epidote, zeolites. Beautiful specimens come from the United States, Canada, Chile, Bolivia, Zambia, the Urals, Sweden etc.
USES: deposits with economically exploitable concentrations are rare; used in the electrical industry and in the manufacture of alloys (brass, bronze).
NOTES: the name is derived from "Cyprus"; in Latin this became *cuprum*, the first two letters of which are used as the symbol for the chemical element. Copper has left its mark on the course of history, giving its name to an era of human civilization (the Copper Age). Very ductile and malleable, it is an excellent conductor of heat and electricity. It fuses at 1978°F (1081°C) and, on a platinum wire, colors the flame green. It is soluble in nitric acid and when treated with ammonia, the solution takes on a blue color. It is not fluorescent.

PLATINUM

COMPOSITION: Pt *native platinum*.
CRYSTALLINE FORM: cubic system, hexoctahedral class. The very rare crystals have a cubic habit with an imperfect form; generally as granules, laminae; rarely as nuggets.
COLOR: silver-gray.
HARDNESS: 4-4.5.
SPECIFIC GRAVITY: 21.47.
LUSTER: metallic.
GEOLOGIC OCCURRENCE: a rare mineral in basic (norites) and ultrabasic (dunites) rocks; usually concentrated in alluvial placers.
ASSOCIATED MINERALS AND LOCALITIES: with chromite, olivine, enstatite, and sulfides of nickel and cobalt. Important deposits are found in South Africa, the Urals, the United States, Canada, Colombia, Peru, and New Zealand.
USES: useful mineral source of the metal; used in jewelry making, as a catalyst, and for high-technology equipment.
NOTES: platinum gives its name to a chemically-related group of minerals (platinum, palladium, iridium, and osmium). It is very rarely found in nature in its pure state; it alloys with other elements such as palladium (up to 37%), copper (up to 13%), iridium, and iron which in some varieties amounts to 30% of the weight and gives the mineral weakly magnetic properties and a lower specific gravity. It fuses at 3182°F (1750°C), is only soluble in aqua regia, and is not fluorescent. It is a good conductor of heat and electricity.

▼ *Nugget of native platinum (ca. x 3), Siberia.*

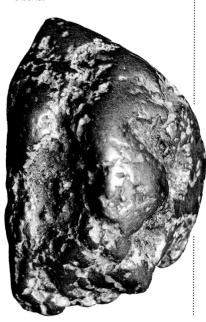

▲ *Gold (ca. x 7), Sandhurst, Victoria, Australia.*

Native silver on ▶ calcite (ca. x 1), Köngsberg, Norway.

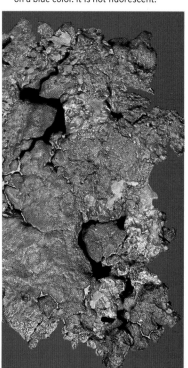

Native ▶ copper (ca. x 1), New Mexico, United States.

PRECIOUS METALS

GOLD AND SILVER

Gold and silver were among the first minerals discovered by mankind. Since time immemorial, man's attention and his greed have focused on the yellow metal.

NOTHING MORE PRECIOUS

Gold (page 86), much rarer than silver (page 86), very quickly became a symbol of power to the extent that a king was not considered to be true royalty without possessing great quantities of it. In addition to the fact that the most highly valued coins were minted from gold, its malleability and brilliance made it very attractive and as a result, it was ostentatiously displayed as an ornament.

Gold brooch with topazes.

According to ancient peoples, gold's value was not limited to this world. The remains of royalty were buried together with the most precious objects that had surrounded them in life. One need only think of Tutankhamen's tomb in which innumerable gold objects were found. Among these was the mummy case made out of a single block of gold and weighing at least 243 lbs. (110 kg).

Many wars waged over the course of history ended with the victor seizing the "spoils of war" consisting of the loser's treasury. With the conquest of Egypt, Alexander the Great took possession of gold in the amount of 60 to 80 tons. The Spanish *conquistadores* did no less, carrying back about 450 tons from the Americas to Europe.

THE GOLD RUSHES

The attraction people feel toward gold and its material and symbolic value has no equal. In the past, every gold mine had always been the property of the powerful. During the settling of the "Wild West," many common folk, dazzled by the idea of easy riches, par-ticipated in the famous "gold rushes." These included the 1848 "rush" in California, and the last gold rush in 1896 in the Klondike of Canada, in which at least 30,000 people took part. Australia, too, had its own "rush" in 1851 in New South Wales.

The enormous power of gold is witnessed by the gold reserves of the great world powers. An immense quantity of gold ingots are stored in these reserves to "balance" the amount of paper currency in circulation in each nation.

Silver, although it is the "poor cousin" to gold, has always been an object of diligent extraction. It is also a by-product of mining operations for other minerals.

PROPERTIES

In contrast with silver which oxidizes easily, gold maintains its characteristic brilliance over time, undiminished by oxidation. Pyrite (fool's gold) is a mineral often mistaken for the noble

Small gold sarcophagus from Tutankhamen's tomb (Cairo Museum, Egypt).

metal by the naïve. It is golden yellow and often found associated with gold in veins. It is distinguished from gold by its greater hardness and its gray-black streak (the color of the mineral when it is ground to a powder).

THE PROSPECTORS' TOOLS

Possibly the best known tools are the *pan* and *batea* which gold prospectors filled with sediment and water collected in rivers. Using a rotary motion, they rinsed away the lighter minerals while the heavier ones, among them grains and rare nuggets of gold, accumulated at the bottom of the pan.

Centerpiece in silver, vermeil, semiprecious stones, and enamel.

MERCURY

COMPOSITION: Hg *native mercury.*

CRYSTALLINE FORM: the only mineral that occurs in the liquid state at room temperature; forms droplets or globules disseminated in rocks with cinnabar; solidifies in the trigonal system, hexagonal scalenohedral class at -38°F (-39°C).

COLOR: silver-white.

HARDNESS: liquid at normal temperatures.

SPECIFIC GRAVITY: 13.6.

LUSTER: metallic.

GEOLOGIC OCCURRENCE: formed in a reducing environment along with cinnabar (HgS), in either bituminous limestones or volcanic rocks; accumulations of mercury are commonly formed in natural cavities that can cause annoying "showers" during some stages of mine excavations.

ASSOCIATED MINERALS AND LOCALITIES: found with cinnabar in Spain, Germany and the United States; the Italian mines are historically noteworthy.

USES: used in the past for the extraction of gold and silver; employed in electrotechnology, in the preparation of explosives; at one time, as a medicine (calomel) and for fungicides.

NOTES: in nature, mercury easily forms alloys with the noble metals such as gold (amalgam), silver (kongsbergite and moschellandsbergite), palladium (potarite), all generically called amalgams. It is an excellent conductor of heat and electricity and it is soluble in nitric acid. At temperatures above 662°F (350°C), it is in the gaseous state.

▼ *Mercury (ca. x 0.8), Idria, Carniola, Slovenia.*

ARSENIC

COMPOSITION: As *native arsenic*.

CRYSTALLINE FORM: trigonal system, hexagonal scalenohedral class. Rare as pseudocubic crystals; in microcrystalline masses, commonly concentric with a mammilary appearance, sometimes stalactitic.

COLOR: white on a fresh cut, gray from surficial alteration.

HARDNESS: 3-3.5.

SPECIFIC GRAVITY: 5.4-5.9.

LUSTER: metallic.

GEOLOGIC OCCURRENCE: found in hydrothermal veins.

ASSOCIATED MINERALS AND LOCALITIES: with arsenides and sulfides of silver, nickel, and cobalt. Masses of remarkable size are found in Siberia; also found in Germany, France, Romania, and Italy.

USES: of no industrial importance; not used as a source of the metal.

NOTES: Arsenic gives its name to a group of chemically analogous metals (arsenic, antimony, bismuth). It is infusible, but it volatilizes at low temperature 842°F (450°C), giving off white fumes with a characteristic garlicky odor. It is a semi-metal with a high reflectance (52.6%) and good electrical conductivity, but unlike true metals, it is very brittle. In some cases it contains antimony, with which it is soluble in the solid state at high temperatures; at low temperatures, it forms the compound AsSb (allemontite) which has a semimetallic character. It may also contain small amounts of copper, silver and nickel.

▼ *Arsenic (ca. x 0.9), Zellerfeld, Germany.*

ANTIMONY

COMPOSITION: Sb *native antimony*.

CRYSTALLINE FORM: trigonal system, hexagonal scalenohedral class. The rare crystals have a pseudocubic, tabular habit and are sometimes twinned; commonly in microcrystalline masses or radial nodules.

COLOR: silver-white.

HARDNESS: 3-3.5.

SPECIFIC GRAVITY: 6.61-6.72.

LUSTER: metallic.

GEOLOGIC OCCURRENCE: in hydrothermal veins as a reduction product of sulfo-arsenides and sulfo-antimonides.

ASSOCIATED MINERALS AND LOCALITIES: with sphalerite, niccolite, stibnite. Small masses are found in Borneo, Canada, the United States, Portugal, Germany, and Sweden.

USES: interest limited to researchers and collectors; the metal is extracted from stibnite (antimony sulfide).

NOTES: the name is derived from the Latin *antimonium*. It belongs to the arsenic group and has the properties of semimetals: good conductivity, high reflectance (74.4%), but also highly brittle. With arsenic, it forms an isomorphic compound called allemontite (AsSb). It fuses at a low temperature 1166°F (630°C), coloring the flame greenish blue, and it is insoluble in acids.

▼ *Antimony (ca. x 0.7), Spain.*

GRAPHITE

COMPOSITION: C *native carbon*.

CRYSTALLINE FORM: hexagonal system, dihexagonal dipyramidal class. Common in small, densely striated lamellae with a hexagonal outline; in foliated masses.

COLOR: dark gray, lead-gray.

HARDNESS: 1-2.

SPECIFIC GRAVITY: 2.1-2.3.

LUSTER: metallic to submetallic.

GEOLOGIC OCCURRENCE: in high-temperature, low- to moderate-pressure metamorphic rocks derived from sediments rich in organic matter; rarely, in pegmatites and hydrothermal veins.

ASSOCIATED MINERALS AND LOCALITIES: with quartz, muscovite, calcite. Remarkable deposits are found in Sri Lanka, Madagascar, Mexico, the United States etc.

USES: in the electronics industry; as a dry lubricant, in the dyestuff and refractories industries; as the lead for pencils.

NOTES: the name is derived from the Greek "to write," in reference to the grayish imprint that the mineral leaves behind when rubbed on paper. It is a polymorph of carbon (the other being diamond) with two polytypes, one dihexagonal and one ditrigonal. It is infusible, and in hot concentrated nitric acid in the presence of KCO_3 it changes into a yellow crystalline substance called "graphitic acid." It is a good conductor of electricity, oxidizes with difficulty, and is greasy to the touch.

CHALCOCITE

COMPOSITION: Cu_2S *copper sulfide*.

CRYSTALLINE FORM: orthorhombic system, rhombic dipyramidal class. Crystals are rare, with a pseudohexagonal habit, striated; commonly in compact microcrystalline masses.

COLOR: lead-gray, blackish with bluish reflections.

HARDNESS: 2.5-3.

SPECIFIC GRAVITY: 5.5-5.8.

LUSTER: metallic.

GEOLOGIC OCCURRENCE: in hydrothermal veins; in oxidation zones of copper deposits.

ASSOCIATED MINERALS AND LOCALITIES: with bornite, covellite, chalcopyrite, pyrite, quartz, cuprite, azurite, malachite, calcite. Beautiful specimens come from South Africa, the United States, Namibia, Chile, Peru, and Mexico.

USES: important mineral source of copper (79.85 weight % when pure).

NOTES: the name is derived from the Greek for "copper," and it is also known as copper glance. It occurs in two polymorphs: one stable below 217°F (103°C) (orthorhombic) and one stable at higher temperatures (dihexagonal). It alters easily to basic copper carbonates. It fuses easily and during heating, releases very irritating sulfur dioxide fumes; colors the flame green. It is soluble in nitric acid.

▼ *Chalcocite (ca. x 1), Kolwezi, Musonoi, Shaba, Zaire.*

Graphite (ca. x 0.7), ▶ *provenance unknown.*

BORNITE

COMPOSITION: Cu_5FeS_4 *copper iron sulfide.*

CRYSTALLINE FORM: cubic system, hexoctahedral class. Rarely as crystals with a cubic, dodecahedral, or octahedral habit; generally in compact microcrystalline masses.

COLOR: reddish, bronze with superficial iridescent violet and blue alteration patinas (peacock ore).

HARDNESS: 3.

SPECIFIC GRAVITY: 5.06-5.08.

LUSTER: metallic.

GEOLOGIC OCCURRENCE: as a primary mineral in basic rocks; in pegmatites; in high-temperature hydrothermal veins; in metasomatic deposits like skarn; as an alteration product in copper deposits.

ASSOCIATED MINERALS AND LOCALITIES: with chalcopyrite, pyrite, quartz, malachite, calcite. Beautiful specimens come from Namibia, the United States, Mexico, Chile, Peru, Australia etc.

USES: among the principal minerals for the extraction of copper with 63.73 weight % (if pure).

NOTES: the name honors the Austrian naturalist Ignaz von Born (1742-1791). Occurs in three polymorphs: one cubic, stable at temperatures greater than 442ºF (228ºC); one tetragonal, stable below this temperature and the most common form in ore deposits; and a transitional trigonal one. It fuses with relative ease, forming a weakly magnetic globule. It is soluble in strong acids with the development of sulfur, and it is not fluorescent.

▼ *Bornite with malachite (ca. x 1.5), Montecatini in Val di Cecina, Pisa, Italy.*

ARGENTITE (ACANTHITE)

COMPOSITION: Ag_2S *silver sulfide.*

CRYSTALLINE FORM: cubic system, hexoctahedral class. Cubic, octahedral crystals; in filamentary, arborescent crystalline aggregates or in granular masses.

COLOR: lead-gray; brilliant when fresh cut, it tarnishes in the air.

HARDNESS: 2-2.5

SPECIFIC GRAVITY: 7.2-7.4.

LUSTER: metallic

GEOLOGIC OCCURRENCE: in low-temperature hydrothermal veins; in some lead and zinc deposits.

ASSOCIATED MINERALS AND LOCALITIES: with proustite, pyrargyrite, galena, cerussite, native silver. Beautiful specimens come from Mexico, Bolivia, Peru, Honduras, the United States, Norway, Germany etc.

USES: the chief source of silver (87.05 weight %, when pure).

NOTES: the compound Ag_2S occurs as three polymorphs: acanthite (monoclinic) is stable at temperatures lower than 354ºF (179ºC); the cubic hexoctahedral phase is stable between 354º and 1087ºF (179 and 586ºC); stable at higher temperatures is a cubic phase whose structure is unknown; crystals with cubic symmetry are designated by the name "argentite." It is a very soft, sectile mineral. It fuses easily, forming a globule of metallic silver, and during heating it gives off very irritating sulfide fumes. It is soluble in acids and is not fluorescent.

▼ *Argentite (ca. x 2), Sardinia, Italy.*

CHALCOPYRITE

COMPOSITION: $CuFeS_2$ *copper iron sulfide.*

CRYSTALLINE FORM: tetragonal system, scalenohedral class. Rare as crystals with a disphenoidal habit and a pseudotetrahedral morphology; common in compact microcrystalline masses, sometimes with a reniform or botryoidal appearance.

COLOR: brass-yellow with greenish tones commonly with alteration patinas in various colors, lending it a beautiful iridescence.

HARDNESS: 3.5-4.

SPECIFIC GRAVITY: 4.1-4.3.

LUSTER: metallic.

GEOLOGIC OCCURRENCE: in pegmatites and high-temperature hydrothermal veins.

ASSOCIATED MINERALS AND LOCALITIES: with pyrite, bornite, molybdenite, chalcocite, gold. There are large deposits in Canda, the United States, Chile, Zambia, and the Urals.

USES: chief mineral source of copper (34.6 weight %); gold and silver are also extracted as by-products.

NOTES: the name is derived from the Greek for "copper" and "fire," because when the mineral is struck with a hammer, it emits sparks. Its appearance and color are similar to pyrite with which it is commonly associated, but the two minerals can be distinguished by their streaks: green for chalcopyrite, black for pyrite. About 80% of the world's copper output comes from the processing of chalcopyrite. It fuses easily, giving off toxic fumes, and colors the flame green. It is slowly soluble in nitric acid with the separation of sulfur, and it is not fluorescent. (*See box, p. 92.*)

▼ *Chalcopyrite (ca. x 1.2), Tri-State District, Kansas, United States.*

ENARGITE

COMPOSITION: Cu_3AsS_4 *copper arsenic sulfide*

CRYSTALLINE FORM: orthorhombic system, pyramidal class. Rare as prismatic, tabular crystals, sometimes vertically striated; normally in radial lamellar aggregates and in microcrystalline masses.

COLOR: gray, iron-black.

HARDNESS: 3.

SPECIFIC GRAVITY: 4.4-4.5.

LUSTER: metallic.

GEOLOGIC OCCURRENCE: in medium-temperature hydrothermal veins.

ASSOCIATED MINERALS AND LOCALITIES: with bornite, tetrahedrite, covellite. Deposits of remarkable size are found in Chile, Peru, Mexico, Bolivia, the United States, Namibia, and the Philippines.

USES: industrially important mineral for the extraction of copper (48.4 weight %) and arsenic.

NOTES: the name is derived from the Greek for "distinct," in reference to the mineral's perfect cleavage. It fuses easily, is soluble in nitric acid, forming flakes of sulfur in the solution, and it is not fluorescent. Sometimes tennantite pseudomorphs of enargite, called "green enargite," are formed.

▼ *Enargite (ca. x 0.7), Peru.*

MILLERITE

COMPOSITION: NiS *nickel sulfide*.

CRYSTALLINE FORM: trigonal system, hexagonal scalenohedral class. Radial, acicular crystals, commonly in tufts with a felted apearance, or in incrustations with a velvety look.

COLOR: brass-yellow to bronze.

HARDNESS: 3-3.5.

SPECIFIC GRAVITY: 5.3-5.7.

LUSTER: metallic.

GEOLOGIC OCCURRENCE: in low-temperature hydrothermal veins; as an alteration product of other nickel minerals.

ASSOCIATED MINERALS AND LOCALITIES: with chalcopyrite, siderite, hematite, calcite, quartz. Beautiful specimens come from Germany, Great Britain, Italy, Bohemia, the United States etc.

USES: despite the fact that it contains a large weight percent of nickel, it does not occur in economically exploitable concentrations and so is a minor source of nickel.

NOTES: Millerite was named in honor of the English mineralogist W. H. Miller (1801-1880). At temperatures above 745°F (396°C), it becomes a polymorph with a structure similar to pyrrhotite. It is a good conductor of electricity and has a high reflectance (about 55%); elongated crystals are slightly elastic. It fuses easily, forming a magnetic globule. It is soluble in aqua regia and is not fluorescent.

▼ *Acicular millerite (ca. x 2), Harz, Germany.*

GALENA

COMPOSITION: PbS *lead sulfide*.

CRYSTALLINE FORM: cubic system, hexoctahedral class. Crystals are cubic, less commonly cubic octahedrons; generally in compact microcrystalline or sparry masses.

COLOR: lead-gray.

HARDNESS: 2.5.

SPECIFIC GRAVITY: 7.4-7.59.

LUSTER: metallic.

GEOLOGIC OCCURRENCE: in medium-temperature hydrothermal veins; in metasomatic deposits like skarns; lead mineralizations are found associated with calcsilicates and limestones, rare in lavas.

ASSOCIATED MINERALS AND LOCALITIES: with sphalerite, argentite, bornite, pyrite, quartz, fluorite, calcite. Beautiful specimens come from the United States, Australia, Mexico, Germany, England etc.

USES: the most important mineral source of lead (when pure, up to 86.6 weight %); silver is obtained as a by-product.

NOTES: the Latin word *galena* was used to designate lead. It fuses easily, 2039°F (1115°C), forming a yellow residue made up of lead monoxide. It is soluble in dilute nitric acid, with the formation of lead sulfate and minute flakes of sulfur, and some specimens exhibit fluorescence with the development of hydrosulfuric acid. Very brittle with perfect cleavage, it has good reflectance (35%). Through the action of atmospheric agents, it commonly alters to lead carbonate (cerussite), sulfate (anglesite), and more rarely, to oxides. (*See box, p. 92.*)

▼ *Galena (ca. x 0.7), Seravezza.*

NICCOLITE

COMPOSITION: NiAs *nickel arsenide*.

CRYSTALLINE FORM: hexagonal system, dihexagonal dipyramidal class. Crystals, with a pyramidal or tabular habit, are rare; usually in compact microcrystalline masses with a botryoidal appearance.

COLOR: copper-red, generally with dark or light green alteration patinas.

HARDNESS: 5-5.5.

SPECIFIC GRAVITY: 7.66-7.78.

LUSTER: metallic.

GEOLOGIC OCCURRENCE: in high-temperature hydrothermal veins; as an accessory in basic rocks (norite gabbros).

ASSOCIATED MINERALS AND LOCALITIES: with pyrrhotite, chalcopyrite, arsenopyrite, pentlandite. Beautiful specimens come from Argentina, Japan, Germany etc.

USES: exploited for the extraction of nickel.

NOTES: the name is derived from the Latin *nicolum* (nickel), in reference to its composition. Also called nickeline and originally known as kupfernickel, it is a brittle mineral with good reflectance (about 50%). Varieties with about 6% antimony are called arite. It fuses easily and during heating gives off arsenic-bearing fumes with a strong garlicky odor. It dissolves in nitric acid, coloring the solution green, and it is not fluorescent.

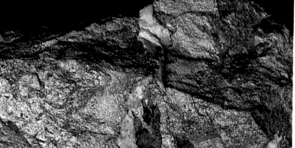

◄ *Massive niccolite (ca. x 1.5), Mansfeld, Germany.*

PYRRHOTITE

COMPOSITION: $Fe_{1-x}S$ *iron sulfide*.

CRYSTALLINE FORM: hexagonal system, dihexagonal dipyramidal class. The rare crystals are prismatic, tabular, pyramidal; generally in compact masses.

COLOR: bronze to reddish, sometimes with iridescence.

HARDNESS: 3.5-4.5.

SPECIFIC GRAVITY: 4.58-4.65.

LUSTER: metallic.

GEOLOGIC OCCURRENCE: primary mineral in basic and ultrabasic igneous rocks, especially norites; in high-grade metamorphic rocks; in high-temperature hydrothermal veins.

ASSOCIATED MINERALS AND LOCALITIES: with pentlandite, pyrite, marcasite, chalcopyrite, sphalerite. Beautiful specimens come from Canada, the United States, Mexico, Bolivia, Brazil, and South Africa.

USES: minor mineral source of iron; pyrrhotite masses with inclusions of pentlandite, $(Fe, Ni)_9S_8$, are very important due to the presence of nickel, cobalt, and platinum.

NOTES: the name is derived from the Greek for "reddish," referring to its color. Pyrrhotite is distinguished from pyrite by its magnetism. It fuses easily, forming a strongly magnetic, black residue. It is soluble in hydrochloric acid, with the development of sulfuric acid, and it is not fluorescent.

Pyrrhotite ▶ crystals (ca. x 1), Kisbanja, Romania.

COVELLITE

COMPOSITION: CuS *copper sulfide.*

CRYSTALLINE FORM: hexagonal system, dihexagonal dipyramidal class. Lamellar crystals, sometimes joined in rose-shaped aggregates; usually in compact micro-crystalline or earthy masses.

COLOR: indigo blue, commonly iridescent in crystals.

HARDNESS: 1.5-2.

SPECIFIC GRAVITY: 4.60-4.76.

LUSTER: submetallic to resinous.

GEOLOGIC OCCURRENCE: in hydrothermal veins; rarely, as a sublimate of volcanic gases.

ASSOCIATED MINERALS AND LOCALITIES: with pyrite, chalcopyrite, bornite. Beautiful specimens come from the United States, Bolivia, Chile, Italy etc.

USES: industrially important mineral source of copper (66.4 weight %, when pure).

NOTES: the name honors the Italian naturalist Nicola Covelli (1790-1829), who found the mineral on the slopes of Mount Vesuvius. Beautiful, very famous specimens used to come from the Sardinian mines (Alghero), but they are now closed down. Easily fusible, it gives off sulfur-dioxide fumes, and while burning, colors the flame blue. It is soluble in hydrochloric acid and is not fluorescent.

▼ *Covellite (ca. x 0.6), Calabona, Sassari, Italy.*

PYRITE

COMPOSITION: FeS *iron sulfide.*

CRYSTALLINE FORM: cubic system, diploidal class. Cubic crystals with characteristic striations on three mutually perpendicular faces, or pentagonal dodecahedral. This latter form is also called the pyritohedron since it is typical of pyrite; octahedral crystals less common; in compact masses.

COLOR: brass-yellow, golden yellow.

HARDNESS: 6-6.5.

SPECIFIC GRAVITY: 4.9-5.

LUSTER: metallic.

GEOLOGIC OCCURRENCE: widespread mineral from a variety of geologic environments: sedimentary (precipitated from natural waters); metamorphic (stable up to granulite facies); igneous (associated with basic rocks); the largest deposits are of medium-temperature hydrothermal origin.

ASSOCIATED MINERALS AND LOCALITIES: with many minerals, in particular sphalerite, galena, gold (auriferous pyrite). The most important deposits are found in the United States, Japan, Spain, Germany, and Italy.

USES: used for the production of sulfuric acid; gold, copper, nickel, and iron can be extracted as by-products.

NOTES: the name is derived from the Greek for "fire," because the mineral produces sparks when struck with a hammer. It is the most abundant sulfide in the earth's crust. It is easily altered by the action of atmospheric agents, forming oxides and hydroxides of iron (limonite) that in some cases form large masses (gossans) in the oxidation zones of sulfide deposits. Pyrite sometimes replaces fossil organic remains. Insoluble in hydrochloric acid, pulverized pyrite is decomposed by nitric acid. It is fusible, giving off sulfurous fumes. (*See box, p. 92.*)

HAUERITE

COMPOSITION: MnS$_2$ *manganese sulfide.*

CRYSTALLINE FORM: cubic system, diploidal class. Occurs as well-formed crystals with an octahedral habit, rarely cubic.

COLOR: chestnut-brown or blackish.

HARDNESS: 4.

SPECIFIC GRAVITY: 3.46.

LUSTER: metallic to adamantine.

GEOLOGIC OCCURRENCE: in sediments of evaporative origin including the cap-rock of salt domes; in iron and manganese nodules found in the ocean depths.

ASSOCIATED MINERALS AND LOCALITIES: with gypsum, calcite, native sulfur. Beautiful specimens come from Italy, the United States, Russia etc.

USES: interest limited to researchers and collectors.

NOTES: although it belongs to the pyrite group, hauerite does not exhibit metallic properties; its properties (rock-like appearance, color, notable magnetic susceptibility) distinguish it from other members of the series. It is soluble in hydrochloric acid and fuses with relative ease; in the closed tube it produces a sulfur sublimate, while in the open tube it yields sulfur dioxide.

▼ *Hauerite (ca. x 2), Raddusa, Catania, Italy.*

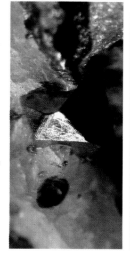

COBALTITE

COMPOSITION: CoAsS *cobalt sulfarsenide.*

CRYSTALLINE FORM: orthorhombic system, rhombic pyramidal class. Pseudoisometric, rarely as beautiful crystals with a pentagonal dodecahedral (pyritohedral) habit.

COLOR: white-gray, commonly slightly rosy (from the alteration of cobalt).

HARDNESS: 5.5.

SPECIFIC GRAVITY: 6.33.

LUSTER: metallic.

GEOLOGIC OCCURRENCE: as an accessory mineral in metalliferous deposits rich in cobalt and nickel.

ASSOCIATED MINERALS AND LOCALITIES: with chalcopyrite, pyrite. Beautiful specimens come from Norway, Sweden, Italy, the Caucasus, Canada etc.

USES: interest limited to researchers and collectors.

NOTES: the name is derived from the German *kobold* for goblin or sprite, alluding to the fact that cobalt deposits are commonly distinguished by a characteristic odor. It contains about 35.5% cobalt but, despite being very widespread, it does not occur in economically exploitable concentrations. It fuses easily, and is soluble in nitric acid with the separation of arsenic sulfides and oxides. It is isomorphous with gersdorffite (nickel sulfarsenide), and it is not flourescent.

▼ *Cobaltite (ca. x 7.5), Modum, Norway.*

◄ *Pyrite cubes (ca. 2.5), Gavorrano, Grosseto, Italy.*

MARCASITE

COMPOSITION: FeS₂ *iron sulfide.*

CRYSTALLINE FORM: orthorhombic system, dipyramidal class. Rare as beautiful isolated crystals with a prismatic, tabular habit; it can form twins called "cockcombs"; in nodules and in radial fibrous masses.

COLOR: golden yellow, brass-yellow.

HARDNESS: 6-6.5.

SPECIFIC GRAVITY: 4-6-4.9.

LUSTER: metallic.

GEOLOGIC OCCURRENCE: in low-temperature hydrothermal veins; in sedimentary rocks, as a chemical precipitate in a reducing environment; alteration product in reducing zones of pyrrhotite deposits.

ASSOCIATED MINERALS AND LOCALITIES: with bornite, galena, pyrite, chalcocite, quartz. Beautiful specimens come from the United States, England, Germany, etc.

USES: employed in the production of sulfuric acid; formerly utilized for costume jewelry because of its remarkable brilliance.

NOTES: the name is of Arabic origin. It alters very easily in the air, forming whitish blotches of melanterite (hydrated iron sulfate) and later disintegrates to a pulverulent product. It is very similar to pyrite from which it can be distinguished by its slightly lighter color and its greenish reflections. A colloidal variety of marcasite is called melnikovite marcasite. It fuses with relative ease and is soluble in hydrochloric acid and nitric acid in which it decomposes, forming flakes of sulfur. It is not fluorescent.

▼ *Marcasite (ca. x 1), Austria.*

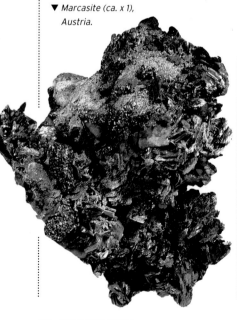

MIXED SULFIDE DEPOSITS

In general, the most commonly used metals are extracted from their sulfur compounds, called "sulfides."
Some of these have a similar origin and are found in the same deposit; in some places they are accompanied by native elements.

The open-pit copper mine in Bingham Canyon is the biggest excavation in the world.

Among the most common sulfides are those of lead (galena, see page 90), zinc (sphalerite, see page 34), copper (chalcopyrite, see page 89), and iron (pyrite, see page 91) which are often found together and can form very extensive deposits. Mixed sulfide deposits can originate in very different environments, but always require the presence of a heat source and of mineralizing fluids such as those produced by igneous intrusions. To simplify matters, we can divide them into those associated with skarns and those with porphyry copper deposits.

SKARNS

A skarn is a product of metasomatic alteration of the rocks surrounding an igneous intrusion. Deposits of this type commonly develop in carbonate rocks, which are among the most reactive relative to infiltrating mineralizing fluids. The transformed rocks, or skarns, are essentially composed of calcsilicate minerals among which are hedenbergite, grossular garnet, vesuvianite, epidote, actinolite, wollastonite, diopside, and anorthite.

Marbles, which represent metamorphosed limestones that have not been chemically altered, are also commonly associated with skarns. In formations with low iron contents, sphalerite and galena are common minerals. In other cases, where iron is prevalent, there may be an abundance of pyrite and chalcopyrite. In general, these deposits have a high concentration of useful minerals but are of limited size. The useful minerals amount, at the most, to tens of millions of tons, and their extraction generally takes place in mines with the classic tunnel construction.

PORPHYRY COPPER

Deposits of the *porphyry copper* type were first identified at the beginning of the last century in the United States, Chile, and Peru. But because of the relatively low percentage of useful minerals, they were not put into production until the beginning of this century when technologic advances made mining them profitable. Porphyry copper deposits with low concentrations of copper or molybdenum disseminated throughout large masses of rocks are worked in open-pit mines (see photo above). In general, there is a superficial portion in which secondary minerals occur that are derived from the alteration of primary, deeper rocks. In the more fractured areas, small masses with a good concentration of useful minerals may be present. However, the majority are scattered in enormous masses so that the minimum acceptable quantity is about 20 million tons of rock with a 0.1 weight percent of copper. A large deposit contains a billion tons of rock with a 0.8 to 3 weight percent of copper. A good deposit can produce about 30 million tons of copper. Interest in these deposits has recently increased since it has been discovered that they also contain discreet amounts of gold and silver.

ARSENOPYRITE

COMPOSITION: FeAsS *iron arsenic sulfide*

CRYSTALLINE FORM: monoclinic system, prismatic class. Elongated, striated crystals with a prismatic habit; twinning generally results in a psuedo-orthorhombic habit with a lance-tip (*fer-de-lance*) shape or, with multiple twinnings, a stellate shape; in compact or granular masses.

COLOR: tin-white, steel-gray.

HARDNESS: 5.5-6.

SPECIFIC GRAVITY: 5.92-6.22.

LUSTER: metallic.

GEOLOGIC OCCURRENCE: found in high-temperature hydrothermal veins.

ASSOCIATED MINERALS AND LOCALITIES: with chalcopyrite, gold, quartz, orthoclase. The largest deposits are found in Sweden, Norway, Mexico, Canada, and the United States.

USES: the most important mineral for the extraction of arsenic (contains 46 weight %); gold, silver, tin and cobalt can be extracted as by-products.

NOTES: because of its similarity to pyrite, arsenopyrite formerly was called "arsenical pyrites"; the present name comes from a contraction of these two words, but it is also known as mispickel. When struck with a hammer, it emits sparks. It fuses easily, giving off white fumes with a garlicky odor. It is soluble in nitric acid with separation of sulfur; is not fluorescent.

▼ *Arsenopyrite (ca. x 1), Panasqueira, Portugal.*

GLAUCODOT

COMPOSITION: (Fe, Co)AsS *iron cobalt arsenic sulfide.*

CRYSTALLINE FORM: monoclinic system, prismatic class. Occurs as prismatic crystals; in micrcrystalline masses with a radial fibrous structure.

COLOR: tin-white, light gray, sometimes with pink spots from the presence of erythrite.

HARDNESS: 5.

SPECIFIC GRAVITY: 5.92-6.14.

LUSTER: metallic.

GEOLOGIC OCCURRENCE: in pneumatolytic and high-temperature hydrothermal veins; rarely, in metamorphic rocks derived from basic volcanics (greenschists).

ASSOCIATED MINERALS AND LOCALITIES: with arsenopyrite, pyrite, chalcopyrite, galena. Very beautiful specimens come from Sweden, Romania, Norway, Chile, and Tasmania.

USES: useful mineral for the extraction of cobalt (average 25 weight %).

NOTES: Glaucodot, gudmundite (FeSbS), and arsenopyrite make up the arsenopyrite series. When subjected to heat, it decomposes, giving off whitish fumes with a garlicky odor. It is soluble in nitric acid with development of sulfur, and is not fluorescent.

▼ *Glaucodot (ca. x 1.5), Haakonsboda, Sweden.*

MOLYBDENITE

COMPOSITION: MoS_2 *molybdenum sulfide.*

CRYSTALLINE FORM: hexagonal system, dihexagonal dipyramidal class. Rare as tabular crystals with a hexagonal outline; common in lamellar aggregates with a foliated appearance.

COLOR: lead-gray.

HARDNESS: 1-1.5.

SPECIFIC GRAVITY: 4.62-4.73.

LUSTER: metallic.

GEOLOGIC OCCURRENCE: as an accessory mineral in many igneous rocks (from granites to basalts); in pegmatites and pneumatolytic veins; in metasomatic deposits, like skarn.

ASSOCIATED MINERALS AND LOCALITIES: with chalcopyrite, cassiterite, pyrite, scheelite, garnets, quartz. Beautiful specimens come from Australia, the United States, Bolivia, Norway etc.

USES: principal mineral for the extraction of molybdenum (60 weight %); used for special alloys and as a highly heat-resistant dry lubricant.

NOTES: a very brittle mineral, molybdenite is greasy to the touch, and leaves a gray-green mark when rubbed on a sheet of paper. It is infusible, partially soluble in nitric acid, dissolves in aqua regia, and it is not fluorescent.

▼ *Molybdenite (ca. x 0.5), Australia.*

SYLVANITE

COMPOSITION: $AuAgTe_4$ *gold silver telluride.*

CRYSTALLINE FORM: monoclinic system, prismatic class. Occurs as stubby crystals, more commonly in dendritic or arborescent aggregates; in microcrystalline or lamellar masses.

COLOR: yellowish white, faded yellow.

HARDNESS: 1.5-2.

SPECIFIC GRAVITY: 8.161.

LUSTER: metallic.

GEOLOGIC OCCURRENCE: in low-temperature hydrothermal veins.

ASSOCIATED MINERALS AND LOCALITIES: fairly rare, found with native gold, calaverite, other tellurides and gangue minerals. Very beautiful specimens come from Romania, Australia, the United States etc.

USES: useful mineral for the extraction of gold (24 weight %), silver (13 weight %), and tellurium.

NOTES: Sylvanite belongs to the calaverite, $(Au,Ag)Te_2$, series. It is a very soft, heavy mineral with a good reflectance (46-57%). Crystalline individuals joined in dendritic aggregates resemble writing, hence the nickname "graphic gold." It fuses easily, giving rise to a yellowish white, metallic globule of gold and silver. Soluble in nitric acid, it forms a yellow residue (gold), while in hot concentrated sulfuric acid, it decomposes and colors the solution dark red (tellurium). It is not fluorescent.

Sylvanite (ca. x 2), ▶ *Nagyag, Romania.*

SKUTTERUDITE

COMPOSITION: $CoAs_3$ *cobalt arsenide.*

CRYSTALLINE FORM: cubic system, diploidal class. Rare as crystals with a cubic or octahedral habit, or a combination of the two forms; usually in granular masses.

COLOR: tin-white, steel-gray with iridescent patinas.

HARDNESS: 5.5-6.

SPECIFIC GRAVITY: 6.1-6.9.

LUSTER: metallic.

GEOLOGIC OCCURRENCE: in medium- and high-temperature hydrothermal veins.

ASSOCIATED MINERALS AND LOCALITIES: arsenopyrite, native silver and bismuth, cobaltite, niccolite, calcite. Beautiful specimens come from Norway, Spain, Germany, Morocco, the United States, and Canada.

USES: industrially important mineral for the extraction of cobalt, arsenic and nickel.

NOTES: the name is derived from the Norwegian mining locality Skutterud, wherethe mineral was discovered. Skutterud also refers to a group of isomorphous minerals made up of skutterudite, chloanthite ($NiAs_3$), and chathamite ((Co, Ni, Fe) As_3). Smaltite is a variety of skutterudite that is particularly rich in nickel, with a weight percentage of arsenic below stoichiometric proportions. Skutterudite fuses easily, forming a magnetic globule, and during heating it gives off arsenical fumes with a garlicky odor. It is soluble in hot nitric acid, coloring the solution pink and it is not fluorescent.

BOURNONITE

COMPOSITION: $PbCuSbS_3$ *lead copper antimony sulfide.*

CRYSTALLINE FORM: orthorhombic system, dipyramidal class. Rare as prismatic, stubby, striated crystals, sometimes with cross-shaped twinning; generally in granular aggregates.

COLOR: lead-gray, steel-gray, iron-black.

HARDNESS: 2.5-3.

SPECIFIC GRAVITY: 5.8-5.86.

LUSTER: metallic.

GEOLOGIC OCCURRENCE: found in medium-temperature hydrothermal veins.

ASSOCIATED MINERALS AND LOCALITIES: with galena, tetrahedrite, chalcopyrite, pyrite, sphalerite, quartz. Beautiful specimens come from Germany, Bolivia, Mexico, the United States etc.

USES: mineral of secondary importance in the industrial sector for the extraction of lead, copper, and antimony.

NOTES: the name honors the French crystallographer Count J.-L. de Bournon (1751-1825). It is isostructural with the more rare $PbCuAsS_3$ (seligmannite) with which it can form partial isomorphous mixtures. Bournonite is easily fusible, and it is soluble in nitric acid, coloring the solution green and leaving a residue of sulfur, antimony and lead powder. It is not fluorescent.

◀ *Skutterudite (ca. x 1), Huelva, Spain.*

TETRAHEDRITE

COMPOSITION: $Cu_{12}(SbS_3)_4S$ *copper sulfantimonide.*

CRYSTALLINE FORM: cubic system, hextetrahedral class. Very highly modified crystals with a tetrahedral habit; generally in compact or granular masses.

COLOR: steel-gray, iron-black.

HARDNESS: 3-4.5.

SPECIFIC GRAVITY: 4.6-5.1.

LUSTER: metallic.

GEOLOGIC OCCURRENCE: in low- or medium-temperature hydrothermal veins.

ASSOCIATED MINERALS AND LOCALITIES: with galena, bornite, pyrite, chalcopyrite, quartz. Beautiful specimens come from Namibia, Bolivia, Chile, Peru, the United States, Romania, and Sweden.

USES: an industrially important mineral for the extraction of copper (about 46 weight %) and of other elements such as silver, mercury, and antimony.

NOTES: the name refers to the typical tetrahedral form of themineral's crystals. It is isostructural with tennantite [$Cu_{12}(AsS_3)_4S$] with which it forms mixtures in which bismuth may also be present, replacing antimony. As a result, many minerals may be formed among which are annivite with bismuth, freibergite with silver (in some specimens up to 18%), and schwazite with mercury. It fuses easily, decomposes in nitric acid with separation of sulfur and oxides of arsenic and antimony, and it is not fluorescent.

▲ *Tetrahedrite (ca. x 3), Baia-Mela, Romania.*

◀ *Bournonite (ca. x 1.5), Horhausen, Germany.*

GEOCRONITE

COMPOSITION: $Pb_5Sb_2S_8$ *lead antimony sulfide*

CRYSTALLINE FORM: monoclinic system, prismatic class. Appears very rarely as stubby, prismatic crystals; rare in granular masses.

COLOR: lead-gray to bluish gray.

HARDNESS: 2.5.

SPECIFIC GRAVITY: 6.3-6.5.

LUSTER: metallic.

GEOLOGIC OCCURRENCE: as an accessory in lead and tin mineralizations in hydrothermal veins.

ASSOCIATED MINERALS AND LOCALITIES: with galena, and quartz. Beautiful specimens come from Sweden, Switzerland, Italy etc.

USES: interest limited to researchers and collectors.

NOTES: Geocronite is isostructural and partially isomorphous with jordanite ($Pb_5As_2S_8$); in these minerals, the ratios among the various elements are not known with certainty and in both arsenic and antimony can be substituted for each other. It fuses easily, is soluble in hot hydrochloric acid with the development of hydrosulfuric acid and separation of lead chloride, and it is not fluorescent.

▼ *Geocronite (ca. x 1), Seravezza, Buca della Vena, Lucca, Italy.*

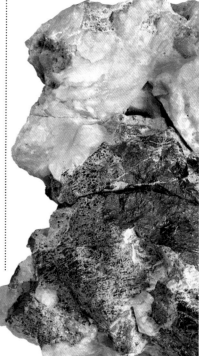

JAMESONITE

COMPOSITION: $Pb_4FeSb_6S_{14}$ *lead iron antimony sulfide.*

CRYSTALLINE FORM: monoclinic system, prismatic class. Crystals are rare, with a prismatic, acicular habit; generally in aggregates with a felted appearance, made up of slender intertwined crystals; in fibrous masses.

COLOR: grayish black.

HARDNESS: 2.5.

SPECIFIC GRAVITY: 5.48–5.72.

LUSTER: metallic.

GEOLOGIC OCCURRENCE: in medium-temperature hydrothermal veins.

ASSOCIATED MINERALS AND LOCALITIES: with tetrahedrite, pyrite, stibnite, gold, quartz. Beautiful specimens come from Romania, Bohemia, Mexico, the United States, and Canada.

USES: mineral of secondary importance for the extraction of lead.

NOTES: the name honors the Scotsman Robert Jameson (1774–1854) who first described the mineral. It fuses easily and is soluble in nitric acid with separation of Sb_2O_3 (valentinite) and lead sulfate. In the past, acicular masses with a feathery appearance called plumosite were considered to be a variety of jamesonite, but X-ray analysis has shown that they are really a type of boulangerite ($Pb_5Sb_4S_{11}$).

▼ *Jamesonite (ca. x 1.4), Bottino, Lucca, Italy.*

STIBNITE

COMPOSITION: Sb_2S_3 *antimony sulfide.*

CRYSTALLINE FORM: orthorhombic system, dipyramidal class. Acicular, prismatic crystals, commonly striated and elongated in the shape of a small cane; in parallel or radial fibrous aggregates; in compact microcrystalline masses or with a felted look.

COLOR: lead-gray with reddish or yellowish pulverulent alteration patinas (kermesite or other secondary antimony minerals).

HARDNESS: 2.

SPECIFIC GRAVITY: 4.61–4.65.

LUSTER: metallic.

GEOLOGIC OCCURRENCE: in low-temperature hydrothermal veins; as a chemical precipitate from hot springs.

ASSOCIATED MINERALS AND LOCALITIES: with realgar, orpiment, cinnabar, calcite. Beautiful specimens come from China, Peru, Mexico, and Bolivia where the largest deposits are found.

USES: principal mineral for the extraction of antimony (71.7 weight %, when pure); used in metal alloys, in the fireworks industry, and for electric batteries. Antimony salts are used in the textile industry and for glassware.

NOTES: the name seems to have been derived from the Greek word for antimony; it is also known under the names antimonite or gray antimony. Crystals are flexible, but not elastic and it is isomorphous with bismuthinite (Bi_2S_3). It is easily fusible 1022°F (550°C), to the point that the heat of a wax match will suffice when the mineral is in very small flakes; colors the flame blue. It is soluble in concentrated hydrochloric acid, and, mixed with potassium hydroxide, it yields a yellow-orange solution.

▼ *Stibnite (ca. x 1), Felsobajna, Romania.*

BISMUTHINITE

COMPOSITION: Bi_2S_3 *bismuth sulfide.*

CRYSTALLINE FORM: orthorhombic system, dipyramidal class. Prismatic, elongated crystals with striated crystalline faces; in compact granular or radial aggregates.

COLOR: lead-gray, tin-white, dull yellowish.

HARDNESS: 2.

SPECIFIC GRAVITY: 6.75–6.81.

LUSTER: metallic.

GEOLOGIC OCCURRENCE: an accessory in mineralizations of high-temperature hydrothermal veins.

ASSOCIATED MINERALS AND LOCALITIES: with arsenopyrite, chalcopyrite, gold, quartz. Beautiful specimens come from Bolivia, Peru, Australia, the United States, Canada, Mexico etc.

USES: principal mineral for the extraction of bismuth.

NOTES: Bismuthinite, or bismuth glance, is isomorphous with stibnite (Sb_2S_3), which it resembles both in appearance and in physical properties. Bismuthinite fuses easily, but not in a wax match flame (test used to distinguish it from stibnite). It is soluble in nitric acid with separation of minute flakes of sulfur.

Bismuthinite (ca. 0.8), ▶
La Roja, Peru.

TENORITE

COMPOSITION: CuO *copper oxide.*

CRYSTALLINE FORM: monoclinic system, prismatic class. Occurs as minute laminae with a pseudohexagonal outline; in pulverulent, earthy masses.

COLOR: black, iron-gray, steel.

HARDNESS: 3.5.

SPECIFIC GRAVITY: 5.8–6.5.

LUSTER: metallic.

GEOLOGIC OCCURRENCE: in the oxidation zones of copper deposits; as a sublimate of volcanic fumaroles.

ASSOCIATED MINERALS AND LOCALITIES: with cuprite, native copper, cotunnite, melanothallite, paratacamite. Beautiful crystallizations come from Mount Vesuvius, Germany, and the United States.

USES: the variety most often found in practicable deposits is the black melaconite, a good secondary mineral for the extraction of copper (79.8 weight %).

NOTES: the name honors botanist M. Tenore. The paratenorite variety is a polymorphous modification with tetragonal symmetry. It is infusible, soluble in dilute hydrochloric and nitric acids, and, it is not fluorescent.

▼ *Tenorite (ca. x 2), Mount Vesuvius, Naples, Italy.*

CHROMITE

COMPOSITION: FeCr$_2$O$_4$ *iiron chromate.*
CRYSTALLINE FORM: cubic system, hexoctahedral class. Very rare as crystals, always small and with an octahedral or dodecahedral habit; generally in microcrystalline masses or scattered granules.
COLOR: gray to gray-black or black-chestnut-brown.
HARDNESS: 5.5.
SPECIFIC GRAVITY: 4.5-4.8.
LUSTER: submetallic to metallic.
GEOLOGIC OCCURRENCE: in basic or ultrabasic igneous rocks; in serpentine rocks; concentrated in detrital deposits such as placers; in meteorites.
ASSOCIATED MINERALS AND LOCALITIES: with olivine, pyroxenes, uvarovite. Beautiful specimens come from Namibia, South Africa, the Philippines, Cuba, the Urals.
USES: principal mineral for the extraction of chromium, used in the iron and steel industry, in the preparation of special paints and for the tanning of hides.
NOTES: the name refers to the element chromium (from the Latin *chroma*), which is contained in the mineral. Chromite also refers to one of the three series of the spinel group, the others being the spinel series and the magnetite series. It is infusible, insoluble in acids, weakly magnetic, and it is not fluorescent.

▼ *Chromite (ca. x 0.8), Bohemia, Czech Republic.*

MAGNETITE

COMPOSITION: Fe^{+2}Fe$^{+3}_2$O$_4$ *iron oxide.*
CRYSTALLINE FORM: cubic system, hexoctahedral class. Crystals have an octahedral, rhombic dodecahedral habit with striated faces; generally in compact and granular masses.
COLOR: black, sometimes iron-gray.
HARDNESS: 5.5-6.5.
SPECIFIC GRAVITY: 5.17-5.18.
LUSTER: metallic.
GEOLOGIC OCCURRENCE: ubiquitous mineral; an accessory in basic and ultrabasic igneous rocks (can be concentrated in very extensive layers); in contact metamorphic rocks; in metasomatic deposits like skarn; in alluvial and marine sands, where it forms large accumulations.
ASSOCIATED MINERALS AND LOCALITIES: with andradite, almandine, chlorite, apatite. Large masses are found in South Africa, the United States, Sweden, the Urals etc.
USES: principal mineral for the extraction of iron (71.5 weight %, when pure).
NOTES: the name is derived from Magnesia, a region of ancient Macedonia. It loses its magnetism when heated to 1072°F (578°C); some varieties contain chromium (chromomagnetite) or titanium (titano-magnetite). Magnetite also refers to to one of the three series of minerals (the others being spinel and chromite) that make up the spinel group. It fuses at 2901°F (1594°C), with separation of hematite. It is soluble in concentrated hydrochloric acid and is not fluorescent.

▼ *Magnetite (ca. x 1.4), Traversella, Turin, Italy.*

HEMATITE

COMPOSITION: Fe$_2$O$_3$ *iron oxide.*
CRYSTALLINE FORM: trigonal system, ditrigonal scalenohedral class. Crystals are rhombohedral, stubby (oligist iron), lamellar, sometimes joined in rose-shaped aggregates (micaceous hematite or iron rose); in compact masses with a radial fibrous structure and a botryoidal appearance (kidney ore); in earthy, microcrystalline masses with an oölitic or pisolitic structure (red ocher).
COLOR: brown, brilliant to dull red; iron-gray, almost black, with blue, green, and red iridescence.
HARDNESS: 5-6.
SPECIFIC GRAVITY: 5.
LUSTER: submetallic to metallic.
GEOLOGIC OCCURRENCE: as an accessory mineral in igneous rocks; in pegmatites and hydrothermal veins; in low-grade metamorphic rocks; common in sedimentary rocks from diagenetic alteration or metasomatism of iron-rich sediments.
ASSOCIATED MINERALS AND LOCALITIES: with magnetite, siderite, diopside, epidote, quartz. Beautiful specimens come from Venezuela, Brazil, Canada, the United States, India, Italy, etc.
USES: important for the extraction of iron (69.9 weight %); used as a pigment (red ocher or burnt sienna).
NOTES: the name is derived from the Greek for "blood red," in relation to the bright red color of its powder, which distinguishes it from magnetite and ilmenite (black powder). Microcrystalline hematite scattered in sediments gives these rocks a reddish coloration. The pseudomorphous variety after magnetite is called martite. It is infusible and soluble in concentrated hydrochloric acid; in the reducing flame, it becomes magnetic. It is not flourescent.

Hematite (ca. x 1), ▶
Island of Elba,
Livorno, Italy.

ILMENITE

COMPOSITION: FeTiO$_3$ *titanium iron oxide.*
CRYSTALLINE FORM: trigonal system, rhombohedral class. Crystals are rhombohedral, tabular; in rose-shaped lamellar aggregates; in compact masses.
COLOR: iron-black, black-red.
HARDNESS: 5-6.
SPECIFIC GRAVITY: 4.72.
LUSTER: submetallic to metallic.
GEOLOGIC OCCURRENCE: as an accessory in plutonic rocks (gabbros, norites) and in metamorphic rocks (gneiss, granulites); the magnesiferous variety is an important constituent of the nodules and xenoliths included in kimberlites; in marine sands.
ASSOCIATED MINERALS AND LOCALITIES: with magnetite, hornblende, micas, rutile, diopside. Beautiful specimens come from Norway, Russia, India, Canada etc.
USES: for the extraction of titanium.
NOTES: the name is derived from the Ilmen Mountains in Russia, where it was first discovered. Weakly magnetic, it sometimes exhibits an alteration product, leucoxene (rutile and pseudorutile), which distinguishes it from magnetite; it can be differentiated from hematite by the black color of its powder. It is infusible, soluble with difficulty in concentrated hydrochloric acid, and it is not fluorescent.

▲ *Ilmenite (ca. x 0.7), Norway.*

PYROLUSITE

COMPOSITION: MnO_2 *manganese oxide.*
CRYSTALLINE FORM: tetragonal system, ditetragonal dipyramidal class. Very rarely as prismatic crystals; generally in radial fibrous aggregates and in compact, earthy, pulverulent masses.
COLOR: iron-black, dark gray.
HARDNESS: 6-6.5.
SPECIFIC GRAVITY: 5.06.
LUSTER: submetallic to metallic.
GEOLOGIC OCCURRENCE: as a chemical precipitate in a sedimentary environment (lacustrine, lagoons); alteration product of manganiferous minerals.
ASSOCIATED MINERALS AND LOCALITIES: with other manganese oxides (wad), hematite, barite, calcite. Large masses are found in India, Brazil, South Africa, Georgia, Germany, and Bohemia.
USES: important mineral for the extraction of manganese; used in the iron industry and in glassworks.
NOTES: the name is derived from the Greek for "fire" and "to wash," in relation to its use as a bleaching agent for glass. The hardness and specific gravity listed above refer to the crystals, as microcrystalline masses have varying values: 1-6 for hardness, and 4.4-5 for specific gravity. The earthy variety, very soft and greasy to the touch, leaves a black streak when rubbed across a rough surface. It is infusible, soluble in hydrochloric acid, giving off acrid chlorine fumes, and it is not fluorescent.

URANINITE

COMPOSITION: UO_2 *uranium oxide.*
CRYSTALLINE FORM: cubic system, hexoctahedral class. Crystals have an octahedral, cubic, or rarely, a rhombic dodecahedral habit; common in compact cryptocrystalline masses with a pitchy appearance (pitchblende variety), sometimes with botryoidal or concretionary surfaces; pseudomorphic after organic material such as wood and bone.
COLOR: black, gray, green, chestnut-brown.
HARDNESS: 5-6.
SPECIFIC GRAVITY: 8-10.6.
LUSTER: greasy submetallic.
GEOLOGIC OCCURRENCE: in pegmatites, medium- and high-temperature hydrothermal veins; detrital mineral (in the matrix of some conglomerates) or autogenous mineral in sedimentary rocks such as sandstones.
ASSOCIATED MINERALS AND LOCALITIES: with pyrite, galena, marcasite, chalcopyrite, cassiterite, gold. Beautiful specimens come from Canada, the United States, South Africa, Portugal, Bohemia etc.
USES: fundamental mineral for the extraction of uranium; used in nuclear reactors for the production of electrical energy and in the preparation of fissionable material for military purposes; radium is obtained as a by-product.
NOTES: Uraninite is highly radioactive. In 1898, while analyzing some pitchblende, a type of uraninite, Marie and Pierre Curie discovered two other radioactive elements: polonium and radium.

COLUMBITE

COMPOSITION: $FeNb_2O_6$ *niobium iron oxide.*
CRYSTALLINE FORM: orthorhombic system, dipyramidal class. Small, prismatic, stubby crystals, commonly in heart-shaped twins and striated; in massive aggregates and scattered grains.
COLOR: iron-black, black-brown, sometimes with reddish tints.
HARDNESS: 6.
SPECIFIC GRAVITY: 5.2.
LUSTER: submetallic.
GEOLOGIC OCCURRENCE: in pegmatites which are particularly rich in silicates and phosphates of lithium; commonly concentrated in placers.
ASSOCIATED MINERALS AND LOCALITIES: with spodumene, lepidolite, beryl, cryolite, corundum, quartz. Beautiful specimens come from Australia, Zaire, Madagascar, Brazil, Canada, Norway etc.
USES: principal mineral for the extraction of niobium and tantalum (present as a replacement for niobium) used in the metallurgic industry and for the construction of high-technology apparatus.
NOTES: Columbite was found for the first time in America in about 1600, and its name commemorates the discoverer of the New World (the original specimen can still be found in the collections of the British Museum). It is moderately magnetic, infusible, soluble only in sulfuric acid, and it is not fluorescent.

GOETHITE

COMPOSITION: $FeO(OH)$ *hydrous ferrous oxide.*
CRYSTALLINE FORM: orthorhombic system, dipyramidal class. Very rarely as elongated crystals with a prismatic habit and vertical striations; generally in stalactitic masses with a radial structure formed by acicular crystals; in earthy and porous masses (yellow ocher variety).
COLOR: gray-black, brown-red, bright yellow (yellow ocher).
HARDNESS: 5-5.5.
SPECIFIC GRAVITY: 3.3-4.3.
LUSTER: submetallic.
GEOLOGIC OCCURRENCE: typical mineral in the oxidation zones (gossan) of ferriferous accumulations; essential component of limonitic deposits.
ASSOCIATED MINERALS AND LOCALITIES: with psilomelane, manganite, hematite, calcite, quartz. Beautiful specimens come from Canada, the United States, Cuba, Germany, and Great Britain.
USES: important mineral in the industrial sector for the extraction of iron (62 weight %, when pure); some yellow ochers are used as dyes.
NOTES: the name honors the German poet J. W. von Goethe (1749-1832). When subjected to prolonged heating, it becomes magnetic; heated in the closed tube, it loses water and turns to hematite. It is infusible, soluble in hydrochloric acid. and it is not fluorescent.

Uraninite (ca. x 2), Czechoslovakia.

Goethite (ca. x 1.5), ▶ Morocco.

Columbite (ca. x 1), ▶ Raahe, Finland.

◀ *Pyrolusite (ca. x 1), Illfeld, Harz, Germany.*

IGNEOUS ROCKS

IGNEOUS ROCKS

Top: Capo d'Orso, Sardinia. Wind erosion has molded a block of granite into a bizarre shape.

Below: Classification diagrams and nomenclature of volcanic rocks (extrusive).

Texture, an important physical characteristic of rocks, can be used to distinguish between intrusive and extrusive igneous rocks with some degree of ease in the field. The term "texture" refers to the shape, size, and arrangement of the crystals in a rock.

Intrusive rocks are characterized by a holocrystalline texture; i.e., they are made up of crystals visible to the naked eye. These crystals are generally equidimensional, although there are some exceptions. For example, in some granites orthoclase is larger than the other minerals and the rock has a porphyritic texture. This milky-looking mineral is commonly pink or even dark red from inclusions of iron oxides. Porphyritic texture can also be found in some diorites (rocks with a generally darker color than granites) in which the larger crystals are plagioclases.

Acidic extrusive rocks are light-colored because of the prevalence of quartz and feldspars. They have, for the most part, a granular, holocrystalline, porphyritic texture, sometimes with flow lines indicative of the lava's movement. Between the larger crystals, there is a microcrystalline or glassy groundmass in which the single crystals are too small to be recognizable.

Basic extrusive rocks are generally very dark in color and are characterized by a fine-grained texture. The crystals are invisible to the naked eye and cannot generally be distinguished even with a hand lens. These rocks may be partially glassy, due to sudden cooling (hyaline texture). Some have porphyritic and fluidal textures. Volcanic rocks may also have small cavities produced by the degassing of the mass or by the alteration of some minerals.

Let us now try to understand how igneous rocks can be distinguished from one another. The ways in which rocks form and the variety of appearances they exhibit (for example, altered or not, fine-grained or coarse-grained, etc.) often make identifying them complicated. Therefore, we will resort to a few simplifications, utilizing only the principal minerals which are generally visible to the naked eye or with the aid of a magnifying glass.

As the composition of a rock reflects the magma from which it formed, using minerals as a means of identifying rocks not only helps us to understand the differences between specific rocks, but also provides us with information about their origins.

The important minerals used to classify igneous rocks include the alkali

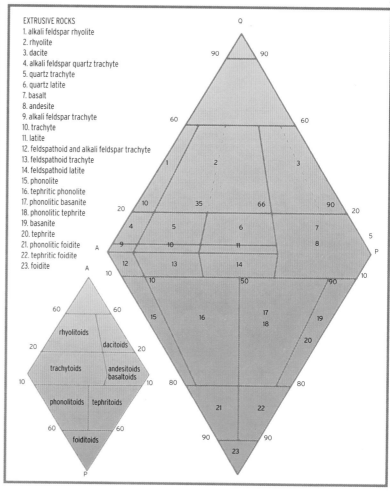

EXTRUSIVE ROCKS
1. alkali feldspar rhyolite
2. rhyolite
3. dacite
4. alkali feldspar quartz trachyte
5. quartz trachyte
6. quartz latite
7. basalt
8. andesite
9. alkali feldspar trachyte
10. trachyte
11. latite
12. feldspathoid and alkali feldspar trachyte
13. feldspathoid trachyte
14. feldspathoid latite
15. phonolite
16. tephritic phonolite
17. phonolitic basanite
18. phonolitic tephrite
19. basanite
20. tephrite
21. phonolitic foidite
22. tephritic foidite
23. foidite

Below left: Classification diagrams and nomenclature of plutonic rocks (intrusive).

At the vertices: Q = quartz; A = alkaline feldspars (including albite); P = plagioclases; F = feldspathoids.

Below: intrusive igneous rocks and their volcanic equivalents. Note the variation in color, from light (granite, rhyolite) to dark (gabbro, basalt), which relates to the reduction in the quantity of light minerals (quartz and feldspars).

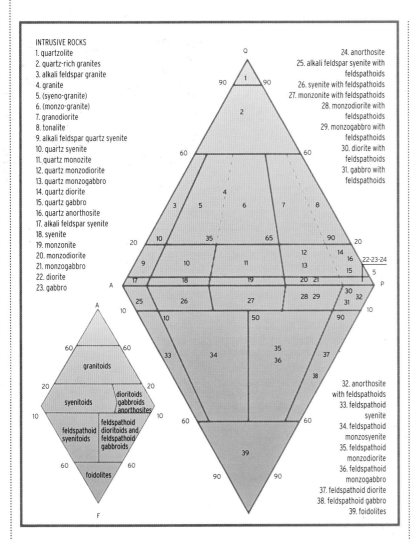

INTRUSIVE ROCKS
1. quartzolite
2. quartz-rich granites
3. alkali feldspar granite
4. granite
5. (syeno-granite)
6. (monzo-granite)
7. granodiorite
8. tonalite
9. alkali feldspar quartz syenite
10. quartz syenite
11. quartz monozite
12. quartz monzodiorite
13. quartz monzogabbro
14. quartz diorite
15. quartz gabbro
16. quartz anorthosite
17. alkali feldspar syenite
18. syenite
19. monzonite
20. monzodiorite
21. monzogabbro
22. diorite
23. gabbro

24. anorthosite
25. alkali feldspar syenite with feldspathoids
26. syenite with feldspathoids
27. monzonite with feldspathoids
28. monzodiorite with feldspathoids
29. monzogabbro with feldspathoids
30. diorite with feldspathoids
31. gabbro with feldspathoids
32. anorthosite with feldspathoids
33. feldspathoid syenite
34. feldspathoid monzosyenite
35. feldspathoid monzodiorite
36. feldspathoid monzogabbro
37. feldspathoid diorite
38. feldspathoid gabbro
39. foidolites

intrusive rock's acidity by observing its color: light-colored rocks or those slightly speckled with dark colors (mafic minerals) are acidic; very speckled will be intermediate to basic; a predominance of dark coloration will indicate basic to ultrabasic. For extrusive rocks, the same rule can generally be applied, but because extrusive rocks are so fine-grained, their color may be deceptive.

Ultramafic rocks composed almost exclusively of dark minerals (mafic > 90%), such as the amphiboles, pyoxenes, and olivine, have been left out of this classification. These rocks exhibit a typical green coloration when olivine prevails and have black or blackish spots when the amphiboles and pyroxenes are predominant.

feldspars (orthoclases, sanidine, microcline, and albite), the plagioclases, and the feldspathoids. The alkali feldspars, albite, and quartz are acidic; the more calcium-rich plagioclases are acidic-intermediate to basic; and the feldspathoids are basic.

The commonly utilized diagrams are shaped like a rhombus composed of two equilateral triangles (see figure). Quartz and the other three groups of minerals are placed at the vertices of the rhombus. Quartz is located at the vertex opposite to the feldspathoids, as there is no compatability between them. In other words, quartz and feldpathoids, theoretically, cannot exist in the same rock.

In order to facilitate the identification of both intrusive and extrusive rocks in the field, one can use a simplified diagram in which rocks are separated into large, readily identifiable groups. Given the minerals present in a rock, the diagram allows one to easily distinguish

the acidic (oversaturated with silica) from the basic and ultrabasic rocks (saturated to undersaturated). The oversaturated rocks (acidic) are all above the horizontal line separating the two triangles; the undersaturated rocks are below it. Finally, the saturated rocks are (in theory) along the line between the two triangles. In practice, they correspond to the intermediate rocks which can be either slightly over- or undersaturated and, therefore, occupy a narrow band straddling the dividing line.

For simplicity, in this diagram the various types of plagioclases and feldspathoids have not been differentiated, and the ferromagnesian (mafic) minerals (also very useful in the identification of rocks) have been left out. An increase in the content of mafic minerals (always dark in color) entails a shift in composition towards the basic and ultrabasic rocks.

We can get a good indication of an

INTRUSIVE ROCKS

White granite (S. Fedelino, Novara, Italy).

Diorite (Sondalo, Sondrio, Italy).

Gabbro, norite variety (Emilian Apennines, Italy).

EXTRUSIVE ROCKS

Rhyolite (Euganean Hills, Padua, Italy).

Trachyte (Euganean Hills, Padua, Italy).

Basalt

ALKALI FELDSPAR GRANITE

CHEMISTRY: acidic.

ESSENTIAL MINERALS: quartz, potassic feldspars (orthoclase, microcline), albite.

ACCESSORY MINERALS: biotite, pyrite, zircon, monazite, tourmaline, garnets, sometimes with hornblende.

APPEARANCE: medium- to coarse-grained; massive structure with rare miarolitic cavities; hypidiomorphic granular texture grading to idiomorphic and to porphyritic.

COLOR: light to rosy, sometimes with bluish hues.

GEOLOGIC ENVIRONMENT: in differentiated masses associated with plutons solidified at great depth.

OCCURRENCE: relatively common rock; the intrusions in Norway (Swedish red), Finland, Portugal, and Canada are famous.

USES: polished slabs used as ornamental stones.

NOTES: The granites appear in numerous varieties, both in grain and color, which correspond to variations in solidification history and chemistry. Alkali feldspar granites exhibit a very light coloration with rare small dark spots (mafic minerals), and, they are distinguished from granites by a higher alkali feldspar content (> 90%) with respect to the total feldspar content. The term alkali granite refers to granitic rocks with alkalic amphiboles or pyroxenes as accessories. The red color of some alkali feldspar granites is due to the oxidation of iron contained in small quantities in the feldspars.

▼ *Alkali feldspar granite (ca. x 1), Norway.*

GRANITE

CHEMISTRY: acidic.

ESSENTIAL MINERALS: quartz, potassic feldspars (orthoclase, microcline), plagioclases (albite-oligoclase), biotite.

ACCESSORY MINERALS: magnetite, apatite, pyrite, zircon, allanite (epidote), tourmaline; sometimes with muscovite, hornblende, garnets, rarely pyroxenes (augite).

APPEARANCE: medium- to coarse-grained; massive structure; hypidiomorphic granular texture, grading to porphyritic from potassic feldspar development.

COLOR: variable, according to the mafic (dark) mineral content; white, light gray, rosy, faded yellow, rarely greenish from alteration.

GEOLOGIC ENVIRONMENT: in large batholiths in Precambrian shields; in ancient crystalline basements and in orogenic chains; more rarely in dikes and sills of various sizes intruded in sedimentary rocks.

OCCURRENCE: the most abundant rocks in the earth's crust; large masses (batholiths) are found in Scandinavia, the United States, Canada, Africa, India, and Siberia.

USES: as rough-cut ashlars and polished slabs, used in the building industry.

NOTES: In modern classification schemes, these rocks are subdivided into two groups: the syenogranites and the monzogranites. The former contain greater amounts of alkali feldspars and lesser amounts of mafic minerals (dark), though both belong to the leucocratic rocks; they are, therefore, light in color with a few small dark spots. The monzogranites, on the other hand, have a greater number of dark spots. The term "adamellite," now discouraged, was synonymous with monzogranite. The red color of some granites is due to the oxidation of iron, contained in small quantities in the feldspars. Because of their hardness, granites were difficult to work with in the past and were only used in rough-cut blocks. Because of granite's high resistance to alteration, many granite monuments are beautifully preserved.

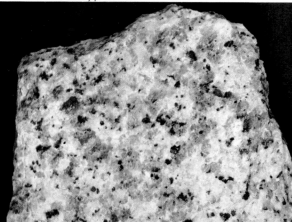

◀ *White granite (ca. x 1), Montorfano, Novara, Italy.*

GRANODIORITE

CHEMISTRY: acidic.

ESSENTIAL MINERALS: quartz, oligoclase (more rarely andesine) and lesser quantities of alkali feldspars, hornblende, biotite.

ACCESSORY MINERALS: augite, apatite, magnetite, sphene.

APPEARANCE: medium- to fine-grained; massive structure; hypidiomorphic granular texture; some exhibit a flow foliation or xenoliths visible in the marginal sections of the plutons; miarolitic cavities and orbicular structures are rarely present.

COLOR: light gray with small dark spots.

GEOLOGIC ENVIRONMENT: in plutons; in differentiated border zones of batholiths.

OCCURRENCE: in orogenic chains; beautiful examples are found in the United States, Japan, Norway, Romania, and Austria.

USES: used both as a building material and, in polished slabs, as an ornamental stone.

NOTES: Among the most widespread intrusive rocks, granodiorite can be confused with the granites; a precise classification can only be obtained by a petrographical analysis. Therefore, in field classifications, the term "granitoid" is used.

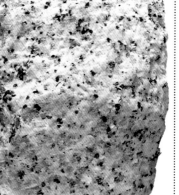

▲ *Granodiorite (ca. x 0.6), Vosges, France.*

TONALITE

CHEMISTRY: acidic.

ESSENTIAL MINERALS: sodic plagioclases (oligoclase, andesine), quartz, hornblende, biotite.

ACCESSORY MINERALS: orthoclase, apatite, sphene, magnetite, zircon, monoclinic and orthorhombic pyroxene, allanite (epidote).

APPEARANCE: coarse-grained; massive structure, rarely fluidal; hypidiomorphic granular texture, sometimes grading to porphyritic from hornblende development.

COLOR: blackish gray.

GEOLOGIC ENVIRONMENT: associated with large batholiths of granitic composition

OCCURRENCE: found in Italy, the United States, Canada, Norway, etc.

USES: utilized in the builiding industry both in the rough state (for ashlars) and as polished slabs.

NOTES: The name is derived from the Tonale pass near the Adamello massif. Tonalites have a light-colored background with numerous dark spots; they are difficult to distinguish from granodiorites and, in the field, are included under the term "granitoids." A precise classification can only be obtained by a petrographical analysis. The words trondhjemite and plagiogranite are synonomous and indicate leucocratic tonalites (with a low mafic mineral content).

▼ *Tonalite (ca. x 0.7), Scotland, Great Britain.*

ALKALI FELDSPAR SYENITE

CHEMISTRY: intermediate-acidic.

ESSENTIAL MINERALS: alkali feldspars (orthoclase).

ACCESSORY MINERALS: quartz, plagioclases (albite), aegirine, nepheline, sphene, pyrrhotite, allanite (epidote), anorthoclase.

APPEARANCE: coarse-grained; massive structure grading to fluidal; hypidiomorphic granular texture.

COLOR: light to dark gray with bluish hues.

GEOLOGIC ENVIRONMENT: in small plutons, laccoliths, sills; in border zones of syenites and monzonites; derived from local differentiations apparently unconnected to tectonic events.

OCCURRENCE: found in Norway, Greenland, Canada, etc.

USES: utilized as an ornamental and building stone.

NOTES: Alkali feldspar syenites are very similar to syenites and quartz syenites. They are distinguished from the former by a higher alkali feldspar content (A > 90% of A+P) and from the latter by a lower quartz content (Q < 5%). Distinguishing among these rocks in the field or in macroscopic specimens is not always easy; for an initial classification, the term "syenitoid" is often used to indicate the syenitic and monzonitic rocks. Correct classification can only be obtained by a petrological analysis.

▼ *Alkali feldspar syenite (ca. x 0.6), Bohemia.*

QUARTZ SYENITE

CHEMISTRY: acidic-intermediate.

ESSENTIAL MINERALS: alkali feldspars, plagioclases, biotite, quartz.

ACCESSORY MINERALS: pyroxene, amphibole, sphene, muscovite, apatite, corundum.

APPEARANCE: medium-grained; massive structure, miarolitic cavities and fluidal structures common; hypidiomorphic texture commonly grading to porphyritic.

COLOR: pink, white, light gray with dark green or black dots conferred by mafic minerals.

GEOLOGIC ENVIRONMENT: in small intrusive bodies; common in differentiated zones of plutons.

OCCURRENCE: found in Germany, the United States, Norway, etc.

USES: in polished slabs, used as an ornamental stone.

NOTES: Quartz syenite is very similar to syenite but has a higher quartz content (5 to 20%). Although a trained eye can detect the quartz grains by their translucent, semitransparent appearance and their light gray color, in the field the term "syenitoid" is used to indicate both syenitic and monzonitic rocks. The correct terminology can only be assigned after petrological analysis.

◄ *Quartz syenite (ca. x 0.8), United States.*

SYENITE

CHEMISTRY: intermediate.

ESSENTIAL MINERALS: alkali feldspars, plagioclases (andesine-labradorite), biotite, pyroxene, amphibole.

ACCESSORY MINERALS: quartz, nepheline, sphene, muscovite, apatite, corundum.

APPEARANCE: medium-grained; massive structure, miarolitic cavities and fluidal structures common, orbicular structures rare; hypidiomorphic texture, commonly grading to porphyritic.

COLOR: pink, white, light gray, pale violet, pale green, with dark green or black spots conferred by mafic minerals.

GEOLOGIC ENVIRONMENT: in small intrusive bodies; common in differentiated zones of basic plutons or stratiform intrusions.

OCCURRENCE: found in Germany, the United States, Norway, and Italy, where Oropa syenite, used as a decorative stone, is well-known: it is part of a granitic mass, differentiated in its border zones to syenite and in its peripheral zones to monzonite.

USES: as polished slabs.

NOTES: The name is derived from the Egyptian city Aswan, known in the past as Syene. In Syene there is an outcrop of amphibole granite which was used widely as an ornamental stone and is easily confused with syenite; the rocks are very similar, making a petrological analysis necessary for assigning the correct name. Syenite is recognizable by its quartz content which must be lower than 5%. Distinction in the field or with macroscopic specimens is not always easy, and, for an initial classification, the term "syenitoid" is often used to indicate both syenitc and monzonitic rocks.

▼ *Syenite (ca. x 1), Balma, Vercelli, Italy.*

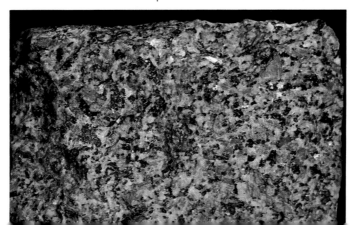

QUARTZ MONZONITE

CHEMISTRY: intermediate.

ESSENTIAL MINERALS: alkali feldspars (microcline, orthoclase), plagioclases (oligoclase, more rarely andesine), quartz, biotite, hornblende.

ACCESSORY MINERALS: augite, apatite, magnetite, ilmenite, zircon.

APPEARANCE: coarse-grained; massive structure, commonly fluidal; miarolitic cavities common; hypidiomorphic granular texture with idiomorphic plagioclase; potassic feldspar may appear as large crystals which include mafic minerals or plagioclases.

COLOR: gray; less commonly pink, reddish.

GEOLOGIC ENVIRONMENT: in differentiated border zones of small plutonic masses (sills, laccoliths) with a dioritic or granodioritic composition.

OCCURRENCE: found in Italy, Norway, and the United States.

USES: used rarely in the building industry.

NOTES: Quartz monzonite is very similar to monzonite but has a higher quartz content (5 to 20%), giving the rock a lighter color. In the field, these rocks are also easily confused with syenite and can only be distinguished from it by petrological analysis; hence the term "syenitoids" is used. The term "adamellite," now discouraged, also included the quartz monzonites.

▼ *Quartz monzonite (ca. x 0.7), Italy.*

MONZONITE

CHEMISTRY: intermediate.
ESSENTIAL MINERALS: alkali feldspars (microcline, orthoclase), plagioclases (oligoclase, labradorite, more rarely andesine), augite, hornblende, biotite.
ACCESSORY MINERALS: quartz (less than 5%), apatite, magnetite, ilmenite, zircon, hypersthene (orthorhombic pyroxene), and olivine.
APPEARANCE: coarse-grained; massive structure, commonly fluidal; miarolitic cavities common; hypidiomorphic granular texture with idiomorphic plagioclase; potassic feldspar may appear as large crystals which include mafic minerals or plagioclases.
COLOR: dark gray; less commonly pink, greenish, reddish.
GEOLOGIC ENVIRONMENT: in differentiated border zones of small plutonic masses (sills, laccoliths) with a dioritic or granodioritic composition.
OCCURRENCE: found in Italy, Norway, the United States, etc.
USES: used rarely in the building industry.
NOTES: The name is derived from the mountainous massif the Monzoni, in the Fassa valley in Italy. It is a leucocratic to mesocratic rock generally composed of equal quantities of plagioclases and alkali feldspars. These minerals give the rocks a light background color in which many dark spots (mafic minerals) are visible. In the varieties richer in mafic minerals, olivine and hypersthene are also present as accessories, giving the rock a darker color.

▼ *Monzonite (ca. x 0.8), Italy.*

QUARTZ MONZODIORITE

CHEMISTRY: intermediate.
ESSENTIAL MINERALS: sodic plagioclases (andesine), alkali feldspars, quartz, pyroxenes, hornblende, biotite.
ACCESSORY MINERALS: apatite, magnetite, ilmenite, zircon.
APPEARANCE: coarse-grained; massive structure; hypidiomorphic granular texture with idiomorphic plagioclase.
COLOR: light gray, greenish.
GEOLOGIC ENVIRONMENT: in differentiated border zones of small plutonic masses with a dioritic or granodioritic composition.
OCCURRENCE: found in Norway, the United States, etc.
USES: used rarely in the building industry.
NOTES: Quartz monzodiorite is a leucocratic rock mesocratic, composed of far more abundant plagioclases than alkali feldspars and of quartz (5–20%); these minerals give the rock a light background color in which dark mafic minerals are visible. It belongs to a group of very similar rocks which in field classifications are called dioritoids. The type of plagioclase (andesine) distinguishes quartz monzodiorite from the quartz monzogabbros which contain labradorite.

▲ *Quartz monzodiorite (ca. x 0.5), California, United States.*

MONZODIORITE

CHEMISTRY: intermediate.
ESSENTIAL MINERALS: plagioclases (andesine), alkali feldspars, pyroxenes, hornblende, biotite.
ACCESSORY MINERALS: quartz (less than 5%), apatite, magnetite, ilmenite, zircon.
APPEARANCE: coarse-grained; massive structure; hypidiomorphic granular texture with idiomorphic plagioclase.
COLOR: light gray, greenish.
GEOLOGIC ENVIRONMENT: in differentiated border zones of small plutonic masses (sills, laccoliths) with a dioritic or granodioritic composition.
OCCURRENCE: found in Italy, Norway, the United States, etc.
USES: used rarely in the building industry.
NOTES: Monzodiorite is a leucocratic to mesocratic rock composed of far more abundant plagioclases than alkali feldpars; these minerals give the rock a light background color in which numerous dark mafic minerals are visible. The type of plagioclase (andesine) and the rock's lower mafic mineral content distinguish it from the monzogabbros which contain labradorite. In the past, these rocks were called syenodiorites. They belong to a group of very similar rocks called dioritoids.

▼ *Monzodiorite (ca. x 0.7), Italy.*

QUARTZ MONZOGABBRO

CHEMISTRY: intermediate.
ESSENTIAL MINERALS: labradoritic plagioclase, alkali feldspars, quartz, pyroxenes, hornblende, biotite.
ACCESSORY MINERALS: apatite, magnetite, ilmenite, zircon.
APPEARANCE: coarse-grained; massive structure; hypidiomorphic granular texture with idiomorphic plagioclase.
COLOR: light gray, greenish.
GEOLOGIC ENVIRONMENT: in differentiated border zones of small plutonic masses (sills, laccoliths) with a dioritic composition.
OCCURRENCE: found in Norway, the United States, etc.
USES: used rarely in the building industry.
NOTES: Quartz monzogabbro is an essentially mesocratic rock, sometimes leucocratic, composed of far more abundant plagioclases than alkali feldspars and of quartz (5–20%); these minerals give the rock a light background color in which dark spots (mafic minerals) are visible. The type of plagioclases (labradorite) distinguishes quartz monzogabbro from the quartz monzodiorites which contain andesine. It belongs to a group of very similar rocks known as anorthosites, gabbroids, or dioritoids.

▼ *Quartz monzogabbro (ca. x 0.8), California, United States.*

THE TEXTURE OF IGNEOUS ROCKS

The texture of an igneous rock is determined by the way it crystallized.
Both macroscopic and microscopic features can be used to distinguish between extrusive and intrusive rocks.

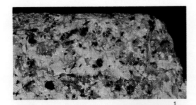
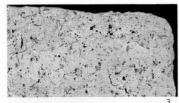

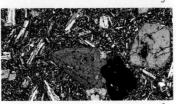

Pink granite (ca. x 20), Baveno, Novara, Italy; below, in a thin section under the polarizing microscope.

Rhyolite (ca. x 15), Euganean Hills, Padua, Italy; below, in a thin section under the polarizing microscope.

Olivinic basalt (ca. x 20), Mount Etna, Italy; below, in a thin section under the polarizing microscope.

The methods employed in the study of igneous rocks differ depending on whether the chemical composition or the mineralogical composition is being analyzed. Optical analysis, by means of a microscope, is a very common, quick method adopted by petrologists to identify minerals and study the texture of a rock: mineral forms, shapes, sizes, and spatial relationships. With the aid of a polarizing microscope, the diagnostic properties of minerals and their relationships can be observed and both the composition of the original magma and its crystallization conditions can be investigated.

INTRUSIVE ROCKS

Intrusive rocks solidify below the earth's surface. Because they cool slowly, intrusive rocks have a granular or holocrystalline texture; the crystals are well-developed, interpenetrated, and very commonly large enough that they are visible to the naked eye. *Figures 1* and *2* depict a macroscopic granite specimen and its thin section as seen under the microscope. The flesh-colored spots in *fig. 1* and the large crystal with mesh-like inclusions in *fig. 2* are orthoclase. The light gray crystals with a glassy look in *fig. 1* and the inclusion-free, fractured, white crystals in *fig. 2* are quartz. The small dark spots in the hand sample and the chestnut-brown crystals in thin section are biotite. Basic intrusive rocks can have the same texture as acidic intrusives, but they differ in mineralogical composition. In the former, quartz crystals are absent, and plagioclases and sometimes feldspathoids are present in generally lower quantities than the mafic minerals (the most common of which are biotite, pyroxenes, amphiboles, and olivine)

EXTRUSIVE ROCKS

Extrusive magmas are made up of a totally or partially molten mass. They solidify quickly as a result of a sudden drop in temperature. The variously colored extrusive rocks are composed of microcrystals and, in some cases, larger crystals which are distinguishable from the minute groundmass. This texture is called "porphyritic." It commonly occurs in acidic magmas, as these have slightly lower temperatures and are more viscous than the basic ones. Under these conditions, well-developed crystals may form before the magma has erupted on the surface and are then carried along with the lava when it erupts.

Rhyolite *(fig. 3)* is an acidic extrusive rock which corresponds to granite in composition. Because of its generally fine grain, it has a homogeneous appearance and, like granite, its light color is due to its essentially quartzo-feldspathic mineral composition. Note in the thin section *(fig. 4)*, a large, light-colored, well-formed phenocryst of sanidine with fractures and inclusions inside it and smaller quartz phenocrysts, lighter in appearance, immersed in a microcrystalline groundmass of the same composition.

Basalts are mafic rocks that crystallize from a very fluid basic magma that readily spreads out on the surface. The molten mass cools quickly, forming a dense interlacing of small crystals among which there may be phenocrysts of larger mafic minerals formed in the volcanic chimney. The lava may also completely solidify as a glass (hyaline texture) or as a mixture of glass and crystals (hypocrystalline). Basalt *(fig. 5)* has a homogeneous appearance and is black in color, either from the abundance of mafic minerals or because of its very fine grain which makes even the plagioclases look dark. In the thin section *(fig. 6)*, one can see phenocrysts of vividly colored pyroxene and polysynthetically twinned plagioclases in a groundmass of plagioclase crystals and glass.

MONZOGABBRO

CHEMISTRY: intermediate-basic.
ESSENTIAL MINERALS: plagioclases (labradorite), alkali feldspars, pyroxenes, hornblende, biotite.
ACCESSORY MINERALS: quartz (< 5%), apatite, magnetite, ilmenite, zircon.
APPEARANCE: coarse-grained; massive structure; hypidiomorphic granular texture with idiomorphic plagioclase.
COLOR: light gray, greenish.
GEOLOGIC ENVIRONMENT: in differentiated border zones of small plutonic masses (sills, laccoliths) with a dioritic composition.
OCCURRENCE: found in Italy, Norway, the United States, etc.
USES: used rarely in the building industry.
NOTES: Monzogabbro is a mesocratic to leucocratic rock, composed of much more abundant plagioclases than alkali feldspars; these minerals give the rock a light background color in which numerous dark spots (mafic minerals) are visible. The type of plagioclase (labradorite) and the greater quantity of mafic minerals distinguish it from the monzodiorites which contain andesine. In the past, these rocks were called syenogabbros. They belong to a group of very similar rocks known as anorthosites, gabbroids, or dioritoids.

▼ *Monzogabbro (ca. x 0.8), Italy.*

QUARTZ ANORTHOSITE

CHEMISTRY: intermediate.
ESSENTIAL MINERALS: plagioclases (labradorite, bytownite), quartz.
ACCESSORY MINERALS: pyroxenes, olivine, amphibole, chromite, magnetite, ilmenite, garnet.
APPEARANCE: coarse-grained; massive structure; granular texture with generally elongated plagioclase crystals.
COLOR: gray; in polished slabs, dark gray with bluish reflections.
GEOLOGIC ENVIRONMENT: in small stratiform plutons associated with gabbros; in very large intrusive bodies (batholiths); in Precambrian metamorphic rocks (Baltic shield, Canadian shield).
OCCURRENCE: found in Canada, the United States, Brazil, South Africa, and Scandinavia.
USES: widely used as an ornamental stone because of its beatiful iridescent bluish gray reflexions, highlighted by polishing.
NOTES: Quartz anorthosite is very similar to anorthosite, but has a higher quartz content (5-20%). It is distinguishable by its semitransparent, glassy appearance, dark gray color, and lack of iridescent plagioclase.

ANORTHOSITE

CHEMISTRY: intermediate-basic.
ESSENTIAL MINERALS: plagioclases (labradorite, bytownite).
ACCESSORY MINERALS: pyroxenes, olivine, amphibole, chromite, magnetite, garnet.
APPEARANCE: coarse-grained; massive structure; granular texture with generally elongated plagioclase crystals.
COLOR: white or light gray; in polished slabs, dark gray with bluish reflections.
GEOLOGIC ENVIRONMENT: in small generally stratiform plutons associated with gabbros; in very large intrusive bodies (batholiths); in Precambrian metamorphic rocks (Baltic shield, Canadian shield).
OCCURRENCE: found in Canada, the United States, Brazil, South Africa, and Scandinavia.
USES: in the preparation of refractory materials; labradoridite (composed almost exclusively of labradorite plagioclase) is widely used as an ornamental stone because of its beautiful iridescent bluish gray reflections, highlighted by polishing.
NOTES: Anorthosite is an almost monomineralic leucocratic rock, made up of over 90% plagioclases. Its mineralogical composition is similar to that of gabbro and diorite. It is distinguished from them by its low mafic mineral content. Anorthosite is also known under the name plagioclasite. The light colored parts of the moon are composed of 4 to 4.5 billion-year-old anorthosite with high calcium plagioclase (almost pure anorthite).

QUARTZ DIORITE

CHEMISTRY: intermediate.
ESSENTIAL MINERALS: plagioclases (somet zoned, polysynthetic twins common), quartz, hornblende.
ACCESSORY MINERALS: orthoclase, monoclinic or orthorhombic pyroxene, biotite, sphene, allanite (epidote), zircon, ilmenite, magnetite, hypersthene (orthorhombic pyroxene).
APPEARANCE: medium- to coarse-grained; massive structure, commonly grading to fluidal; orbicular structures present rarely; hypidiomorphic texture, to porphyritic from plagioclase or hornblende development.
COLOR: more or less dark gray.
GEOLOGIC ENVIRONMENT: in border zones of batholiths with a granitic or granodioritic composition.
OCCURRENCE: in the Harz massifs of central Germany, also in Romania, Scandinavia, and the United States.
USES: in the building industry as a building covering stone and as polished slabs.
NOTES: Quartz diorite can be mistaken for tonalite, quartz gabbro, and quartz anorthosite. It is distinguished from tonalite by its lower quartz content (< 20%), from quartz anorthosite by its higher mafic mineral content (in both cases, a darker color ensues), and from quartz gabbro by its andesinic plagioclase, rather than labradoritic.

DIORITE

CHEMISTRY: intermediate.
ESSENTIAL MINERALS: plagioclases (some zoned, polysynthetic twins common), hornblende (amphibole).
ACCESSORY MINERALS: quartz (< 5%), orthoclase, monoclinic or orthorhombic pyroxene, biotite, sphene, allanite (epidote), zircon, ilmenite, magnetite, hypersthene (orthorhombic pyroxene).
APPEARANCE: medium- to coarse-grained; massive structure, commonly grading to fluidal; orbicular structures are rarely present; hypidiomorphic texture, to porphyritic from plagioclase or hornblende development.
COLOR: gray to blackish; rarely greenish; black in polished slabs.
GEOLOGIC ENVIRONMENT: in masses, sills and dikes; commonly in border zones of batholiths with a granitic or granodioritic composition; along the edges of gabbro masses.
OCCURRENCE: in the Härz massifs of central Germany, also in Romania, Scandinavia, and the United States.
USES: in the building industry, as a building covering stone and as polished slabs.
NOTES: Diorite is a mesocratic rock whose name is derived from the Greek for "to distinguish." Its mineralogical composition makes it similar to gabbro and anorthosite: it is differentiated from the former by its andesinic plagioclase rather than labradoritic, and from the latter by its higher mafic mineral content which gives it a darker color. A petrological analysis is usually required to distinguish diorite from gabbro.

▲ Quartz anorthosite (ca. x 0.7), provenance unknown.

Anorthosite ▶ (ca. x 1.2), Canada.

▲ Quartz diorite (ca. x 0.5), Vosges, France.

Diorite (ca. x 0.7), ▶ Monte Varallo, Vercelli, Italy.

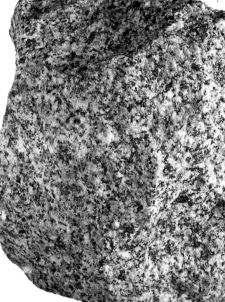

QUARTZ GABBRO

CHEMISTRY: intermediate.

ESSENTIAL MINERALS: calcic plagioclases (generally labradorite, less commonly bytownite or anorthite), monoclinic pyroxenes (diallage = hyperaluminous augite), quartz.

ACCESSORY MINERALS: ilmenite, hematite, apatite, chromite, pyrope, spinel, rutile.

APPEARANCE: medium- to coarse-grained; massive structure; hypidiomorphic granular texture.

COLOR: light green to greenish gray, rarely reddish, according to its ferromagnesian mineral content and the size of these minerals.

GEOLOGIC ENVIRONMENT: associated with basic-intermediate intrusive bodies.

OCCURRENCE: found in Minnesota and Sudbury, Ontario

USES: of limited use because it is easily altered and is not very durable.

NOTES: Quartz gabbros, fundamentally dark colored, are easily confused with gabbros, monzogabbros, and quartz monzogabbros; therefore, the term gabbroid is used in the field. They can also be mistaken for quartz monzodiorites from which they are distinguished by the presence of labradorite plagioclase rather than andesine.

▲ Quartz gabbro (ca. x 0.6), Minnesota, United States.

GABBRO

CHEMISTRY: intermediate-basic.

ESSENTIAL MINERALS: calcic plagioclases (essentially labradorite, less commonly bytownite or anorthite), monoclinic pyroxenes (diallage = hyperaluminous augite).

ACCESSORY MINERALS: olivine, quartz, ilmenite, hematite, apatite, chromite, pyrope, spinel, rutile

APPEARANCE: medium- to coarse-grained; massive structure; often displays bands of concentrations of particular minerals; hypidiomorphic granular texture.

COLOR: light green to greenish gray, rarely reddish, according to its ferromagnesian mineral content and the size of these minerals.

GEOLOGIC ENVIRONMENT: as batholiths, in large intrusive bodies concordant with the country rocks (lopoliths); in dikes.

OCCURRENCE: common in ophiolitic associations, typical of ancient and current areas of extension (large crustal fractures). In large orogenic chains.

USES: of limited use because it is easily altered and is not very durable.

NOTES: Gabbros are a group of rocks with a fairly variable composition (within certain limits). Based on the type of mafic mineral, gabbros are classified into: gabbro *s.s.* (plagioclases and monoclinic pyroxenes), norite (plagioclases and orthorhombic pyroxenes), troctolite (plagioclases and olivine), gabbronorite (plagioclases with equal amounts of monoclinic and orthorhombic pyroxenes); hornblende gabbro is characterized by the presence of amphibole. The term diabase is used to indicate fine-grained gabbros (microgabbros). Common to this group of rocks is plagioclase with a labradoritic-bytownitic composition.

▲ Gabbro (ca. x 0.90), Tuscany, Italy.

FELDSPATHOID SYENITE

CHEMISTRY: basic.

ESSENTIAL MINERALS: feldspathoids (nepheline predominant, sodalite), alkali feldspars, sodic amphiboles, sodic pyroxenes (aegirine).

ACCESSORY MINERALS: cancrinite, plagioclases (oligoclase-andesine), analcime, biotite, fluorite, eudialyte (rare silicate).

APPEARANCE: coarse-grained; massive structure, sometimes fluidal; granular texture with idiomorphic crystals of varying sizes, sometimes porphyritic.

COLOR: light gray; rarely rosy or greenish, due to the color of the feldspars and feldspathoids.

GEOLOGIC ENVIRONMENT: in small masses associated with syenites.

OCCURRENCE: not very widespead; outcrops in Scandinavia, Canada, and Greenland are noteworthy.

USES: associated with rubidium, cesium, thorium, and uranium mineral deposits

NOTES: The term feldspathoid syenite indicates a group of plutonic rocks, rich in light minerals, included in the leucocratic alkali rock category. They are made up of feldspathoids and predominant alkali feldspars and contain a maximum of 35% mafic minerals; to define the type of rock more precisely, the predominant feldspathoid is generally specified (sodalite syenite, nepheline syenite, etc.). In some varieties the characteriastic feldspathoid is leucite, which being an unstable mineral in intrusive conditions, alters to an aggregate of orthoclase and nepheline while maintaining its original crystalline form (pseudo-leucite). A pseudo-leucite syenite is found in Scotland and is called borolanite. The term foyaite is now employed for nepheline syenite with a trachytic (fluidal) texture.

◄ Feldspathoid syenite (ca. x 1), Bancroft, Canada.

SHONKINITE

CHEMISTRY: basic.

ESSENTIAL MINERALS: abundant augite, alkali feldspars and feldspathoids (generally nepheline).

ACCESSORY MINERALS: olivine, hornblende, biotite, aegirine.

APPEARANCE: coarse-grained; massive structure; granular texture with allotriomorphic crystals.

COLOR: blackish gray.

GEOLOGIC ENVIRONMENT: in small masses associated with alkali syenitic complexes.

OCCURRENCE: fairly rare rock; found in the United States.

USES: of exclusively scientific interest.

NOTES: The name derives from Shonkin, the Native American name for the Highwood Mountains in the United States. In the modern nomenclature of igneous rocks, the term shonkinite corresponds to a variety of feldspathoid syenite, dark in color (melanocratic) from the presence of abundant augite.

▲ Shonkinite (ca. x 0.8), Predazzo, Trento, Italy.

FELDSPATHOID MONZODIORITE

CHEMISTRY: basic.
ESSENTIAL MINERALS: feldspathoids, acidic plagioclases, pyroxenes, amphiboles, biotite.
ACCESSORY MINERALS: alkali feldspars, olivine, apatite, magnetite, sphene.
APPEARANCE: fine- to medium-grained; massive structure; hypidiomorphic granular texture.
COLOR: gray to blackish.
GEOLOGIC ENVIRONMENT: in masses, in differentiated zones of plutons with a gabbroic composition.
OCCURRENCE: found in the United States, Canada, Bohemia, and Scotland.
USES: of exclusively scientific interest.
NOTES: Feldspathoid monzodiorite is very similar to the monzonites and diorites; it is distinguished from them by its higher feldspathoid content (10–60%). These rocks are named according to the dominant feldspathoid. The ratios between feldspathoids and feldspars (light minerals) and between the latter and the dark (mafic) minerals is relatively variable. In addition, plagioclases are more prevalent than potassic feldspars. Feldspathoid monzodiorites are distinguished from feldspathoid monzogabbros by the type of plagioclase present in the rock: andesine in the former and labradorite in the latter.

▼ *Feldspathoid monzodiorite (ca. x 1), Rongstock, Czechoslovakia.*

FELDSPATHOID MONZOGABBRO

CHEMISTRY: basic.
ESSENTIAL MINERALS: feldspathoids, calcic plagioclases, pyroxenes, amphiboles, biotite.
ACCESSORY MINERALS: olivine, apatite, magnetite, sphene.
APPEARANCE: fine- to medium-grained; massive structure; hypidiomorphic granular texture.
COLOR: dark gray to blackish.
GEOLOGIC ENVIRONMENT: in masses, in differentiated zones of plutons with a gabbroic composition.
OCCURRENCE: sometimes found associated with monzogabbros.
USES: of exclusively scientific interest.
NOTES: The name feldspathoid monzogabbro indicates a group of plutonic rocks characterized by the presence of feldspathoids; their color may be darker than that of feldspathoid monzodiorites. The ratios between feldspathoids and feldspars (light minerals) and between fledspars and the darker mafic minerals is relatively variable. In addition, plagioclases are more prevalent than potassic feldspars. Feldspathoid monzogabbros are distinguished from feldspathoid monzodiorites by the presence of labradorite rather than andesine plagioclase. For both groups there exist corresponding nepheline-rich varieties: both called essexite, from Essex County, Massachusetts (United States).

▼ *Feldspathoid monzogabbro (ca. x 0.7), Italy.*

FELDSPATHOID GABBRO

CHEMISTRY: basic.
ESSENTIAL MINERALS: augite, labradorite, nepheline.
ACCESSORY MINERALS: olivine, biotite, hornblende.
APPEARANCE: fine- to medium-grained; massive structure; hypidiomorphic granular texture, sometimes porphyritic.
COLOR: more or less dark gray.
GEOLOGIC ENVIRONMENT: generally present as sills.
OCCURRENCE: rare.
USES: of exclusively scientific interest.
NOTES: Feldspathoid gabbros are a group of alkaline plutonic rocks characterized by the presence of feldspathoids. The feldspathoids may be more abundant than the (primarily plagioclase) feldspars. Feldspathoid gabbros are distinguished from feldspathoid diorites by the presence of labradorite rather than andesine plagioclase. Teschenite and theralite are two varieties of feldspathoid gabbros that contain analcime and nepheline respectively.

Feldspathoid gabbro ▶ (ca. x 0.6), provenance unknown.

ITALITE

CHEMISTRY: basic.
ESSENTIAL MINERALS: pseudo-leucite, monoclinic pyroxene (aegirine).
ACCESSORY MINERALS: alkali amphiboles.
APPEARANCE: coarse-grained; holocrystalline granular texture.
COLOR: whitish or faded yellow.
GEOLOGIC ENVIRONMENT: in small masses intruded at shallow depth.
OCCURRENCE: very rare rock.
USES: of exclusively scientific interest.
NOTES: Italite takes its name from Italy, as it was found in the Alban Hills. In modern classifications of igneous rocks, italite is defined as a leucocratic variety of leucitite, belonging to the foidolite group: plutonic rocks characterized by a high feldspathoid content (> 60% of the light minerals). The prevailing feldspathoid is used to specify the rock's name more accurately (e.g., nephelinite, leucitite, etc.).

▼ *Italite (ca. x 1.8), Italy.*

MISSOURITE

CHEMISTRY: basic.
ESSENTIAL MINERALS: pseudo-leucite, monoclinic pyroxene.
ACCESSORY MINERALS: olivine, biotite, and metallic minerals.
APPEARANCE: coarse-grained; holocrystalline granular texture.
COLOR: gray to black.
GEOLOGIC ENVIRONMENT: in small masses intruded at shallow depth.
OCCURRENCE: very rare rock; found in the United States.
USES: of exclusively scientific interest.
NOTES: The name is derived from the Missouri River in the United States. In modern classifications of igneous rocks, missourite is a melanocratic (dark) variety of foidolite. Foidolites are plutonic rocks characterized by a high feldspathoid content (> 60% of the light minerals). The predominant feldspathoid is used to specify the rock's name more accurately (e.g., nephelinite, leucitite, etc.).

DUNITE

CHEMISTRY: ultrabasic.
ESSENTIAL MINERALS: olivine.
ACCESSORY MINERALS: spinel, monoclinic pyroxene, garnet, metallic minerals.
APPEARANCE: medium- to fine-grained; massive structure; variations in color and grain are common; idiomorphic granular texture, sometimes saccharoidal.
COLOR: light green to yellowish.
GEOLOGIC ENVIRONMENT: in stratiform masses, sometimes very extensive, in differentiated zones of cumulate sequences; in small masses in ophiolite associations.
OCCURRENCE: found in New Zealand, the Urals, Turkey, Italy, and the United States.
USES: of exclusively scientific interest.
NOTES: The name is derived from Mount Dun in New Zealand. Dunite belongs to the ultramafic rock family (characterized by a dark mineral content over 90%), they are divided into peridotites (with more than 40% olivine) and pyroxenites and hornblendites (both with less than 40% olivine). Dunite is a special peridotite, composed of over 90% olivine with spinel as an accessory. Olivinite is a type of dunite that has magnetite as the accessory mineral.

LHERZOLITE

CHEMISTRY: ultrabasic.
ESSENTIAL MINERALS: olivine, monoclinic and orthorhombic pyroxene.
ACCESSORY MINERALS: spinel, pyrope (garnet), hornblende, chromite.
APPEARANCE: medium-grained; massive structure; idiomorphic granular texture; in ophiolite sequences, it exhibits obvious tectonic deformation structures linked to solid-state flow.
COLOR: dark green to black.
GEOLOGIC ENVIRONMENT: in masses, sometimes of great dimensions, at the base of ophiolite associations; in border zones of basic and ultrabasic plutons; within Precambrian metamorphic basements.
OCCURRENCE: found in Italy, Switzerland, Cyprus, the Urals, the United States, Canada, Cuba, New Caledonia, and South Africa.
USES: of exclusively scientific interest.
NOTES: Lherzolite belongs to the ultramafic rock family (characterized by a dark mineral content greater than 90%,) they are divided into peridotites (olivine greater than 40%), and pyroxenites and hornblendites (both with less than 40% olivine). Lherzolite is a peridotite composed of olivine and subordinate amounts of monoclinic and orthorhombic pyroxene. In ophiolite associations, it is commonly affected by serpentinization processes.

PYROXENITE

CHEMISTRY: ultrabasic.
ESSENTIAL MINERALS: pyroxenes (both monoclinic and orthorhombic).
ACCESSORY MINERALS: olivine, hornblende, chromite, magnetite, pyrrhotite, plagioclase, biotite, ilmenite, garnet, apatite.
APPEARANCE: coarse-grained; massive structure, sometimes has stratified structures from gravitational selection of the crystals; hypidiomorphic to idiomorphic granular texture.
COLOR: dark green to brownish, to black.
GEOLOGIC ENVIRONMENT: in small dike-like masses; in small strata within ultrabasic complexes of ophiolite sequences.
OCCURRENCE: found in Italy, Spain, Morocco, and the United States.
USES: some pyroxenites contain mineable concentrations of metals such as platinum and palladium.
NOTES: The term pyroxenite refers to a family of ultramafic rocks (characterized by a dark mineral content over 90%) made up essentially of pyroxenes. Based on the mafic mineral's symmetry, pyroxenites are differentiated as follows: bronzitite (with orthorhombic pyroxene); diallagite (with monoclinic pyroxene); websterite (composed of almost equal parts of monoclinic and orthorhombic pyroxene).

▼ *Missourite (ca. x 1), Italy.*

▼ *Dunite (ca. x 1), New Zealand.*

▼ *Pyroxenite (ca. x 1), Malgola, Trento, Italy.*

◄ *Lherzolite (ca. x 1), Lherz, France.*

HORNBLENDITE

CHEMISTRY: ultrabasic.

ESSENTIAL MINERALS: hornblende.

ACCESSORY MINERALS: olivine, orthorhombic or monoclinic pyroxene, magnetite, chromite, pyrrhotite.

APPEARANCE: medium- to fine-grained; massive structure; idiomorphic granular texture.

COLOR: dark green to black.

GEOLOGIC ENVIRONMENT: in small masses within peridotites or in differentiated zones of plutons with a gabbroic composition; rarely as dikes associated with syenitic-nephelinic rocks.

OCCURRENCE: found in Spain, Italy, the United States, Cuba, and the Urals.

USES: of exclusively scientific interest.

NOTES: The name hornblendite refers to the extremely high percentage (> 90%) of hornblende in the rock. With the presence of other mafic minerals, hornblendite grades into peridotites or pyroxenites. Pedrosite, a variety of hornblendite, is rich in alkali amphiboles (riebeckite) and also contains magnetite, and, rarely, albite and analcime. It is found in the Kola Peninsula (Russia).

▼ *Hornblendite (ca. x 1), Valcamonica, Brescia, Italy.*

CARBONATITE

CHEMISTRY: ultrabasic.

ESSENTIAL MINERALS: calcite or dolomite.

ACCESSORY MINERALS: various carbonates, nepheline, phlogopite, olivine, apatite, monazite, barite, pyrochlore, fluorite, perovskite (calcium titanium oxide), oxides and sulfides of iron, niobium, and tantalum.

APPEARANCE: coarse- to medium-fine-grained; massive structure; idiomorphic to allotriomorphic texture, generally without phenocrysts.

COLOR: light gray, yellowish.

GEOLOGIC ENVIRONMENT: linked to nepheline-syenitic intrusions; in dikes and in lava flows.

OCCURRENCE: found in Sweden, Norway, Russia, Canada, the United States, and South Africa.

USES: of exclusively scientific interest.

NOTES: The term carbonatite is used to indicate a family of igneous rocks composed in large part of carbonate minerals (greater than 50%). The carbonatites are of either plutonic or volcanic origin.

▼ *Carbonatite (ca. x 0.5), Oka, Canada.*

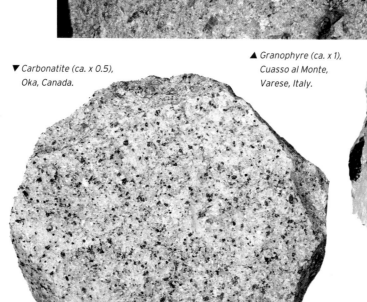

GRANOPHYRE

CHEMISTRY: acidic.

ESSENTIAL MINERALS: potassic feldspar (orthoclase, microcline), quartz.

ACCESSORY MINERALS: plagioclases (albite-oligoclase), biotite, amphibole, aegirine, muscovite, apatite, zircon, molybdenite.

APPEARANCE: medium-fine-grained; massive structure, sometimes with miarolitic cavities; porphyritic texture with quartz and alkali feldspar intergrowths.

COLOR: light colored, commonly reddish.

GEOLOGIC ENVIRONMENT: in small dikes and pockets in border and peripheral zones of granitic plutons intruded at shallow depth.

OCCURRENCE: very widespread, like the granites with which it is commonly associated.

USES: for paving roads and as a low-quality ornamental stone.

NOTES: In the past, the term granophyre was used to indicate a variety of granite porphyry with a microcrystalline groundmass, but it now refers to porphyritic rocks with a granitic composition characterized by a micrographic or micropegmatitic texture. Felsite is the term used for very compact, microcrystalline rocks with a granitic composition lacking micrographic structures.

▲ *Granophyre (ca. x 1), Cuasso al Monte, Varese, Italy.*

PEGMATITE

CHEMISTRY: acidic.

ESSENTIAL MINERALS: quartz, alkali feldspars (orthoclase, microcline, albite), muscovite, biotite, lepidolite (lithium mica).

ACCESSORY MINERALS: tourmaline, beryl, topaz, zircon, apatite, cassiterite, columbite.

APPEARANCE: very coarse-grained; massive structure with crystal-lined cavities; pegmatitic structure with crystals of sometimes gigantic proportions.

COLOR: generally light.

GEOLOGIC ENVIRONMENT: in dikes or small masses associated with granitic or syenitic intrusions rich in volatile constituents; they are commonly associated with aplites.

OCCURRENCE: found in many localities such as Brazil, the United States, Great Britain, the Urals, Madagascar, India, etc.

USES: pegmatites contain many kinds of relatively large, well-formed mineral crystals, highly sought-after by collectors; some are found in sufficient concentrations for extraction on an industrial scale.

NOTES: The name is derived from the Greek for "bond" and was proposed by the French mineralogist Haüy as a synonym for granites with a graphic texture. Pegmatites are the product of the crystallization of residual fluids in a granitic intrusion particularly rich in silicates and volatile constituents with extremely low viscosity and relatively low temperature, permitting the growth of very large crystals.

▼ *Pegmatite (ca. x 0.5), Piona, Como, Italy*

SHALLOW INTRUSIVE ROCKS

Dike rocks are formed as a consequence of the cooling and crystallization of magma inside crustal fractures. During the geodynamic evolution of a mountain belt, large fractures occur in a variety of rock types.

It is thought that magma is injected into preexistent fractures or propagates new fractures along zones of weakness to create rock bodies of various shapes. If the intruded magma crystallizes at shallow depths (less than 13,124–16405 ft or approximately 4–5 km), the corresponding rocks are called hypabyssal. This group also includes those rocks that crystallize inside volcanic chimneys.

MODE OF OCCURRENCE

Hypabyssal rocks have specific names associated with the form they assume within the country rock. Dikes are tabular bodies that cross cut structures in the surrounding country rocks. They have thicknesses that range from inches to hundreds of yards. Sills, on the other hand, are formed by magma injected parallel to the stratification of the country rock. Both dikes and sills extend laterally for great distances; in some cases they are miles long. Laccoliths are large bodies that generally have well-defined boundaries with a roughly circular outline. Like sills, they are emplaced concordant with the surrounding rocks, but in this case have a domed upper surface. Necks are fossil volcanic conduits (chimneys) exposed by erosion of the surrounding volcanic cone. They are vertical, subcylindrical bodies with diameters from dozens to hundreds of yards. While dikes are relatively

common, sills, laccoliths, and necks are less so.

COMPOSITION AND TEXTURE

Dike rocks are crystallized under conditions that are intermediate between those of intrusive rocks and extrusive rocks. They are easily recognizable because they generally differ markedly

In the drawing: C = a volcanic conduit known as a neck, once it is exposed by erosion; D= dikes; L = laccolith; S = sills.

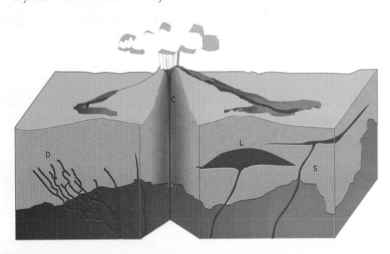

from the rocks they intruded and because of their particular form. Dike rocks that appear light in color because they are composed essentially of quartz and feldspar are called aplites, whereas dark ones, such as lampophyres, are composed mostly of mafic minerals.

These rocks are described more specifically by a compound name that indicates both their characteristic texture and their composition. For example, "granite porphyry" and "diorite porphyry" refer to rocks which have a porphyritic texture (large crystals in a fine-grained groundmass), and acidic and intermediate compositions, respectively. Microgabbro refers to a mafic dike rock with microscopically small, equidimensional crystals of plagioclase feldspar and pyroxene. A granophyre is a rock with a granitic composition in which quartz and alkali feldspar are intergrown on a microscopic scale.

Pegmatites are coarse-grained rocks that generally form in crustal fractures by crystallization of a vapor or a vapor-rich magma. They have an acidic composition and are, therefore, rich in quartz and feldspar. They are especially interesting because in many

The monumental Ship Rock rises over 1,300 ft. (400 m) above the sediments on the New Mexico desert floor. It was formed by the solidification of molten material inside a conduit that originally fed magma through the crust to a volcano. It was gradually exposed by erosion of the cone.

cases they include a rich variety of collectible minerals including tourmaline, fluorite, muscovite, topaz, beryl, and apatite. In some cases, the crystals reach metric dimensions.

Necks, as already indicated, are much less common, although some of them are of great importance. Among the most picturesque ones are Ship Rock in New Mexico, which emerges like a cathedral from the surrounding sediments, and an unnamed one in Le Puy, France. The most famous volcanic pipe is the one in Kimberley, South Africa. This city gives its name to an ultrabasic igneous rock (kimberlite) which is believed to have originated directly from the mantle and fills an extremely old and large subsurface volcanic chimney. Another special feature of this rock in its occurrence at Kimberley is the valuable mineral it contains: diamond.

APLITE

CHEMISTRY: acidic.

ESSENTIAL MINERALS: quartz, alkali feldspars (orthoclase, microcline, albite), muscovite, biotite.

ACCESSORY MINERALS: tourmaline.

APPEARANCE: fine- to extremely fine-grained; massive or zonal structure; allotriomorphic granular or saccharoidal texture.

COLOR: white or light gray.

GEOLOGIC ENVIRONMENT: in dikes and small masses associated with granitic intrusions; constitute the border zones of pegmatite dikes.

OCCURRENCE: rock present in almost all granitic massifs.

USES: of exclusively scientific interest; negatively affects the extraction of granites from quarries.

NOTES: The name is derived from the Greek for "simple" and is used to indicate leucocratic microgranites in vein-like emplacements. Aplites are very similar in origin and composition to pegmatites, but have a finer-grained, commonly sugary or saccharoidal texture.

GRANITE PORPHYRY

CHEMISTRY: acidic.

ESSENTIAL MINERALS: potassic feldspar (orthoclase, microcline), quartz.

ACCESSORY MINERALS: biotite, plagioclases (albite, oligoclase), amphibole, muscovite, apatite, zircon, xenotime (yttrium phosphate), cassiterite, molybdenite.

APPEARANCE: medium-fine-grained; massive to zonal structure, miarolitic cavities common; porphyritic texture with quartz and feldspar phenocrysts and with a microcrystalline, microfelsitic groundmass.

COLOR: light gray, rosy, red.

GEOLOGIC ENVIRONMENT: in dikes; in the border zones of shallow granitic intrusions.

OCCURRENCE: fairly common rock present in almost all granitic plutons, whether in ancient shields or orogens.

USES: used rarely in the building industry.

NOTES: Granite porphyry is compositionally similar to the granites, but it has a porphyritic texture with a fine-grained granular groundmass with larger feldspar phenocrysts.

DIORITE PORPHYRY

CHEMISTRY: intermediate.

ESSENTIAL MINERALS: plagioclases (andesine, labradorite), hornblende, biotite, quartz.

ACCESSORY MINERALS: monoclinic pyroxene, epidote, magnetite, apatite, sphene, rutile, zircon, monazite, pyrite, orthoclase, hematite.

APPEARANCE: medium-grained; massive or zonal structure with quartz xenoliths; porphyritic texture with phenocrysts of plagioclase (tabular and zonal), biotite, or hornblende with a microcrystalline or intersertal groundmass.

COLOR: gray, green, bluish, reddish from the presence of hematite.

GEOLOGIC ENVIRONMENT: in dikes of variable size associated with plutons with a granitic-granodioritic composition.

OCCURRENCE: in localities where granitic-dioritic plutons crop up.

USES: for pavements and facings; a variety called "antique red porphyry" was used extensively by the ancient Egyptians and Romans for use in monuments.

NOTES: The name is derived from the Greek for "purple red," relating to the color of the altered rock. The term porphyry is used for mesocratic rocks of varying chemistry with a porphyritic texture (e.g., monzonite porphyry, etc.)

MINETTE

CHEMISTRY: basic.

ESSENTIAL MINERALS: biotite, diopside, orthoclase.

ACCESSORY MINERALS: olivine, augite, hornblende, plagioclase, calcite, chlorite, rarely quartz.

APPEARANCE: medium- to fine-grained; zonal structure, vesicular; porphyritic texture with phenocrysts of mafic minerals.

COLOR: when freshly cut, it is blackish in color with a characteristic glitter due to flakes of biotite; it is commonly altered, whereupon it appears brownish and is friable.

GEOLOGIC ENVIRONMENT: in dikes and small masses associated with intrusions.

OCCURRENCE: relatively common; found in France, Germany, Norway, Finland, and the United States.

USES: of exclusively scientific interest.

NOTES: The name is derived from the Minette Valley in the Vosges (France). The word *minette* has its origins in mining: the altered rock was at one time used to stop up the bungholes of explosive charges in mines. It is a variety of lamprophyre, characterized by phenocrysts chiefly of biotite and a groundmass composed of orthoclase and small amounts of plagioclase.

▼ *Aplite in gneiss (ca. x 1)., Val Masino, Sondrio, Italy.*

▲ *Granite porphyry (ca. x 1), Erzgebirge, Germany.*

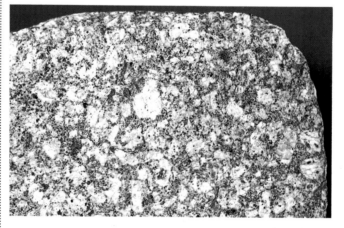

Porphyry ► *(ca. x 1), Valtellina, Sondrio, Italy.*

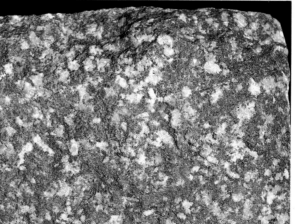

▼ *Minette (ca. x 1), Vosges, France.*

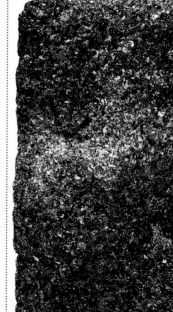

KERSANTITE

CHEMISTRY: basic.
ESSENTIAL MINERALS: magnesium-rich biotite (phlogopite), augite, plagioclase (oligoclase or andesine).
ACCESSORY MINERALS: olivine, diopside, hornblende, orthoclase, rarely quartz.
APPEARANCE: medium- to fine-grained; zonal structure, vesicular; porphyritic texture with phenocrysts of mafic minerals.
COLOR: dark with a characteristic glitter due to flakes of biotite.
GEOLOGIC ENVIRONMENT: in dikes and small masses associated with intrusions.
OCCURRENCE: found in France, Germany, Great Britain, etc.
USES: of exclusively scientific interest.
NOTES: The name has replaced an older term, kersanton, from a locality of the same name in Brittany (France). As is common to all lamprophyres, when olivine is present it is found only as a phenocryst and never in the groundmass. In addition, it almost always alters to carbonates or serpentine. Other typical characteristics of lamprophyres are the occurrence of mafic minerals as phenocrysts and the absence of orthorhombic pyroxenes.

▼ *Kersantite (ca. x 1), Brest, France.*

SPESSARTITE

CHEMISTRY: basic.
ESSENTIAL MINERALS: amphibole (green or brown hornblende), plagioclase (andesine), augite.
ACCESSORY MINERALS: biotite, olivine, diopside, orthoclase, rarely quartz.
APPEARANCE: medium- to fine-grained; zonal structure, vesicular; porphyritic texture with phenocrysts of mafic minerals.
COLOR: dark.
GEOLOGIC ENVIRONMENT: in dikes and small masses associated with intrusions.
OCCURRENCE: found in France, Germany, Great Britain, etc.
USES: of exclusively scientific interest.
NOTES: The name is derived from the Bavarian locality Spessart (Germany). It is a variety of lamprophyre characterized by hornblende phenocrysts and a groundmass with more plagioclase than orthoclase grains.

◄ *Spessartite (ca. x 0.4), provenance unknown.*

CAMPTONITE

CHEMISTRY: ultrabasic.
ESSENTIAL MINERALS: sodic amphiboles (kaersutite, barkevikite), titanaugite, olivine.
ACCESSORY MINERALS: titaniferous biotite, plagioclase (andesine or labradorite), orthoclase, zeolites.
APPEARANCE: medium- to fine-grained; zonal structure, vesicular; porphyritic texture with phenocrysts of mafic minerals.
COLOR: dark.
GEOLOGIC ENVIRONMENT: in dikes and small masses associated with intrusions.
OCCURRENCE: found in France, Germany, Great Britain, etc.
USES: of exclusively scientific interest.
NOTES: The name is derived from its U.S. locality, Campton Falls in New Hampshire. It is a variety of lamprophyre in which plagioclase is the prevalent feldspar. A typical characteristic of lamprophyres is that they are profoundly altered so that secondary minerals such as carbonates, chlorite, and zeolites, probably resulting from autometasomatism, are common.

▼ *Camptonite dike in granite (ca. x 0.5), Predazzo, Trento, Italy.*

KIMBERLITE

CHEMISTRY: ultrabasic.
ESSENTIAL MINERALS: serpentinized olivine, orthorhombic and monoclinic pyroxene, carbonates.
ACCESSORY MINERALS: phlogopite, chromite, pyrope, ilmenite, monticellite, rutile, perovskite, diamond.
APPEARANCE: porphyritic texture.
COLOR: black, blue, greenish.
GEOLOGIC ENVIRONMENT: in volcanic pipes.
OCCURRENCE: found from central to southern Africa (from the Congo to South Africa), in the United States, Canada, and Brazil.
USES: of no industrial use; the primary source of diamonds.
NOTES: The name is derived from the South African locality, Kimberley. It is an intensively studied rock and the primary one in which natural diamonds occur. The emplacement of kimberlite in deep pipes has resulted in diamond extraction by open-pit mining; the pit is in the shape of a cone with the vertex at the bottom.

▼ *Kimberlite (ca. x 1), South Africa.*

ALKALI FELDSPAR RHYOLITE

CHEMISTRY: acidic.

ESSENTIAL MINERALS: quartz, sanidine, albite, aegirine-augite and/or riebeckite.

ACCESSORY MINERALS: anorthoclase, orthoclase, aenigmatite (inosilicate), magnetite, hematite, zircon.

APPEARANCE: very fine-grained; massive structure; commonly has porphyritic hypocrystalline texture with phenocrysts in a glassy, microcrystalline, pilotaxitic, or perlitic groundmass.

COLOR: whitish, faded yellow, greenish, reddish, black.

GEOLOGIC ENVIRONMENT: in domes and lava flows; flows are not as extensive as basaltic ones since the rhyolitic magma has a higher viscosity; some pyroclastic rocks also have a rhyolitic composition.

OCCURRENCE: found in many areas of acidic volcanism.

USES: it is used locally for making architectural bosses and ashlars.

NOTES: Alkali feldspar rhyolites are the extrusive equivalents of alkali feldspar granites. Several named varieties exist, among which are pantellerite and comendite. The term alkali rhyolite is used when over 90% of the feldspars present are alkali feldspars; also, sodic amphiboles and/or pyroxenes are present in the rock.

▼ *Alkali feldspar rhyolite: comendite variety (ca. x 1), San Pietro Island, Cagliari, Italy.*

RHYOLITE

CHEMISTRY: acidic.

ESSENTIAL MINERALS: quartz, sanidine.

ACCESSORY MINERALS: oligoclase, biotite, albite, magnetite, tridymite, diopside.

APPEARANCE: fine-grained; variable structure: massive, with flow structures, or vesicular on surfaces of flows; generally porphyritic hypocrystalline texture with idiomorphic phenocrysts in a totally glassy, spherulitic, or perlitic groundmass; groundmass in some cases is microcrystalline or pilotaxitic.

COLOR: whitish or light gray; glassy varieties are dark.

GEOLOGIC ENVIRONMENT: in domes, dikes, and small size lava flows; some pyroclastic rocks also have a rhyolitic composition.

OCCURRENCE: very common in volcanic areas with acidic magmatism; among the numerous localities that can be cited are the United States, Japan, Ethiopia, Romania, and Italy.

USES: harder than granite, therefore difficult and costly to work; because of its resistance to weathering by atmospheric agents, it can be used for making architectural bosses and ashlars.

NOTES: The name is from the Greek word meaning "to flow." It is the extrusive equivalent of granite. Rhyolite is also known as liparite. Some light colored rhyolites that are especially fine-grained might be confused with some sedimentary rocks. In such cases, the distinctive vesicular and spherulitic structures of rhyolite can be very helpful. The term quartz porphyry was employed in the past, especially by petrologists of the European school, for porphyritic rocks with a rhyolitic composition and older than the Tertiary Period.

▼ *Rhyolite (ca. x 1), Euganean Hills, Padua, Italy.*

RHYODACITE

CHEMISTRY: acidic.

ESSENTIAL MINERALS: quartz, plagioclase (oligoclase, andesine), biotite, hornblende, pyroxenes.

ACCESSORY MINERALS: alkali feldspars, magnetite.

APPEARANCE: fine-grained; massive or with flow structure; generally porphyritic hypocrystalline texture with a microcrystalline, glassy groundmass.

COLOR: light.

GEOLOGIC ENVIRONMENT: in domes or small lava flows; some pyroclastic rocks also have a dacitic composition.

OCCURRENCE: associated with rhyolites and dacites.

USES: used for making architectural bosses and ashlars.

NOTES: The term rhyodacite is used for rocks transitional between rhyolites and dacites. Toscanite, first described on Mount Amiata (Tuscany, Italy) is a variety of rhyodacite; it consists of plagioclase, sanidine, hypersthene, and biotite in a glassy matrix of rhyolitic composition.

▲ *Rhyodacite (ca. x 0.5), Panarea, Lipari Islands, Messina, Italy.*

DACITE

CHEMISTRY: acidic.

ESSENTIAL MINERALS: plagioclase (andesine or labradorite), quartz, biotite, brown hornblende.

ACCESSORY MINERALS: sanidine, pyroxenes (orthorhombic and monoclinic), magnetite, ilmenite, sphene, garnet.

APPEARANCE: fine-grained; massive or with flow structure; generally porphyritic hypocrystalline texture with a glassy or microcrystalline groundmass.

COLOR: various shades of gray.

GEOLOGIC ENVIRONMENT: small lava flows, domes; some pyroclastic rocks also have a dacitic composition.

OCCURRENCE: fairly common; found in volcanic areas with acidic magmatism in numerous localities including Romania, Hungary, France, the Andes, and the Caribbean Islands.

USES: used for making architectural bosses and ashlars.

NOTES: The name is derived from the ancient Roman province Dacia (Romania). Dacites are the extrusive equivalents of granodiorites and tonalites. Dacites that are plagioclase-rich are sometimes called quartz andesite or plagidacites, but these terms are discouraged in modern classification. Dacites are similar to rhyolites, but are distinguished from them by having more plagioclase and pyroxene (diopside); the latter are visible as small dark spots.

▼ *Dacite (ca. x 1), Transylvania, Romania.*

NATURE'S PYROTECHNICS

PRODUCTS OF VOLCANOES

Different types of magma are produced when rocks of differing compositions are melted. Magma produced by melting rocks of the Earth's upper mantle initially has an ultrabasic composition, whereas magma that originates in the continental crust has a more acidic composition.

Eruptions of molten rock on the surface of the Earth produce lava flows that are often very spectacular. The chemical and physical properties of the molten mass results in different types of lava flows. The lava's viscosity, which varies according to its chemical composition, its temperature, and its amount of dissolved gases (water vapor and volatile elements) are all important in determining the eruption characteristics.

RIVERS OF LAVA

Magmas with ultrabasic and basic compositions have relatively low viscosities and are therefore very fluid. They have temperatures on the order of 1832°–2192°F (1000–1200°C) and degassing occurs with relative ease. The eruptions are quiet in nature and the lava spreads out over the land, sometimes covering great distances (flood basalts). Small explosions may occur, throwing small quantities of ash and shreds of lava into the air. The lava remains in a partially molten state even during its journey through the air. Eruptions of this kind are classified as Hawaiian, after the typical style of volcanic activity in those islands.

These lavas exhibit different properties depending on their rate of cooling and their viscosity. They assume characteristic shapes, some of which are designated by indigenous names:

– *pahoehoe lava* (the term may be translated as "ropy lava") is very fluid and its surface cools quickly, solidifying into superficially smooth shapes; as molten lava continues to flow beneath the solid crust, the surface of the lava becomes wrinkled and looks like an irregular tangle of large ropes.

– *aa lava* (the term may be translated as "blocky lava") is generally cooler, has more crystals and has released more of its volatile elements (degassed) than pahoehoe; as a consequence, it is more viscous and flows at a lower velocity. The surface of the flow solidifies to form many rough, angular blocks. These are pushed for-

ward by the underlying molten mass which acts like a bulldozer.

– *pillow lava:* are typical of underwater lava flows formed by the rapid cooling of the extrusive material in contact with the water. They appear as spheroidal or tubular masses covered by a glassy crust.

RAIN OF FIRE

In contrast to basic magmas, acidic magmas with their high silica content and lower temperatures 1472–1832°F (around 800–1000°C) are very viscous. In these magmas, gas release is difficult. The volatiles condense to form bubbles which expand as the magma rises toward the upper part of the volcanic chimney. However, the high viscosity of the magma may inhibit full gas expansion, so that the pressure within the bubbles increases until they burst and the surrounding magma is disrupted. In this case, the eruptions are explosive, ejecting lava and rock fragments into the air.

The erupted fragments of lava and rock are called pyroclasts, and are classified chiefly on the basis of size: *blocks* are angular pyroclasts > 2.5 in. (64 mm) that were in a solid state when ejected; *bombs* are also > 2.5 in. (64 mm) but were ejected in a partially or totally molten state; *lapilli* are pyroclasts between .08 in. and 2.5 in. (2 mm and 64 mm) and come in a

Materials erupted from a volcano (left to right): pahoehoe lava, aa lava, pillow lava, lapilli, blocks, ashes.

variety of shapes; *ashes* are pyroclasts < .08 in. (2 mm) in diameter. Explosive volcanic eruptions are differentiated into various types which, in some cases, take their names from specific volcanoes:

– *strombolian:* named after the volcano on the island of Stromboli in the Lipari Islands. This type is characterized by relatively frequent but moderately intense eruptions of basaltic lava accompanied by the emission of white clouds rich in water vapor but poor in pyroclasts. These eruptions reach maximum heights of a few hundred feet.

– *vulcanian:* named after the volcano on Vulcano Island in the Lipari. The eruptions occur at much longer intervals and eject incandescent, viscous bombs accompanied by clouds of gases that are darkened by large quantities of ash.

– *plinian:* named after the naturalist Pliny the Younger, who described the eruption of Mount Vesuvius (79 A.D.) during which his uncle Pliny the Elder died. This eruption occurs at very long intervals (in some cases, once every few thousand years) and is particulary violent, emitting enormous quantities of gas and volcanic materials.

▼ *Alkali feldspar trachyte with large crystals of sanidine (ca. x 0.6), Bohemia, Czech Republic.*

QUARTZ TRACHYTE

CHEMISTRY: intermediate.
ESSENTIAL MINERALS: sanidine, quartz, sodic plagioclase.
ACCESSORY MINERALS: biotite, amphiboles, magnetite, apatite, zircon, sphene.
APPEARANCE: fine-grained; massive structure; generally, hypocrystalline porphyritic texture with a cryptocrystalline or glassy groundmass.
COLOR: whitish, light gray.
GEOLOGIC ENVIRONMENT: short lava flows, dikes, bodies intruded at shallow depths.
OCCURRENCE: associated with trachytes.
USES: used rarely in the building industry.
NOTES: Quartz trachyte is the extrusive equivalent of quartz syenite. It is distinguished from trachyte by its greater abundance of quartz. Quartz trachyte is very similar in appearance to trachyte, latite, and quartz latite and cannot always be distinguished from these rock types. In such cases, the more general term trachytoid is used and the rock can only be precisely identified by petrological analysis.

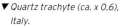

▼ *Quartz trachyte (ca. x 0.6), Italy.*

TRACHYTE

CHEMISTRY: intermediate.
ESSENTIAL MINERALS: sanidine, sodic plagioclase.
ACCESSORY MINERALS: quartz (< 5% by volume), biotite, amphiboles, pyroxenes, magnetite, apatite, zircon, sphene, melanite (garnet).
APPEARANCE: fine-grained; massive structure, or more commonly with flow texture; holocrystalline, rarely hypocrystalline, porphyritic texture; in the groundmass, small lath-shaped crystals of sandine are aligned parallel or almost parallel to one another and indicate direction of flow (trachytic texture).
COLOR: light, from whitish to light gray, faded yellow, rarely greenish.
GEOLOGIC ENVIRONMENT: short lava flows, dikes, bodies intruded at shallow depths; associated with alkali basalts.
OCCURRENCE: found in Hungary, New Zealand, and islands in the Atlantic and Pacific Oceans.
USES: in the building industry for flooring, pavements, and external facings (pink trachyte).
NOTES: The name is derived from the Greek for "rough" and refers to the rock's surface. It is the extrusive equivalent of syenite.

*Quartz latite ▶
(ca. x 0.8),
Italy.*

▼ *Trachyte
(ca. x 1),
Euganean
Hills, Padua,
Italy.*

QUARTZ LATITE

CHEMISTRY: intermediate.
ESSENTIAL MINERALS: plagioclase (andesine), sanidine, quartz, amphiboles (brown hornblende), biotite.
ACCESSORY MINERALS: augite, magnetite, hematite, apatite.
APPEARANCE: fine-grained; massive structure, grading to fluidal; holocrystalline porphyritic texture with a microcrystalline, pilotaxitic, or partially glassy groundmass.
COLOR: generally light, from whitish to gray.
GEOLOGIC ENVIRONMENT: lava flows and dikes, associated with andesitic rocks.
OCCURRENCE: relatively common; found in France, Germany, Italy, the Canaries, and the Azores.
USES: of no industrial use.
NOTES: Quartz latite is the extrusive equivalent of quartz monzonite. It is distinguished from latite by a greater abundance of quartz which, however, cannot be detected by the naked eye. Quartz latite appears very similar to latite, trachyte, and quartz trachytes and cannot always be distinguished from them. In such cases, the term trachytoid is used as a temporary classification before carrying out more thorough analyses.

LATITE

CHEMISTRY: intermediate.
ESSENTIAL MINERALS: plagioclases (andesine), sanidine, augite, amphibole (brown hornblende).
ACCESSORY MINERALS: quartz (< 5% by volume), biotite, magnetite, hematite, apatite.
APPEARANCE: fine-grained; massive structure, grading to fluidal; texture commonly holocrystalline porphyritic with a microcrystalline, pilotaxitic groundmass, rarely with glass.
COLOR: generally light, from white to gray, faded yellow or reddish.
GEOLOGIC ENVIRONMENT: lava flows and dikes, associated with andesitic and basaltic rocks.
OCCURRENCE: relatively commom; found in France, Germany, Italy, the Canaries, and the Azores.
USES: of no commercial use.
NOTES: The name is derived from the Italian region Latium where the variety known as vulsinite occurs. Latite is the extrusive equivalent of monzonite. These rocks appear very similar to trachyte, quartz trachyte, and quartz latite. It is not always possible to tell them apart. In such cases, the term trachytoid is used as an initial general classification.

▼ *Latite (ca. x 1), Roccamonfina, Caserta, Italy.*

ANDESITE

CHEMISTRY: intermediate.

ESSENTIAL MINERALS: plagioclase (andesine), pyroxenes (augite, pigeonite, hypersthene), brown hornblende and/or biotite.

ACCESSORY MINERALS: quartz or olivine, magnetite, hematite, apatite, rarely zircon.

APPEARANCE: medium-fine-grained; massive structure with many variations in color; typically porphyritic texture, holocrystalline, rarely hypocrystalline, with pilotaxitic groundmass.

COLOR: black, brownish or greenish.

GEOLOGIC ENVIRONMENT: lava flows, domes, dikes.

OCCURRENCE: common among the volcanic rocks found along convergent margins of tectonic plates; in the Alpine and Himalayan orogenic mountain chains, the Andes, Japan, Indonesia, Melanesia, Iran, and Turkey.

USES: locally used as building material.

NOTES: The name is derived from the Andes Mountains in South America, where the rock is very common. It is the extrusive equivalent of diorite and is distinguished from basalt by its SiO_2 content (> 52% by weight). It is easily confused with some latites rich in mafic minerals. These rocks are geologically important in active continental margins where an oceanic plate sinks beneath a continental one. For example, numerous volcanoes along the entire perimeter of the Pacific Ocean typically are associated with andesitic lavas.

▼ Andesite (ca. x 1), Euganean Hills, Padua, Italy.

BASALT

CHEMISTRY: basic.

ESSENTIAL MINERALS: plagioclase (labradorite, bytownite), pyroxenes (augite, in some cases titaniferous, pigeonite, hypersthene).

ACCESSORY MINERALS: olivine, amphiboles (brown hornblende), biotite, magnetite, hematite, ilmenite, sphene (common in altered basalts).

APPEARANCE: fine-grained; massive structure, in some cases with meter-scale columnar fractures, commonly vesicular, in some the vesicles are lined with secondary minerals (calcite, zeolites, chalcedony, etc.); under the microscope, basalts commonly exhibit a porphyritic holocrystalline texture with phenocrysts of plagioclase and pyroxenes and a glassy (intersertal) or microcrystalline (intergranular or subophitic) groundmass; glass is rare in alkali basalts.

COLOR: generally black; can be brown or reddish from oxidation.

GEOLOGIC ENVIRONMENT: subaerial lava flows including flood basalts, dikes, sills; submarine lava flows.

OCCURRENCE: it is the most common type of extrusive rock; flood basalts of enormous dimensions are found in India (the famous Deccan traps), Southern Africa (Karroo), Brazil, Paraguay, Ethiopia, etc. Many volcanic islands are made up of basalts (Hawaii).

USES: used rarely in the building industry.

NOTES: The name is probably of Egyptian origin, usually attributed to Pliny. The extrusive equivalent of gabbro, basalt is a rock of great geologic importance since it forms the earth's crust beneath the oceans. It may sometimes include nodules of peridotites. In some basalts olivine is present as large crystals which can be cut as gems. These rocks further classified on the basis of their macroscopic structures and textures: columnar basalt, pillow basalt, pahoehoe basalt, etc. The term spilite is used for altered basaltic lavas where the feldspars are altered to albite.

▼ Basalt (ca. x 1), Mount Etna, Catania, Italy.

PHONOLITE

CHEMISTRY: intermediate.

ESSENTIAL MINERALS: sanidine or anorthoclase, feldspathoids.

ACCESSORY MINERALS: monoclinic pyroxene (aegirine, titaniferous augite), riebeckite, albite, aenigmatite, titaniferous magnetite, apatite, zeolites.

APPEARANCE: fine-grained; structure sometimes massive, commonly with flow texture; holocrystalline porphyritic texture with a generally microlitic, pilotaxitic, or trachytic groundmass.

COLOR: light gray, brown, greenish.

GEOLOGIC ENVIRONMENT: lava flows, domes, dikes.

OCCURRENCE: found associated with alkali basalts in ocean islands (Kerguelen, Canaries, Azores); in Brazil, the United States, Germany, Bohemia, etc.

USES: of limited use in the potassic salts industry and in the building industry.

NOTES: The name is derived from the Greek for "music" and "stone" and refers to the resonance of the rock when struck. It is the extrusive equivalent of feldspathoid syenite.

▼ Basalt (ca. x 1), Mount Etna, Catania, Italy.

TEPHRITIC PHONOLITE

CHEMISTRY: intermediate-basic.

ESSENTIAL MINERALS: feldspathoids (nepheline, leucite, etc.), sanidine, sodic plagioclase (oligoclase).

ACCESSORY MINERALS: monoclinic pyroxene (aegirine, titaniferous augite), riebeckite, anorthoclase, albite, aenigmatite, titaniferous magnetite, apatite, sphene, zeolites.

APPEARANCE: fine-grained; structure sometimes massive, commonly with flow texture; holocrystalline porphyritic texture with a generally microlitic, pilotaxitic, or trachytic groundmass.

COLOR: light gray, brown, green, rosy.

GEOLOGIC ENVIRONMENT: lava flows, domes.

OCCURRENCE: found associated with alkali basalts, especially in continental rift environments.

USES: used rarely in the building industry.

NOTES: Tephritic phonolite is the extrusive equivalent of feldspathoid monzosyenite. These rocks are very similar to the phonolites and, therefore, in field classifications, the generic term phonolitoids is used.

◄ Phonolite (ca. x 1), Bohemia, Czech Republic.

▼ Tephritic phonolite with leucite (ca. x 0.7), Bracciano, Rome, Italy.

THE SLEEPING BEAUTIES

VOLCANOES

Many areas over the surface of the earth are sites of magmatic eruptions.
The lavas quickly and dramtically alter the local topography giving rise to vast plains or great mountains.

In 1963, a volcanic cone emerged from the Atlantic Ocean and, erupting lava for almost 4 years, formed the island of Surtsey (Iceland).

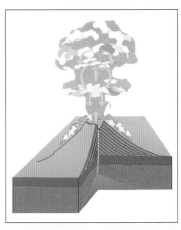

A nuée ardente is a dense incandescent cloud of ash and pumice suspended in hot gases. A less dense cloud erupts vertically (above), and the speed at which suspended particles drop out of it depends on gravity and rarely exceeds 93mph (150 km/h); the denser, lower part of the cloud surges downhill (below) attaining a speed of up to 298 mph (480 km/h).

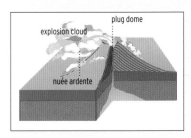

Magmas reach the earth's surface by traveling through deep fractures in the crust which are typically located at plate boundaries. Basaltic magma flows from the faults along mid-ocean ridges and solidifies to form new oceanic crust.

Along the mid-ocean ridges, contact with sea water causes the lava to solidify rapidly rather than flowing far from the crustal fissure. In this way, successive lava flows construct a mountain chain along the zone of fractures and faults from which they emanated. As a result of particularly intense and prolonged eruptions, the submarine mountain chain grows until it surfaces. This is the case for Iceland, justifiably regarded as the "Island of Fire,"; its morphology is characterized by a central depression produced by the system of mid-ocean fractures. Great expanses of basalt are found within continents where the lava flowed laterally over vast areas and successive outflows were superimposed on one another to depths of as much as a few miles. These flows

solidify into what are called "flood basalts," which produce large tablelands that cover as much as a few million square miles of land.

In other cases, along convergent margins and in plate interiors magma erupts at discrete locations to form a cone. When magma travels from a reservoir within the crust or upper mantle (magma chamber) to the earth's surface through essentially vertical conduits, the resultant volcanic activity is called a "central eruption." A cone is produced by the accumulation of flows and pyroclastic material and typically has a central crater. This complex edifice is commonly known as a "volcano."

THE SHAPES OF VOLCANOES
The word is derived from Vulcan, the Roman god of fire. The shapes of volcanoes vary depending on the chemical composition of the lava. Shield volcanoes originate from eruptions of relatively fluid magmas with a basic composition. They have circumferences of tens of miles and heights of

thousands of yards. Mauna Loa on the island of Hawaii is a classic example of this type of volcano. It has rather remarkable dimensions since from its base at the bottom of the ocean to its top, it stands 6.2 miles (10 kilometers) tall with a diameter of 62.10 miles (100 kilometers). The portion of the volcano above sea level is 13,123 ft. (4,000 meters) high.

Magmas with an acidic composition have a greater viscosity than basaltic ones and, consequently, do not flow as freely and spread out.

Composite cone, or stratovolcano, typical of volcanoes in subduction zones.

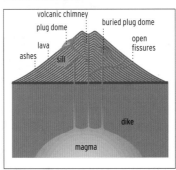

Instead, volcanic structures with fairly steep sides and varying shapes and sizes are formed. For example, lava domes or cupolas are common within volcanic craters. In some cases the material extruded is nearly solid, giving rise to a monolith known as a spine.

Volcanoes of intermediate to acidic compositions typically have explosive activity alternating with effusions of lava. In this case, successions of solidified lava flows and pyroclastic rocks produce imposing volcanoes with fairly steep slopes and the classic cone shape. These structures are called "composite volcanoes" or "stratovolcanoes." Classic examples of this type are Mount Etna, Mount Vesuvius, and Mount Fuji.

PHONOLITIC TEPHRITE

CHEMISTRY: basic-intermediate.

ESSENTIAL MINERALS: feldspathoids (nepheline, leucite, etc.), plagioclase (labradorite), pyroxenes (augite, aegirine).

ACCESSORY MINERALS: brown hornblende, biotite, sodalite, analcime, magnetite, haüynite.

APPEARANCE: fine-grained; vesicular, in some cases with amygdules or cavities lined with zeolites; porphyritic holocrystalline texture with a microcrystalline groundmass.

COLOR: dark gray with whitish spots (feldspathoids, zeolites).

GEOLOGIC ENVIRONMENT: lava flows, dikes

OCCURRENCE: relatively rare rocks; found in associated with provinces of alkaline magmatism such as those in Central Italy, Arizona, Wyoming, Montana, and the east African rift.

USES: used rarely in the building industry.

NOTES: Phonolitic tephrite is the extrusive equivalent of feldspathoid monzogabbro. It is very similar to phonolitic basanite from which it is distinguished on the basis of the olivine content (< 10% by volume). The higher olivine content in the basanites gives them a darker color. Distinguishing phonolitic tephrites, basanites, and tephrites in the field is not easy. Therefore, the general term tephritoid is used to describe all three rock types.

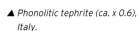

▲ *Phonolitic tephrite (ca. x 0.6), Italy.*

TEPHRITE

CHEMISTRY: basic-ultrabasic.

ESSENTIAL MINERALS: feldspathoids (nepheline, leucite, etc.), plagioclase (labradorite, bytownite), pyroxenes (titanium-rich augite).

ACCESSORY MINERALS: brown hornblende, biotite, sodalite, analcime, magnetite.

APPEARANCE: fine-grained; vesicular, in some cases with amygdules or cavities lined with zeolites; porphyritic holocrystalline texture with a microcrystalline groundmass.

COLOR: dark gray, rarely black with whitish spots (feldspathoids, zeolites).

GEOLOGIC ENVIRONMENT: lava flows, dikes.

OCCURRENCE: not very common rocks; leucite tephrites are found in the lavas of Mount Vesuvius (Italy) and in Uganda; nepheline tephrites in Germany and also in the islands of the Atlantic Ocean (Canaries, Azores, Madeira).

USES: used rarely in the building industry.

NOTES: The name tephrite, probably assigned by Pliny, is derived from the Greek for "ashes." It is the extrusive equivalent of theralite (nepheline gabbro). It is distinguished from basanites by less abundant olivine (< 10% by volume). In the past, tephrites were considered to be a feldspathoid-rich variety of basalt. The principal type of feldspathoid is generally specified as part of the name (e.g., leucite tephrite, etc.), except for nepheline tephrite which is simply called tephrite.

Leucite ▶
tephrite
(ca. x 1),
Latium,
Italy.

BASANITE

CHEMISTRY: ultrabasic.

ESSENTIAL MINERALS: plagioclase (labradorite, bytownite), pyroxenes (titanium-rich augite, aegirine), feldspathoids (nepheline, leucite, etc.), olivine.

ACCESSORY MINERALS: brown hornblende, biotite, sodalite, analcime, magnetite.

APPEARANCE: fine-grained; vesicular in some cases, with amygdules or cavities lined with zeolites; porphyritic holocrystalline texture with, for the most part, a microcrystalline groundmass.

COLOR: dark gray with whitish spots (feldspathoids, zeolites).

GEOLOGIC ENVIRONMENT: lava flows, dikes associated with alkali basalts.

OCCURRENCE: found in the Sahara (Tassili), in Uganda, and the United States; nepheline varieties are found in Germany and in the islands of the Atlantic Ocean (Canaries, Azores, Madeira, St. Helena).

USES: used rarely in the building industry.

NOTES: This very old name is derived from the Greek for "black flint" and is attributed to Theophrastus. Basanite is the volcanic equivalent of olivine theralite (olivine nepheline gabbro). It is distinguished from tephrite by its greater abundance of olivine (> 10% by volume). Basanite is similar in composition and structure to the basalts, so that in the past it was considered to be a porphyritic variety of basalt.

◀ *Leucite*
basanite
(ca. x 1),
Viterbo, Italy.

NEPHELINITE

CHEMISTRY: ultrabasic.

ESSENTIAL MINERALS: nepheline, monoclinic pyroxene (augite, aegirine-augite).

ACCESSORY MINERALS: olivine, sphene, perovskite, melilite, nosean, sodalite, hauyne, zeolites.

APPEARANCE: fine-grained; massive to vesicular; porphyritic holocrystalline texture with a microcrystalline groundmass.

COLOR: light gray, greenish, rosy.

GEOLOGIC ENVIRONMENT: lava flows of modest dimensions associated with alkali basalts.

OCCURRENCE: rare rocks; typical of volcanic areas with alkali magmatism such as those in Germany, France, Italy, and the east African rift.

USES: used rarely in the building industry.

NOTES: The name refers to the rock's composition. It is the extrusive equivalent of ijolite and urtite. In the past, nephelinite was considered to be a nepheline-rich basalt. It is currently classified as a mesocratic variety of foidite, a rock consisting almost entirely of a feldspathoid (90-100% by volume). When 60-90% of the rock volume is made up of feldspathoids, the terms phonolitic nephelinites and tephritic nephelinites are used since the rocks are transitional in mineralogy and chemistry to phonolites and tephrites.

▼ *Nephelinite (ca. x 1), Bohemia, Czech Republic.*

LEUCITITE

CHEMISTRY: basic-ultrabasic.

ESSENTIAL MINERALS: leucite, augite (commonly titaniferous).

ACCESSORY MINERALS: olivine, brown hornblende, biotite, sphene, melanite, melilite, haüyne.

APPEARANCE: fine-grained; massive to vesicular, with amygdules containing zeolites; porphyritic holocrystalline texture with a microcrystalline groundmass.

COLOR: light gray, whitish.

GEOLOGIC ENVIRONMENT: lava flows of modest dimensions.

OCCURRENCE: rare rocks; typical of provinces with alkaline magmatism, such as those in Italy, the east African rift, and Java.

USES: used rarely in the building industry.

NOTES: The name refers to the rock's composition. It is the extrusive equivalent of fergusite. Leucitite is a mesocratic variety of foidite, a rock consisting almost entirely of a feldspathoid (90-100% by volume). When 60 to 90% of the rock volume is made up of feldspathoids, the terms phonolitic leucitites and tephritic leucitites are used since the rocks are transitional to phonolites and tephrites. Ugandite is a melanocratic variety of leucitite composed of monoclinic pyroxene, olivine, and leucite with a glassy groundmass.

▼ *Leucitite (ca. x 1), Acquacetosa, Rome, Italy.*

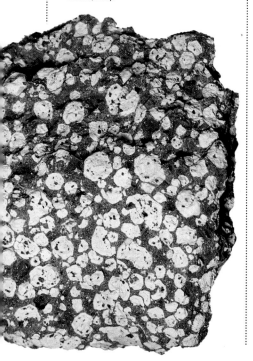

MELILITITE

CHEMISTRY: ultrabasic.

ESSENTIAL MINERALS: melilite, aegerine-augite, olivine.

ACCESSORY MINERALS: perovskite, chromite, picotite, melanite, phlogopite, apatite, sphene.

APPEARANCE: fine-grained; massive to vesicular; holocrystalline porphyritic texture with a microcrystalline or, rarely, glassy groundmass.

COLOR: gray, light brown.

GEOLOGIC ENVIRONMENT: dikes associated with basalts, tephrites, and carbonatites.

OCCURRENCE: found in Italy, specifically as the varieties venanzite and coppaelite; also in Sweden, Madagascar, South Africa, and Canada.

USES: of exclusively scientific interest.

NOTES: These rocks are the extrusive equivalents of the melilitolites. The name refers to the rock's abundance of melilite, a sorosilicate made up by the solid solution of gehlenite and akermanite. Non-ultramafic rocks that contain melilite are known by a modified version of the standard rock name (e.g., melilitic nephelinite).

◄ *Melilitite (ca. x 1), Latium, Italy.*

HYALOCLASTITE

CHEMISTRY: intermediate-basic.

ESSENTIAL MATERIALS: glass.

ACCESSORY MINERALS: none.

APPEARANCE: brecciated; it is made up of very splintered, angular, glassy fragments cemented by minerals such as calcite and zeolite or weakly bound by an argillaceous matrix.

COLOR: generally dark.

GEOLOGIC ENVIRONMENT: in small masses generally associated with pillow lavas (pillow basalts); sometimes stratiform, intercalated with sediments.

OCCURRENCE: not very common rock.

USES: of no practical use.

NOTES: The name is derived from the Greek for "transparent" and "to break" and refers to the rock's appearance. It is generally formed by the contact of lava with water. Commonly, the basaltic glass (sideromelane), the major constituent of this rock, reacts with the water to form palagonite, (a brownish or orange product) chlorite, or other argillaceous minerals. Volcanic glass with an acidic composition is known as obsidian.

▲ *Hyaloclastite (ca. x 1.2), Monti Iblei, Sicily, Italy.*

IGNIMBRITE

CHEMISTRY: acidic-intermediate.

ESSENTIAL MINERALS: glassy shards; quartz, sanidine, albite, biotite in acidic rocks; pyroxene, hornblende, plagioclase in intermediate rocks.

ACCESSORY MINERALS: feldspathoids, fragments of other rocks.

APPEARANCE: a rock with a variable degree of consolidation that consists of a variety of pyroclasts and ash which form a glassy matrix. In some cases, the glassy groundmass is cracked; may have abundant vesicles lined with crystals.

COLOR: light gray, brownish, sometimes reddish or light violet from oxidation.

GEOLOGIC ENVIRONMENT: in flows of remarkable expanse with thicknesses of one to hundreds of feet.

OCCURRENCE: fairly common in the vicinity of volcanoes that erupt expolosively.

USES: in the building industry.

NOTES: The name has its origins in the Latin *ignis* (fire) and *imber* (rain), and refers to the spectacular, catastrophic nuées ardentes. They are very dense masses of incandescent pyroclastic materials mixed with gas that erupt and surge very quickly down the slopes of a volcano destroying everything in their path. The cooled and consolidated deposit is an ignimbrite. In some cases, directional flow textures are present in these rocks.

▼ *Ignimbrite (ca. x 0.7), Viterbo, Italy.*

TUFF

CHEMISTRY: variable (usually acidic-intermediate).

ESSENTIAL MINERALS: plagioclase, sanidine, quartz, micas, feldspathoids, pyroxenes, and amphiboles depending on the composition of the magma; glassy shards and fragments of rocks.

ACCESSORY MINERALS: olivine, zeolites, hematite, magnetite.

APPEARANCE: porous, low-density rock, medium- to fine-grained, partially cemented; common for grain size to decrease from the bottom upwards within a rock unit resulting from one eruption event; generally porphyritic with larger crystals and rock fragments that are generally black or brownish yellow and fine-grained.

COLOR: light or dark gray, rosy, greenish, yellowish, brownish.

GEOLOGIC ENVIRONMENT: in deposits associated with extrusive rocks.

OCCURRENCE: in the vicinity of volcanoes that erupt explosively.

USES: some cemented varieties are used in the building industry.

NOTES: Coarse-grained pyroclastic rocks (> 2.5 inches or 64 mm) are called agglomerates or volcanic breccias; intermediate-grained ones (.08-2.5 inches or 2-64 mm) are called lapilli tuffs. Tuffs have a grain size smaller than .08 inches (2 mm). Vesicles in tuffs are filled with minerals like zeolites and calcite which crystallize out of the vapor and aid in cementing the rock.

PEPERINO TUFF

CHEMISTRY: intermediate.

ESSENTIAL MINERALS: augite, biotite, leucite.

ACCESSORY MINERALS: feldspars, zeolites; lapilli and calcareous clasts are also present.

APPEARANCE: porous, low-density, coherent rock, medium-grained with scattered coarse particles. In some cases, it has the texture of a breccia from the presence of volcanic tephra that is crystal-rich (leucite, augite).

COLOR: dark gray, brownish with small black spots (augite).

GEOLOGIC ENVIRONMENT: vaguely stratiform; in fairly extensive deposits associated with with trachytic or tephritic-leucititic extrusive rocks.

OCCURRENCE: in the vicinity of volcanoes that erupt explosively.

USES: a well-cemented, light, and easily worked variety of tuff; often used as a building material.

NOTES: The name refers to the blackish spots scattered on the rock's surface that call to mind grains of pepper. Pozzolana is a tuff with a trachytic composition used in the preparation of special cements.

OBSIDIAN

CHEMISTRY: variable (usually acidic).

ESSENTIAL MATERIALS: glass.

ACCESSORY MINERALS: iron oxides and various minerals.

APPEARANCE: very compact rock, with typical conchoidal fracture; massive structure, sometimes with concentric cracks and perlitic texture; hyaline (glassy) texture with rare microphenocrysts.

COLOR: glossy black, dark brown, gray or black, speckled with red.

GEOLOGIC ENVIRONMENT: as flows; fragments ejected by explosive volcanoes; as crusts covering rhyolitic and dacitic domes.

OCCURRENCE: relatively common product of acidic volcanism.

USES: in prehistoric times, for making various tools; today it is the basic material for so-called "rock wool."

NOTES: The composition of obsidian is similar to that of granite. Chemical analysis or a measurement of the index of refraction of the glass reveals more about its precise composition.

PUMICE

CHEMISTRY: variable (usually acidic).

ESSENTIAL MATERIALS: glass.

ACCESSORY MINERALS: various silicates, calcite.

APPEARANCE: scoriaceous, in many cases with vesicles oriented parallel to each other, in some cases with amygdules filled with calcite and zeolites; vesicular (or pumiceous) texture, consisting of both solitary and interconnected vesicles, makes pumice remarkably light so that it floats on water.

COLOR: light gray, yellowish, rosy.

GEOLOGIC ENVIRONMENT: in blocks ejected by explosive-type volcanoes; rarely on surfaces of some lava flows.

OCCURRENCE: associated with acidic to intermediate volcanism; abundant in some islands of volcanic origin, such as the Lipari, Thíva (Aegean Sea), and in the islands of Indonesia and Japan.

USES: as a low-quality abrasive; in the building industry as a non-conductor.

NOTES: The name is derived from the Latin *pumex* and is attributed to Theophrastus. The bubbly texture of pumice originates in the release and trapping of gases within viscous acidic magmas traveling through a volcanic chimney; gas is released from the vesicles after the magma has been fragmented into shards near the surface. The historic eruption of Mount Vesuvius in 79 A.D., which caused the destruction of Pompeii, began with the eruption of pumice and lapilli.

◀ Peperino tuff (ca. x 1), Latium, Italy.

Pumice (ca. x 1), ▶ Lipari Islands, Messina, Italy.

Obsidian (ca. x 1), ▶ Lipari Islands, Messina, Italy.

◀ Porphyritic tuff (ca. x 1), Arona, Novara, Italy.

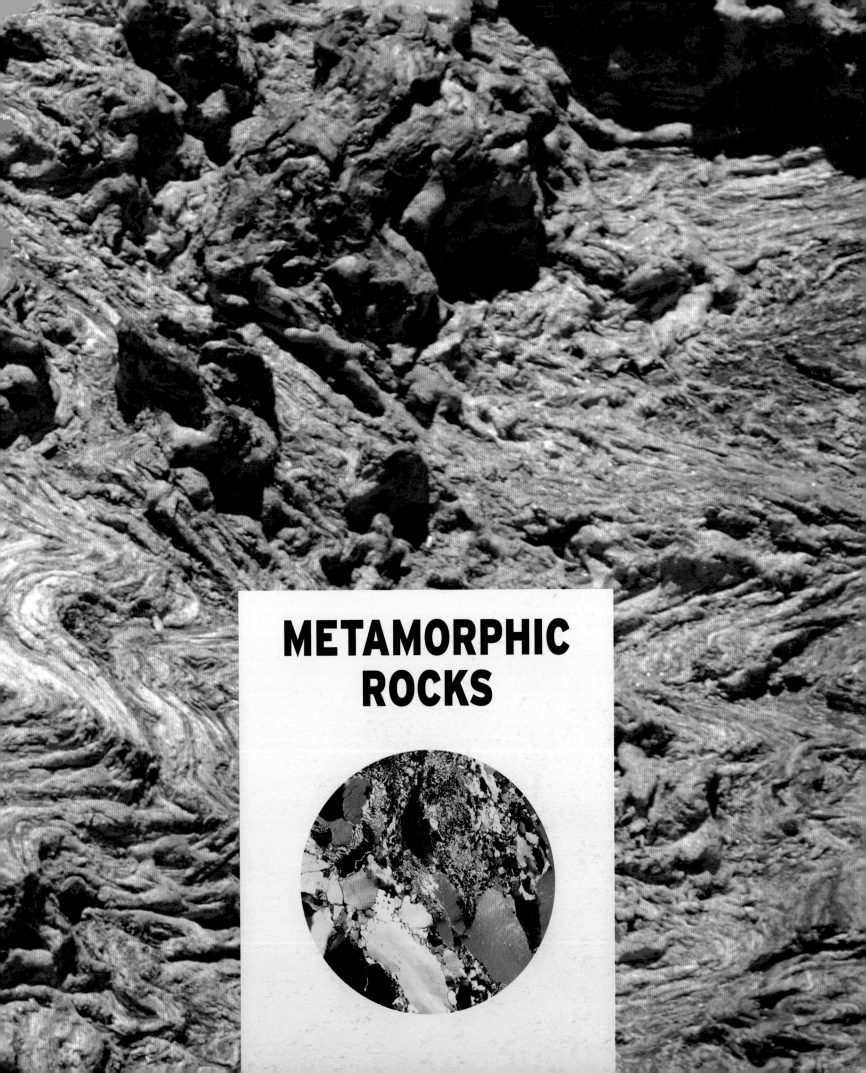

METAMORPHIC ROCKS

METAMORPHIC ROCKS

Top: Sericite quartz schist (Apuan Alps, Tuscany, Italy) in which the schistosity and the intense folding of the rock can be seen.

Bottom: Diagram representing the distribution of temperature in a contact metamorphic aureole.

Metamorphic rocks are formed inside the earth's crust from preexisting rocks, usually by an increase in temperature or pressure. The original mineral grains are transformed in size, arrangement, or type (recrystallization). At a given point within the crust, the pressure is determined by the mass of overlying rocks. Temperature, too, varies with depth. Both measures, temperature and pressure are used to specify a metamorphic environment. In some environments, temperature plays the principal role; in others, pressure is the decisive factor;

and finally, there are cases in which both are fundamental.

We can subdivide metamorphism into two principal types: contact or thermal, and regional. The former results from proximity to a source of heat, typically a magmatic body; the drawing at the bottom is a schematic representation of the distribution of temperature in a

contact metamorphic environment. Regional metamorphism, on the other hand, involves increases in both temperature and pressure and occurs as the result of colliding tectonic plates. The collision produces a mountainous zone of intensely deformed and metamorphosed rocks, known as an orogenic belt.

Subduction zone metamorphism is a special type of regional metamorphism that involves very high pressure. A subducting slab of oceanic crust sinks into the mantle to great depths. However, since the slab is initially cold and is then gradually heated by the surrounding rocks, the temperature in this environment is less than what is generally expected at these depths. Subduction zone metamorphism occurs oceanward of a regional metamorphic orogen.

Another kind of metamorphism takes place on the ocean floor. It develops near mid-ocean ridges where the rocks are affected not only by elevated temperatures, but also by chemical interactions with ocean water. This type of metamorphism was identified only about 25 years ago. Up until that time, it was thought that the ocean floor consisted of a thin layer of oceanic sediments underlain by basalt and other basic and ultrabasic rocks. Then, scientists dredged metamorphosed basalts from near both the Mid-Atlantic Ridge and the Carlsberg Ridge in the Indian Ocean. Further research confirmed that a great deal of the basalt from the ocean floor is characterized by this type of metamorphism. Rocks sampled from deeper parts of the oceanic crust were metamorphosed at higher temperature. Recognition of ocean floor metamorphism is an important aid in understanding the origin of ophiolites. Ophiolites are sections of ancient ocean crust that were thrust up onto continents in subduction zones.

Metamorphic rocks that exhibit schistosity are easily differentiated from igneous and sedimentary rocks. This feature is produced by a directed stress that tends to orient elongate or flat, tabular crystals on a plane perpendicular to the direction of compression. The process involved is similar to stomping on a box to flatten it. Schistosity is less well-developed in very high-grade metamorphic rocks and in those derived from rocks like sandstone and granite that have a granular texture and few micaceous minerals.

In order to better understand the origin and characteristics of metamorphic rocks, a diagram (*see p. 125*) is used to classify the rocks according to the metamorphic environment in which they formed. At the top, temperature increases to the right and is recorded in degrees Celsius; pressure, listed in kilobars (1 bar = 1 atmosphere), increases toward the bottom of the figure. The diagram is valid for conditions where the pressure produced by the weight of the overlying rocks is balanced by that of the fluids present inside the rocks ($P_L = P_{H_2O}$). To the left of the solid line, there is an area showing conditions that do not exist in nature. The conditions for diagenesis are not shown, but are understood to be in the corner just to the left of area 1, the zeolite facies. Diagenesis occurs at relatively low pressures and temperatures,

and does not generally involve significant mineralogical transformations.

Contact (or thermal) metamorphism occupies the uppermost fields (5, 6, and 7), which are characterized by very low pressures and temperatures ranging from about 570° F (300° C) to the melting point of the rocks (1290–1472° F, 700–800° C). These environmental conditions favor the formation of hornfelses, and the fields are named on the basis of a characteristic mineral for each of the three sets of conditions. Schistosity and other foliations are not generally present in these rocks, but they are typically spotted by porphyroblasts: large grains of new minerals that formed with the increased temperature. Hornfels are also recognized by their proximity in the field to a magmatic body.

The remaining fields correspond to regional metamorphic environments. Fields 1, 2, 3, and 4 are characterized by low temperature and pressures ranging from a few hundred bars to 12 kbars. They include low-grade metamorphism as well as subduction zone metamorphism. Zeolites are present in the rocks under conditions of low and medium pressure. At greater depths, jadeite and glaucophane appear. Glaucophane, a blue amphibole, is an indicator mineral for these high-pressure environmental conditions, and, therefore, it is used to name both the rocks that contain glaucophane as well as the conditions under which they formed (fields 4 and 9). With the exception of the rocks in field 1, which commonly have relict textures, the rocks formed at high pressure generally have obvious schistosity and are fine-grained.

Fields 8, 9, 10, and 11 are typical conditions of regional metamorphism. In particular, fields 8 and 9 are characterized by moderate temperatures and by pressures that range from low to very high values. Field 8, also known as the greenschist facies, corresponds to low- and medium-grade metamorphism. Rocks belonging to this facies are schistose, fine- to medium-grained, and rich in micaceous minerals. At the higher pressure of field 9, glaucophane may appear in the rocks.

Fields 10 and 11 are characterized by high temperatures and pressure varying from medium-low to high values; they correspond to the amphibolite facies. Metamorphism in these fields is medium

The arrows indicate the direction of increasing temperature. The highest degree of metamorphism in the contact metamorphic environment occurs where the temperature is at a maximum. The temperature distribution is predominantly a function of distance away from the magmatic body.

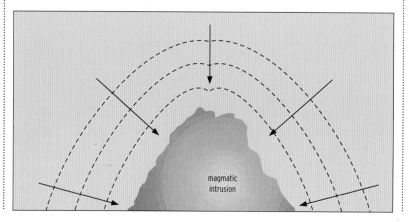

magmatic intrusion

Top: Mont Blanc (seen from Val Veny). This imposing mountain is made up of metamorphic rocks.

Bottom: diagram where the numbered fields correspond to the conditions of temperature and pressure that are typically associated with the various metamorphic facies.

to high grade and the rocks generally are medium- to coarse-grained. In some, schistosity is not well-developed due to the paucity of micaceous minerals. Gneiss and amphibolites are typical rocks of this environment.

The granulite and eclogite facies are excluded from this diagram, since both are formed under anhydrous conditions (low water vapor content in the fluid). Granulites are formed in high-temperature and high-pressure environments ranging inclusively from high-grade metamorphism to the beginning of rock melting (anatexis). They are typically associated with migmatites, towards which they grade.

Finally, eclogites are formed in an anhydrous environment with high or extremely high pressure and a very broad range of temperatures. They are easily altered by reaction with water, which may in part account for their scarcity on the earth's surface. Small masses are found associated with metamorphosed Alpine ophiolites.

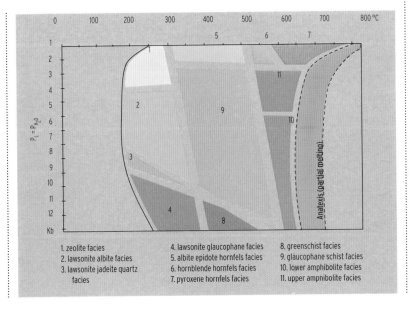

1. zeolite facies
2. lawsonite albite facies
3. lawsonite jadeite quartz facies
4. lawsonite glaucophane facies
5. albite epidote hornfels facies
6. hornblende hornfels facies
7. pyroxene hornfels facies
8. greenschist facies
9. glaucophane schist facies
10. lower amphibolite facies
11. upper amphibolite facies

SLATE

PROTOLITH: pelitic sedimentary rock.
METAMORPHIC ENVIRONMENT: extremely low-grade regional metamorphism, between diagenesis and low-grade metamorphism; contact metamorphism.
ESSENTIAL MINERALS: argillaceous minerals, quartz, muscovite.
ACCESSORY MINERALS: feldspars, carbonaceous substances, calcite, pyrite.
APPEARANCE: extremely fine-grained; schistosity (in this case also called slaty cleavage) with parallel planar surfaces, commonly at an angle to the original bedding surface on which the sediments were deposited; under the microscope, in addition to recrystallization, mechanical reorientation of the argillaceous minerals can be seen: the minerals are flattened along the cleavage planes as a result of applied pressure.
COLOR: gray, bluish, blackish, yellowish or brick-colored.
GEOLOGIC ENVIRONMENT: in beds, in some cases remarkably extensive; grades into phyllites and mica schists at higher metamorphic grades.
OCCURRENCE: fairly common rock in mountain chains or in contact aureoles.
USES: easily separated into thin sheets (fissile) which are resistant to breaking.
NOTES: The fossils present in pelitic rocks are not generally found in slates because of transformations produced by the metamorphic event. The one reproduced in the photograph is a rare example of good preservation. The wide range of colors in these rocks results from the nature of the depositional environment as well as the degree of oxidation.

Metagraywacke (ca. x 0.7), ▶
Lucca, Italy.

▼ *Slate with a fossil (ca. x 0.5), provenance unknown*

METAGRAYWACKE

PROTOLITH: arenaceous sedimentary rock.
METAMORPHIC ENVIRONMENT: low-grade regional metamorphism (lawsonite-albite facies).
ESSENTIAL MINERALS: quartz and other relict minerals from the original sediment; also albite, chlorite, and zeolites.
ACCESSORY MINERALS: prehnite, muscovite, illite, lawsonite.
APPEARANCE: fine-grained; massive; original structures can be recognized; under the microscope, it exhibits complete recrystallization of the matrix.
COLOR: gray, greenish, brownish.
GEOLOGIC ENVIRONMENT: in masses of moderate size; an increase in metamorphic grade produces quartz-feldspar schist.
OCCURRENCE: in mountain chains.
USES: used rarely in local industry.
NOTES: The prefix "meta" indicates that the protolith, in this case graywacke, has been subjected to metamorphism, commonly at low grade.

PHYLLITE

PROTOLITH: pelitic sedimentary rock.
METAMORPHIC ENVIRONMENT: low-grade regional metamorphism (greenschist facies).
ESSENTIAL MINERALS: quartz, light micas (muscovite, phengite, paragonite).
ACCESSORY MINERALS: chlorite, albite, calcite, epidotes, apatite.
APPEARANCE: fine-grained; mica flakes are larger than in slate and aligned along the schistosity to give the rock a luster, but are generally still fine enough not to be resolvable by the naked eye; schistosity occurs as wavy surfaces with a silky luster; sometimes with lenses or nodules of quartz with a medium-coarse grain size, identifiable by the naked eye by their milky appearance (quartz phyllite); under the microscope, equidimensional crystals (granoblastic texture) are evident, sometimes with porphyroblasts of albite.
COLOR: light gray, silver-gray, lead-gray, greenish.
GEOLOGIC ENVIRONMENT: in large masses associated with albite quartzite and albite paragneiss.
OCCURRENCE: very common rock in large mountain chains.
USES: used rarely in the building industry.
NOTES: As metamorphic grade increases, phyllite constitutes rocks transitional between slates and mica schists. They are metamorphic rocks with a broad compositional spectrum and, therefore, are usually described by adding the name of the predominant mineral (quartz, muscovite, etc.). Phyllites that are formed by retrograde metamorphism of higher-grade metamorphic rocks (mica schists, paragneiss) are called phyllonites and exhibit relicts of minerals produced at high temperature.

▼ *Quartz phyllite (ca. x 1), Sondrio, Italy.*

ALBITE PARAGNEISS

PROTOLITH: arenaceous to pelitic sedimentary rock.
METAMORPHIC ENVIRONMENT: low-grade regional metamorphism (greenschist facies).
ESSENTIAL MINERALS: albite, quartz, epidote.
ACCESSORY MINERALS: biotite, chlorite, muscovite, calcite, apatite, tourmaline, chloritoid.
APPEARANCE: medium-grained, with light-colored quartz or feldspar crystals visible to the naked eye; structures of the original rocks are still recognizable; under the microscope, a fine-grained recrystallized matrix is visible, sometimes with porphyroblasts of quartz and albite.
COLOR: light and dark alternating bands.
GEOLOGIC ENVIRONMENT: in very extensive masses, commonly associated with phyllites; at higher metamorphic grades, this rock grades into mica schists and high-temperature paragneiss.
OCCURRENCE: common rock from interiors of eroded mountain chains.
USES: used rarely in the building industry.
NOTES: The major differences between this rock and higher grade paragneisses have to do with mineralogical composition and texture. As the metamorphic grade increases, plagioclase replaces albite and epidote; in addition, as grain size increases, the original textures of the protolith disappear.

▼ *Albite paragneiss (ca. x 1), Val Lozen, Trent, Italy.*

TWO MICA SCHIST

PROTOLITH: pelitic sedimentary rock.
METAMORPHIC ENVIRONMENT: medium-grade regional metamorphism with medium to low pressure (greenschist facies).
ESSENTIAL MINERALS: quartz, plagioclase (oligoclase, andesine), biotite, muscovite, paragonite.
ACCESSORY MINERALS: garnet (almandine), andalusite, cordierite, iron and titanium oxides, apatite, zircon.
APPEARANCE: medium-grained; well-developed schistosity, generally planar but in some cases wavy; fresh fractures exhibit remarkable luster due to the presence of parallel coarse muscovite flakes; granoblastic texture with porphyroblasts of plagioclase and cordierite.
COLOR: light silver-gray, when muscovite predominates; blackish chestnut, when biotite predominates.
GEOLOGIC ENVIRONMENT: in very extensive masses, generally associated with paragneiss; grading into phyllitic rocks at low grade, and to gneiss at high grade.
OCCURRENCE: very common rocks in mountain chains.
USES: used rarely in the building industry.
NOTES: The word "schist" is probably of Germanic origin, from *skivaro* (crushed rocks or wood) or from *schiver* or *schivere* (fragments of crushed rock or wood). In schists, the mica crystals are easily visible with the naked eye; with the phyllites, a magnifying glass is helpful.

▼ *Two mica schist (ca. x 0.7), Saxony, Germany.*

GARNET MICA SCHIST

PROTOLITH: pelitic sedimentary rock.
METAMORPHIC ENVIRONMENT: medium-grade regional metamorphism with medium-high pressure (beginning of the amphibolite facies).
ESSENTIAL MINERALS: garnet (almandine), quartz, biotite, muscovite.
ACCESSORY MINERALS: plagioclase (oligoclase, andesine), iron and titanium oxides, apatite, zircon, epidotes rare.
APPEARANCE: medium-grained with garnet porphyroblasts up to several centimeters in diameter that are easily recognized by their dark red color and nodular appearance; structure is schistose, planar, in some cases wavy or irregular; granoblastic texture with porphyroblasts (garnet) and poikiloblasts (plagioclase with inclusions).
COLOR: silvery to dark with whitish and dark red spots.
GEOLOGIC ENVIRONMENT: in very extensive masses generally associated with paragneiss; grading to phyllites at low grade and to gneiss, migmatites, or granulites at high grade.
OCCURRENCE: very common rocks; they crop out in mountain chains.
USES: rarely used in the building industry.
NOTES: The presence of abundant parallel flakes of mica make mica schists clearly schistose in contrast to paragneisses which exhibit a more massive structure. However, all gradations between these two textures exist in nature, so that gneissic mica schists and schistose paragneisses can be found and are named according to the predominant features. Within the mica schist family, the garnet-rich variety is the most abundant in nature.

▼ *Garnet mica schist (ca. x 1), Bolzano, Italy.*

STAUROLITE GARNET MICA SCHIST

PROTOLITH: pelitic sedimentary rock.
METAMORPHIC ENVIRONMENT: medium grade, medium- to high-pressure regional metamorphism (amphibolite facies).
ESSENTIAL MINERALS: quartz, plagioclase, staurolite, garnet (almandine), biotite.
ACCESSORY MINERALS: iron and titanium oxides, apatite, zircon.
APPEARANCE: medium- to coarse-grained; the structure is schistose with lineations; granoblastic texture with large pophyroblasts of plagioclase (light), andalusite, and staurolite (dark, may have typical cruciform twins).
COLOR: silvery to dark.
GEOLOGIC ENVIRONMENT: in masses associated with paragneiss; grading into phyllite at low grade, and to gneiss at high grade.
OCCURRENCE: in mountain chains.
USES: used rarely in the building industry.
NOTES: The presence of staurolite yields important information about such environmental conditions as temperature, pressure, and fluid composition.

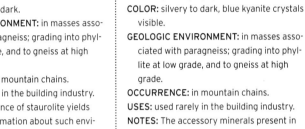

▲ *Staurolite garnet mica schist (ca. x 0.5), France.*

KYANITE MICA SCHIST

PROTOLITH: pelitic sedimentary rock.
METAMORPHIC ENVIRONMENT: medium- to high-grade, high-pressure regional metamorphism (amphibolite facies).
ESSENTIAL MINERALS: quartz, plagioclase (oligoclase, andesine), kyanite, biotite, garnet (almandine).
ACCESSORY MINERALS: staurolite, iron and titanium oxides, apatite, zircon.
APPEARANCE: coarse-grained; discontinuous schistose structure with banding: micaceous layers distinct from the kyanite ones; granoblastic texture, sometimes with kyanite porphyroblasts.
COLOR: silvery to dark, blue kyanite crystals visible.
GEOLOGIC ENVIRONMENT: in masses associated with paragneiss; grading into phyllite at low grade, and to gneiss at high grade.
OCCURRENCE: in mountain chains.
USES: used rarely in the building industry.
NOTES: The accessory minerals present in mica schists yield useful information about the temperature and pressure to which the original rocks were subjected. Sillimanite indicates a medium to high metamorphic grade and, specifically, higher temperatures than either andalusite or kyanite-bearing rocks. Andalusite and cordierite are typical of relatively low pressure conditions; kyanite and garnets, of high pressure. In the latter case, the rocks exhibit less obvious schistosity and are coarse-grained, with textures analogous to those of gneisses.

Kyanite ▶ *mica schist (ca. x 0.7), Apuan Alps, Tuscany, Italy.*

TWO MICA PARAGNEISS

PROTOLITH: arenaceous to pelitic sedimentary rock (in particular, graywackes).

METAMORPHIC ENVIRONMENT: medium- to high-grade regional metamorphism (amphibolite facies).

ESSENTIAL MINERALS: quartz, plagioclase (oligoclase, andesine), orthoclase, biotite, muscovite.

ACCESSORY MINERALS: garnet, staurolite, zircon, calcite, apatite, tourmaline.

APPEARANCE: medium-grained; schistosity barely evident; commonly not homogeneous, but has bands of different mineral assemblages; granoblastic texture with porphyroblasts or poikiloblasts (garnet, staurolite).

COLOR: grayish brown.

GEOLOGIC ENVIRONMENT: in extensive masses associated with mica schists; as metamorphic grade increases, rock grades to migmatites or to granulites, and to phyllites at lower grade.

OCCURRENCE: very common in interiors of eroded mountain chains.

USES: used rarely in the building industry.

NOTES: The prefix "para" before "gneiss" refers to a sedimentary protolith. However, the protolith cannot always be determined unequivocally. Arkosic protoliths (quartzose-feldspathic sandstones) have a chemical composition very similar to that of the granites, and consequently they produce metamorphic rocks that are difficult to distinguish from metagranites.

▼ Two mica paragneiss (ca. x 0.7), Arendal, Norway.

BIOTITE PARAGNEISS

PROTOLITH: arenaceous to pelitic sedimentary rock (in particular, graywackes).

METAMORPHIC ENVIRONMENT: medium- to high-grade regional metamorphism (amphibolite facies).

ESSENTIAL MINERALS: quartz, plagioclase (oligoclase, andesine), alkali feldspar, biotite.

ACCESSORY MINERALS: muscovite, garnet, staurolite, zircon, calcite, apatite, tourmaline.

APPEARANCE: medium-grained; schistosity is scarcely evident; commonly not homogeneous but has bands of different mineral assemblages; lens-shaped areas (gray-green) rich in feldspars are evident. Under the microscope, equidimensional crystals (granoblastic texture) with porphyroblasts are visible.

COLOR: alternating light and dark bands.

GEOLOGIC ENVIRONMENT: in large masses generally associated with mica schists; as metamorphic grade increases, rock grades to migmatites or to granulites, and at low grade, to phyllites.

OCCURRENCE: in interiors of eroded mountain chains.

USES: used rarely in the local building industry; cut into slabs and polished.

NOTES: Accessory minerals are in some cases especially abundant in paragneisses. This mineral is then added to the name of the rock; for example: cordierite-andalusite paragneiss or garnet paragneiss.

▼ Biotite paragneiss (ca. x 0.6), Rhône, France.

GARNET PARAGNEISS

PROTOLITH: arenaceous to pelitic sedimentary rocks.

METAMORPHIC ENVIRONMENT: high-grade regional metamorphism (amphibolite facies).

ESSENTIAL MINERALS: garnet (almandine), microcline, orthoclase, quartz, plagioclase, biotite.

ACCESSORY MINERALS: cordierite, hornblende, augite, apatite, zircon, tourmaline.

APPEARANCE: medium-to coarse-grained; barely discernible schistosity; granoblastic texture; garnets common as large porphyroblasts.

COLOR: generally whitish with reddish garnet grains and dark bands.

GEOLOGIC ENVIRONMENT: in masses of various sizes, commonly associated with migmatites.

OCCURRENCE: very common in highly eroded mountain chains.

USES: used rarely in the local building industry; cut into slabs and polished.

NOTES: Some gneisses are similar in mineralogy to certain varieties of mica schist. However, the gneisses fracture with difficulty to form thick irregular slabs, whereas mica schists are easier to fracture and produce thinner plates.

Garnet paragneiss ▶
(ca. x 0.5),
Sondrio, Italy.

MUSCOVITE QUARTZITE

PROTOLITH: quartz-rich sedimentary rock with small amounts of argillaceous minerals or feldspars.

METAMORPHIC ENVIRONMENT: low-grade regional metamorphism (greenschist facies).

ESSENTIAL MINERALS: quartz, muscovite, chlorite.

ACCESSORY MINERALS: biotite, albite, epidote, garnet (iron-rich), magnetite, tourmaline, apatite.

APPEARANCE: fine-grained; some have schistosity with weak fissility, producing durable thin plates; under the microscope, equidimensional crystals (granoblastic texture), quartz crystals with irregular boundaries, and parallel flakes of muscovite are visible.

COLOR: generally whitish, light gray.

GEOLOGIC ENVIRONMENT: lenticular intercalations of moderate thickness within phyllites, mica schists, and paragneisses; more rarely of great thickness.

OCCURRENCE: fairly common rock in regional metamorphic terrain derived from metamorphism of sedimentary strata.

USES: utilized for flooring and in polished slabs for facings.

NOTES: The presence of other minerals in quartzose sandstones react to form muscovite and chlorite during low-grade regional metamorphism with an intense deformational component. These micas concentrate in thin layers that give the rock a weak fissility. As the mica content increases, the rock grades to quartz mica schist.

▼ Muscovite quartzite (ca. x 0.9), provenance unknown.

QUARTZITE

PROTOLITH: quartz-rich sedimentary rock (orthoquartzites, quartzose siltstones, chert).

METAMORPHIC ENVIRONMENT: variable due to the broad stability field of quartz: found resulting from diagenesis, low-, medium- and high-grade metamorphism and contact metamorphism.

ESSENTIAL MINERALS: quartz.

ACCESSORY MINERALS: muscovite, biotite, alkali feldsparss, plagioclase, garnet (iron-rich), zircon, tourmaline, apatite, graphite, magnetite.

APPEARANCE: fine- to coarse-grained; massive, banded, or schistose (muscovite quartzite, quartz schist) depending on the presence and distribution of the micas; relict sedimentary features may be present; granoblastic texture with irregular quartz crystals, often fractured.

COLOR: commonly whitish, soft gray; dark to black in quartzites rich in graphite, magnetite, biotite.

GEOLOGIC ENVIRONMENT: lenticular intercalations of moderate thickness within phyllites, mica schists, and paragneiss; more rarely in thick beds.

OCCURRENCE: fairly common rock.

USES: quartzite with planar schistosity is split into slabs and used in the building industry; massive varieties are cut into slabs and polished.

NOTES: Schistose structure is common in mica-rich quartzites derived from regional metamorphism where, in addition to recrystallization, there is a deformational component. The thin bands of quartz present in phyllites and in mica schists are not included under the definition for quartzite.

▼ *Quartzite (ca. x 0.9), Tyrol, Austria.*

THE TEXTURES OF METAMORPHIC ROCKS

During its continual, slow movements, the earth's crust transports rocks into different environments from those in which they initially were formed. The minerals that constitute these rocks react to form a new assemblage of minerals that are stable under the changed conditions.

Garnet mica schist (ca. x 1), Val Passeria, Bolzano, Italy; below, in a thin section under the polarizing microscope (ca. x 20; crossed nicols).

Augen gneiss (ca. x 1), Sondrio, Italy; below, in a thin section under the polarizing microscope (ca. x 20; crossed nicols).

Pink marble (ca. x 1), Candoglia, Novara, Italy; below, in a thin section under the polarizing microscope (ca. x 20; polarizer only).

Inside the earth's crust, high temperatures and pressures instigate deformations in the crystalline latice of minerals and initiate chemical reactions between different species. New crystals of preexistent minerals, or even new mineral species can thus be formed. A consequence of these phenomena is the alteration in the texture of the rocks.

The generic term for the texture of metamorphic rocks is "crystalloblastic." The suffix "blasto," of Greek origin, is attached to many terms to describe the variety of textures produced by the growth of new crystals. The intensity of the metamorphic changes results in different textures which are distinguished on the basis of the form, orientation, and size of the crystals.

As with all other rocks, metamorphic rocks are studied as hand samples as well as in thin section with the aid of a polarizing microscope. These sections are slices of rocks that are so thin as to be transparent. Once prepared, they are placed between two pieces of glass in order to protect the section from damage. To observe the minerals and textures a beam of polarized light is transmitted through the section.

EXAMPLES OF TEXTURE

Figure 1 shows a macroscopic specimen of garnet mica schist: the texture is schistose, with micaceous minerals preferentially arranged along planes to produce thin, dark-colored layers and whitish lenses of quartz. Large roundish crystals of brownish-red garnet can also be seen.

In the corresponding thin section (fig. 2), the bright-colored, flat muscovite grains form layers parallel to the flattened, light-colored, quartz-rich lenses (lepidoblastic texture). The large grayish crystals with inclusions are albite (poikiloblasts), and the blackish spots are garnets (porphyroblasts).

In figure 3, a specimen of gneiss is reproduced in which one can see lenticular aggregates of both glassy quartz and large brownish crystals of orthoclase surrounded by thin, dark-colored layers composed of micas.

Taken together, the mineral layers resemble an eye (augen texture).

Figure 4 represents the thin section of the same gneiss; a large grayish crystal of orthoclase (porphyroblast) is very obvious, in which light zones of plagioclase are included. The colored crystals are muscovite; the smaller transparent, clear ones are quartz.

A pink marble is reproduced in figure 5. It was originally a limestone but metamorphism caused the calcite to recrystallize as new equidimensional, fairly large crystals. Each small area that reflects light like a mirror corresponds to one crystal grain.

The thin section (fig. 6) of the pink marble shows interlocking calcite crystal (granoblastic texture). Small, colored flakes of muscovite can be seen intermingled with the calcite grains. The muscovite was produced by the transformation of other minerals, perhaps argillaceous ones, present in the original limestone.

MARBLE

PROTOLITH: carbonate rocks.

METAMORPHIC ENVIRONMENT: low-, medium-, and high-grade regional metamorphism; contact metamorphism.

ESSENTIAL MINERALS: calcite (in some varieties, constitutes essentially the entire rock), dolomite.

ACCESSORY MINERALS: assemblage varies depending on the metamorphic grade: muscovite, quartz, albite, pyrite, epidote, grossular, wollastonite, diopside, tremolite, vesuvianite, sulfur, gypsum, wurtzite, etc.

APPEARANCE: fine- to coarse-grained; structure is massive, in some cases zoned according to grain size or slightly schistose; texture is typically granoblastic, but tends to be more varied in silicate-rich varieties.

COLOR: commonly white; impurities produce spots and veining of various colors: greenish (chlorite), grayish (pyrite), brownish (goethite), reddish (hematite); blackish-gray striations in graphitic marbles.

GEOLOGIC ENVIRONMENT: in large masses or thin intercalations associated with metamorphic rocks from a range of grades.

OCCURRENCE: marbles from Carrara, the Apuan Alps, and some Greek localities (Náxos, Páros, Lávrion) are well-known.

USES: in the building industry; used in the rough state for exterior architecture and in polished slabs for monuments.

NOTES: The word is derived from the Greek for "block" or "white rock." Commercially, the term marble is used improperly to indicate all soft rocks, largely carbonate-rich, that can be easily cut and polished. The term granite is used for hard rocks.

CIPOLINO MARBLE

PROTOLITH: impure limestones and calcareous marls.

METAMORPHIC ENVIRONMENT: low-grade regional metamorphism.

ESSENTIAL MINERALS: calcite, chlorite, and muscovite.

ACCESSORY MINERALS: quartz, albite, pyrite.

APPEARANCE: fine-grained and streaked with alternating bands of different composition; in some cases, schistose; granoblastic texture.

COLOR: whitish (calcite) and greenish (chlorite, muscovite, etc.) bands.

GEOLOGIC ENVIRONMENT: intercalated with other marbles.

OCCURRENCE: in mountain chains with calcareous metamorphic rocks; important deposits are found in the Apuan Alps.

USES: in polished slabs, for both facings and a variety of furnishings.

NOTES: Cipolino, a very common marble, may exhibit bands of varying thickness and color, but always with shades between dark green and gray. The colored streaks in some cases appear to be displaced along small fractures produced during metamorphism; in such cases, the dark bands will look serrated.

▲ Cipolino marble (ca x 1), Apuan Alps, Tuscany, Italy.

◄ White marble (ca. x 0.8), Tháxos, Greece.

VEINED MARBLE

PROTOLITH: limestones.

METAMORPHIC ENVIRONMENT: medium- to high-grade regional metamorphism.

ESSENTIAL MINERALS: calcite, dolomite present at lower grades.

ACCESSORY MINERALS: muscovite, quartz, pyrite, adularia.

APPEARANCE: medium-grained; massive structure; interlocking granoblastic texture.

COLOR: pearly white to light gray; with dark gray, almost regular veining due to microcrystalline pyrite.

GEOLOGIC ENVIRONMENT: intercalated with other marbles.

OCCURRENCE: in mountain chains with calcareous metamorphic rocks; important deposits are found in the Apuan Alps.

USES: in slabs for use in the building industry.

NOTES: The name refers to the dense network of dark veins that criss-cross the light-colored rock; this marble grades into other marbles, such as the ordinary white variety which has a coarser grain size.

▼ Veined marble (ca. x 0.7), Apuan Alps, Tuscany, Italy.

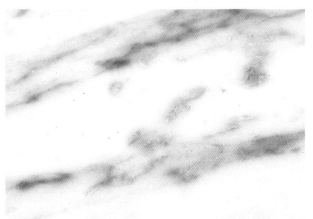

Bardiglio ► marble (ca. x 1), Apuan Alps, Tuscany, Italy.

BARDIGLIO MARBLE

PROTOLITH: limestones.

METAMORPHIC ENVIRONMENT: low-grade regional metamorphism.

ESSENTIAL MINERALS: calcite.

ACCESSORY MINERALS: muscovite, adularia, quartz, albite, pyrite.

APPEARANCE: fine-grained and massive structure; under the microscope, equidimensional crystals (granoblastic texture) are visible.

COLOR: somewhat intense ashy blue, with whitish or dark veining.

GEOLOGIC ENVIRONMENT: intercalated with other marbles.

OCCURRENCE: in mountain chains with calcareous metamorphic rocks; important deposits are found in the Apuan Alps.

USES: in polished slabs, for use as facings or various furnishings.

NOTES: Bardiglio marble, the most common marble, was widely used in the past for mantelpieces and decorative slabs on furniture. It appears homogeneous when observed on a large scale in the field, but on a small scale, as in hand samples, it exhibits substantial variations. The best-known varieties are chapel bardiglio and imperial bardiglio; others are called by local names.

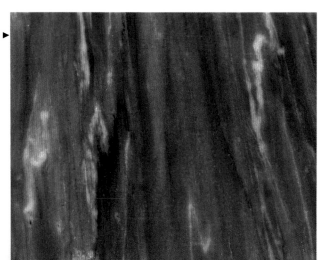

STATUARY MARBLE

PROTOLITH: limestones.
METAMORPHIC ENVIRONMENT: low-grade regional metamorphism.
ESSENTIAL MINERALS: calcite.
ACCESSORY MINERALS: muscovite, quartz, albite, pyrite.
APPEARANCE: fine-grained; massive structure; granoblastic texture.
COLOR: ivory-white with small grayish spots from the presence of minor quantities of microcrystalline pyrite.
GEOLOGIC ENVIRONMENT: intercalated with other marbles.
OCCURRENCE: in mountain chains with calcareous metamorphic rocks. The deposits in the Apuan Alps are unique in character; similar but coarser grained marbles come from Greece (Páros, Náxos, Tháxos).
USES: for ornamental objects and as slabs for valuable flooring.
NOTES: Statuary marble is the most sought-after and costly marble, and it is becoming increasingly rare. The pure, extremely white and translucent variety is currently found on the market only in small blocks. Because it is easy to work, it was widely used in the past for monumental projects, some of which are very famous.

▼ Statuary marble (ca. x 0.8), Apuan Alps, Tuscany, Italy.

PINK MARBLE

PROTOLITH: limestones.
METAMORPHIC ENVIRONMENT: medium-grade regional metamorphism; contact metamorphism.
ESSENTIAL MINERALS: calcite.
ACCESSORY MINERALS: muscovite, plagioclase, pyrite.
APPEARANCE: medium- to coarse-grained; massive structure; texture typically granoblastic.
COLOR: typically very delicate flesh-pink.
GEOLOGIC ENVIRONMENT: intercalated with other marbles.
OCCURRENCE: in mountain chains with calcareous metamorphic rocks. It is a fairly rare variety: the Portuguese deposits are famous; it is also found in Italy (Novara) and a few other localities.
USES: in polished slabs for interior flooring and facings.
NOTES: Like all marbles, it is very delicate and tends to be easily worn down; in fact, areas of floors subject to heavy traffic have an obvious loss of luster and typically have depressions where the rock is worn away. Pink marble with a decidedly rosy color is now rare, and a slightly more opaque shade is more commonly found on the market.

Arabesque ▶ marble (ca. x 0.7), Apuan Alps, Tuscany, Italy.

ARABESQUE MARBLE

PROTOLITH: carbonate rocks.
METAMORPHIC ENVIRONMENT: medium- to high-grade regional metamorphism.
ESSENTIAL MINERALS: calcite.
ACCESSORY MINERALS: dolomite, muscovite, quartz, albite, adularia, pyrite.
APPEARANCE: coarse-grained; it is a breccia made up of marble clasts, flattened parallel to the schistosity, with a calcareous matrix cementing the fragments; under the microscope, the matrix exhibits equidimensional crystals.
COLOR: pure white with gray striations from the presence of microcrystalline pyrite.
GEOLOGIC ENVIRONMENT: intercalated with other marbles.
OCCURRENCE: in mountain chains with calcareous metamorphic rocks. The deposits in the Apuan Alps are unique in character; coarser grained but similar marbles come from Greece (Páros, Náxos, Tháxos).
USES: in polished slabs for interior flooring and facings.
NOTES: Arabesque marble is a breccia in which the individual fragments are cemented by a gray matrix with an irregular reticulated appearance. Numerous varieties exist, with differences in fragment size and in the thickness of the dark network; as a consequence, the general appearance varies substantially.

◀ *Pink marble (ca. x 0.8), Portugal.*

BRECCIATED MARBLE

PROTOLITH: carbonate rocks.
METAMORPHIC ENVIRONMENT: medium- to high-grade regional metamorphism.
ESSENTIAL MINERALS: calcite, dolomite.
ACCESSORY MINERALS: muscovite, quartz, pyrite, iron oxides.
APPEARANCE: heterogeneous grain size; a breccia made up of clasts some of which are larger than 3.9 inches (10 centimeters) in size; the cement that binds the clasts is medium-to fine-grained.
COLOR: variable, depending on the constituent particles: for the most part gray, whitish, yellow, white-gray; the matrix is generally light in color, gray or brown.
GEOLOGIC ENVIRONMENT: intercalated with other marbles; commonly associated with features resulting from tectonic activity (e.g., faults).
OCCURRENCE: in mountain chains with calcareous metamorphic rocks. Important deposits are found in various parts of the world.
USES: in polished slabs for interior flooring and facings.
NOTES: Numerous types of marbled breccia exist. They are differentiated on the basis of both color and grain size of the fragments and the cement. They are generally given imaginative names or names based on the locality where they are quarried: peach blossom breccia, Seravezza breccia, and Medicean breccia, among others.

▼ *Marbled breccia (ca. x 0.7), Apuan Alps, Tuscany, Italy.*

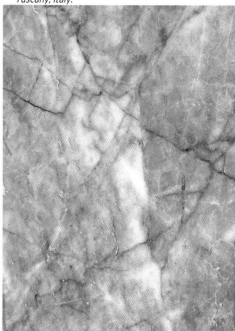

"CALACATTA" (CLASTIC MARBLE)

PROTOLITH: carbonate rocks.

METAMORPHIC ENVIRONMENT: medium- to high-grade regional metamorphism.

ESSENTIAL MINERALS: calcite.

ACCESSORY MINERALS: dolomite, muscovite, quartz, albite, pyrite.

APPEARANCE: wide range of grain size, generally made up of large fragments that are flattened parallel to schistosity and cemented with calcite.

COLOR: pearly white crossed with yellow-green, sometimes golden, striations (cement) from the presence of muscovite and microcrystalline pyrite.

GEOLOGIC ENVIRONMENT: intercalated with other marbles; commonly associated with features resulting from tectonic activity (e.g., faults).

OCCURRENCE: in mountain chains with calcareous metamorphic rocks. Important deposits are found in various parts of the world.

USES: in polished slabs for pavements and facings.

NOTES: "Calacatta" is a breccia made up of clasts of statuary marble, cemented by a micaceous carbonate matrix in which the other accessory minerals are found. It is a fairly rare marble that is best known from the Apuan Alps.

▼ "Calacatta" (ca. x 1), Apuan Alps, Tuscany, Italy.

CALC-SCHIST

PROTOLITH: marls, carbonate or tuffaceous sandstones.

METAMORPHIC ENVIRONMENT: low- to medium-grade regional metamorphism (greenschist facies, beginning of the amphibolite facies).

ESSENTIAL MINERALS: calcite, muscovite and paragonite, chlorite, biotite, quartz.

ACCESSORY MINERALS: actinolite, albite, epidote, graphite, pyrite, ilmenite, tremolite, dolomite.

APPEARANCE: fine-grained, easily degraded rock; superficial cavities result from dissolution of carbonates commonly present in lenses; clearly schistose, with shiny surfaces (schistes lustrés); microscopic observation shows equidimensional crystals (granoblastic texture), typically with parallel flakes of mica and chlorite (lepidoblastic texture).

COLOR: gray, greenish-bluish, brown.

GEOLOGIC ENVIRONMENT: in masses that in some cases are very thick; associated with marbles rich in silicates, phyllites, and mica schists.

OCCURRENCE: generally in Alpine-type settings; in some cases associated with metamorphosed ophiolites.

USES: of exclusively scientific interest.

NOTES: Calc-schists (regional metamorphism) have a mineralogical composition similar to that of the calciphyres (contact metamorphism), from which they can be distinguished by the schistosity of the former and the presence of small lenses of calcite flattened parallel to the schistosity. As the metamorphic grade increases, calc-schists grade to calcsilicates with a more massive structure, in which calcite has partially replaced by reactions that formed calcium silicates.

▼ Calc-schist (ca. x 0.6), Rigoli, Perugia, Italy.

MUSCOVITE SCHIST

PROTOLITH: volcanic and volcaniclastic rocks with an acidic to intermediate composition; arkosic sandstones.

METAMORPHIC ENVIRONMENT: low-grade regional metamorphism (greenschist facies).

ESSENTIAL MINERALS: muscovite, quartz, microcline, albite.

ACCESSORY MINERALS: chlorite, epidote, sphene, iron and titanium oxides.

APPEARANCE: fine-grained; structure is schistose, with brilliant, wavy surfaces, due to metamorphic muscovite; the original structure is essentially obliterated.

COLOR: light gray, greenish gray.

GEOLOGIC ENVIRONMENT: in masses associated with phyllite, phyllitic paragneiss, and semi-schists.

OCCURRENCE: fairly common rock in mountain chains.

USES: of exclusively scientific interest.

NOTES: Muscovite schists are very soft rocks that formerly were mistaken for talc schists. When metamorphism is accompanied by weak deformation, only the groundmass is recrystallized, while the large grains of feldspar and quartz retain their original characteristics. The rock is then called "porphyroid."

▲ Muscovite schist (ca. x 0.6), Taunus, Germany.

METAGRANITE

PROTOLITH: granites.

METAMORPHIC ENVIRONMENT: low-grade regional metamorphism with a slight deformational component (greenschist facies).

ESSENTIAL MINERALS: quartz, microcline, albite, chlorite.

ACCESSORY MINERALS: orthoclase (commonly sericitized), epidote, sphene, iron and titanium oxides.

APPEARANCE: medium- to fine-grained; massive structure; granular texture.

COLOR: whitish, yellowish, light gray; in all cases speckled with darker minerals.

GEOLOGIC ENVIRONMENT: in bodies of various sizes, sometimes associated with granites.

OCCURRENCE: in mountain chains.

USES: used rarely in the building industry.

NOTES: Metagranite is derived from low-grade regional metamorphism of acidic plutonic rocks where original igneous textures are preserved. However, some mineralogical transformations do take place (e.g. plagioclase to albite + epidote). If there is a strong deformational component, a crushed texture results that is called "flaser": in this case, the metagranites are called "flaser granites."

▼ Metagranite (ca. x 0.8), Brossasco, Cuneo, Italy.

ORTHOGNEISS

PROTOLITH: igneous rocks with an acidic or intermediate composition; tuffs and arkosic sandstones.

METAMORPHIC ENVIRONMENT: medium- to high-grade regional metamorphism (amphibolite facies).

ESSENTIAL MINERALS: quartz, microcline, orthoclase, plagioclase, muscovite, biotite.

ACCESSORY MINERALS: hornblende, augite, apatite, zircon, ilmenite, garnet.

APPEARANCE: medium- to coarse-grained; minerals are commonly aligned but schistosity is not pronounced; alternating bands of light and dark discontinuous lenticular layers up to an inch thick (gneissic texture); granoblastic texture, grading to porphyroblastic (augen gneiss).

COLOR: generally whitish with dark bands.

GEOLOGIC ENVIRONMENT: in masses of various dimensions, associated with migmatites and granitic rocks.

OCCURRENCE: very common in highly eroded orogenic chains; especially in Archaean provinces.

USES: some gneisses are used in the building industry, in either rough blocks or polished slabs.

NOTES: Gneiss probably derives its name from an old mining term of Saxon origin, in use since 1500. In the past, the term gneiss had a very broad meaning, without any precise relationship to the true nature of the rock. We owe the current definition of this term to the German, Abraham Gottlob Werner.

▼ Two mica garnet orthogneiss (ca. x 0.8), Sondrio, Italy.

AUGEN GNEISS

PROTOLITH: igneous rocks with an acidic or intermediate composition; tuffs and sedimentary rocks (arkosic sandstones).

METAMORPHIC ENVIRONMENT: medium- to high-grade regional metamorphism (amphibolite facies).

ESSENTIAL MINERALS: quartz, microcline, orthoclase, plagioclase, muscovite, biotite.

ACCESSORY MINERALS: hornblende, augite, apatite, zircon, ilmenite, garnet.

APPEARANCE: medium- to coarse-grained; foliated (gneissic texture), with large whitish crystals of feldspar surrounded by dark bands; the feldspars are easily visible in a hand sample and look similar to eyes (augen); matrix has a granoblastic texture.

COLOR: whitish, commonly in large spots with dark bands.

GEOLOGIC ENVIRONMENT: in masses of various dimensions, commonly associated with migmatites and granitic rocks.

OCCURRENCE: very common in highly eroded orogenic chains and especially in Archaean provinces; a typical augen gneiss is the "ollo de sapo" found in Spain and Portugal.

USES: some kinds of gneiss are used in the building industry, either in rough blocks or polished slabs; it is no longer used for pavement.

NOTES: One way the augen texture develops is from the metamorphism of granites with a porphyritic texture. The large potassium feldspar crystals (microcline), and more rarely plagioclase, more or less maintain their original dimensions during the metamorphic event and have micaceous bands wrapped around them. When this rock is unweathered, it is much sought-after.

▼ Augen gneiss (ca. x 1), Sondrio, Italy.

GLAUCOPHANE SCHIST

PROTOLITH: volcanics and volcaniclastic rocks with a basic composition.

METAMORPHIC ENVIRONMENT: subduction metamorphism (blueschist or lawsonite-glaucophane facies).

ESSENTIAL MINERALS: glaucophane, crossite, or riebeckite; lawsonite.

ACCESSORY MINERALS: calcite, aragonite, magnetite, chlorite, albite, quartz, jadeite, epidote, garnet, kyanite.

APPEARANCE: medium- to coarse-grained; schistosity not well developed in all cases; in some rocks, bands of different mineral assemblages are visible; nematoblastic to lepidoblastic texture.

COLOR: dark blue to bluish green with light violet shading.

GEOLOGIC ENVIRONMENT: in bodies of various sizes.

OCCURRENCE: commonly found in mountain chains.

USES: of exclusively scientific interest.

NOTES: Glaucophane schist is formed in subduction zones along active continental margins under high pressure.

▲ Glaucophane schist (ca. x 1), Norway.

PRASINITE (CHLORITE-AMPHIBOLE-EPIDOTE-ALBITE SCHIST)

PROTOLITH: basalts and basic volcaniclastic rocks or clast-rich sediments with a basic composition.

METAMORPHIC ENVIRONMENT: low-grade regional metamorphism (greenschist facies).

ESSENTIAL MINERALS: chlorite, actinolite, albite, epidote.

ACCESSORY MINERALS: calcite, quartz, sphene, rutile, muscovite, ilmenite.

APPEARANCE: fine-grained; structure is massive or with bands of different composition; in some cases, relict structures of the original rocks are visible; under the microscope, one can discern parallel acicular crystals of amphibole (nematoblastic texture), with small roundish grains of white albite around which are thin layers of darker-colored minerals (ocellar texture).

COLOR: various shades of green with faded yellow or bluish tones.

GEOLOGIC ENVIRONMENT: in masses associated with phyllites, mica schists, and calc-schists.

OCCURRENCE: found in Alpine-type settings.

USES: of exclusively scientific interest.

NOTES: Prasinite is formed by retrograde metamorphism of glaucophane schist after a sharp decrease in pressure, as well as by low-grade regional prograde metamorphism; a large number of the prasinites found in Alpine areas are of the latter type. The term prasinite is typically used in Europe; elsewhere they are called "chlorite-amphibole-epidote-albite schist."

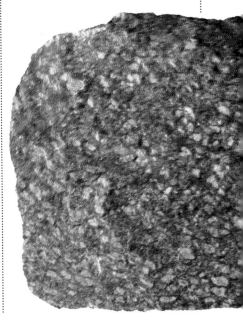

Prasinite (ca. x 1), Val Varaita, Cuneo, Italy. ►

EPIDOTE AMPHIBOLITE

PROTOLITH: basalts, volcaniclastic rocks and clastic sedimentary rocks with a basic composition.

METAMORPHIC ENVIRONMENT: medium- to low-grade regional metamorphism (amphibolite facies, greenschist facies).

ESSENTIAL MINERALS: hornblende (sodium-rich), epidote (clinozoisite, pistacite, epidote), albite, plagioclase (oligoclase).

ACCESSORY MINERALS: quartz, biotite, chlorite, magnetite, hematite, sphene, calcite.

APPEARANCE: wide range of grain sizes; schistose to almost massive structure, with easily visible bands of differing grain size, color, and composition; under the microscope, parallel, prismatic, sometimes very elongated amphibole crystals are visible (nematoblastic texture).

COLOR: light green, sometimes with yellowish green or dark green bands.

GEOLOGIC ENVIRONMENT: generally found in masses of remarkable size, commonly associated with low grade metabasites.

OCCURRENCE: common rocks; found in mountain chains.

USES: used rarely in the building industry.

NOTES: Epidote amphibolite is formed at lower temperatures than amphibolite. It constitutes an intermediate stage between amphibolite and low-grade metabasite (metamorphosed basic igneous rocks). Epidote amphibolite is differentiated from prasinite by a greater abundance of hornblende in the former, as well as a generally coarser grain size.

▼ *Epidote amphibolite (ca. x 0.6), Italy.*

MICHELANGELO'S "WHITE"

MARBLES

The term "marble" is commonly used commercially to indicate all soft, polishable stones. However, "marble" (from the Greek for "stone" or "boulder") as a scientific term refers only to carbonate rocks, either calcitic or dolomitic, that have undergone recrystallization through metamorphism.

Veined Carrara white

Siena yellow.

Marbles are generally found intercalated with mica schists, phyllites, and paragneisses and are extremely common in rocks from the Triassic period, as well as the Precambrian and early Paleozoic eras. During recrystallization, all of the sedimentary structures are obliterated; only rarely are fossil remains found in these rocks.

Marble par excellence, the classic, highly desirable variety, is statuary marble. It is an especially beautiful, silky marble without visible impurities. As many of us have observed firsthand, the works created with this rock truly emanate warmth, as if, at any moment, the sculpted beings might begin to move.

In order to obtain a pure marble of this kind, the original limestone must be of exceptional quality: the purer the initial limestone, the whiter the resulting marble. Also, the lower the temperature of metamorphism, the smaller the new crystals in the marble will be. In fact as a general rule, the crystals tend to grow larger with increasing metamorphic grade.

Aesthetically pleasing white marbles can be found throughout the world. However, there are no deposits of statuary marble comparable to those in the Apuan Alps (Carrara, Italy). Always used to create special works of art, the white, saccharoidal, microcrystalline, Carrara marble is perhaps the most widely known carbonate rock in the world. In the past, Apuan marbles were used to decorate the homes of Roman patricians. Michelangelo personally selected the blocks of marble for his sculptures from quarries near

Carrara. This very precious marble also embellishes many of the most important Italian monuments, including the exterior walls of Santa Maria del Fiore and Giotto's famous campanile.

With the exception of an occasional rare block, the statuary marble preferred by Michelangelo no longer exists. The current statuary marble

has thin, soft, irregular lines but is nonetheless highly valued. Its use is generally limited to floorings and facings and, like all marbles, it is very delicate. It is a soft rock (calcite has a hardness of three on the Mohs scale) that is very easily worn down, especially when used for flooring. This drawback, which causes flooring tiles to lose their luster, is only partially remedied by treatment with protective resins.

OTHER CONSTITUENTS

Even the purest marbles contain accessory minerals such as quartz, light micas, graphite particles, iron oxides, and sulfides, among others. When present in extremely small quantities and as small grains, these minerals are practically invisible. However, in larger quantities they impart different colors to the marble. For example, bardiglio has a somewhat intense, fairly homogeneous, gray-blue color and, depending on the specimen, there may also be small, elongated, white spots. The coloration is due to very small crystals of pyrite scattered in a fairly uniform fashion throughout the rock. Siena yellow has intense tones that alternate with lighter tones; the color results from the presence of very small crystals of iron oxides. Differences in concentration of these crystals influence the shade of color, and names have been coined to distinguish the different varieties (e.g., broccatello, veined yellow, plain yellow, and ivory yellow). In other cases, the impurities are locally concentrated to form striking irregular or elongated spots. For example, ordinary white or veined white marble abounds with dark veins that result from the presence of very small crystals of pyrite.

Michelangelo's "Genius of Victory," 1532–1534., Palazzo Vecchio, Sala dei 500, Florence, Italy.

AMPHIBOLITE

PROTOLITH: basalts, basic volcaniclastic rocks.

METAMORPHIC ENVIRONMENT: medium- to high-grade regional metamorphism (amphibolite facies).

ESSENTIAL MINERALS: amphibole (hornblende, anthophyllite, cummingtonite), plagioclase (andesine to bytownite).

ACCESSORY MINERALS: biotite, diopside, iron-rich garnet, apatite, magnetite, ilmenite, sphene.

APPEARANCE: medium- to coarse-grained; massive structure to slight schistosity; in some high- grade amphibolites, banded with alternating hornblende-rich and plagioclase-rich layers; granoblastic, diablastic, or nematoblastic texture.

COLOR: dark green to blackish (amphibole), sometimes with whitish spots (plagioclase).

GEOLOGIC ENVIRONMENT: in large masses associated with mica schists and paragneiss.

OCCURRENCE: common in eroded mountain belts; found in France, England, Germany, Scandinavia, and the United States.

USES: used rarely in the building industry.

NOTES: Orthoamphibolite is derived from metamorphism of igneous rocks, whereas paraamphibolite is derived from sedimentary ones. Clues such as the geologic environment and the adjacent metamorphic rocks are used to help distinguish the two types in the field.

▼ *Amphibolite (ca. x 1), Val Malenco, Sondrio, Italy.*

GARNET AMPHIBOLITE

PROTOLITH: basalts, volcaniclastic rocks and sediments rich in clasts with a basic composition.

METAMORPHIC ENVIRONMENT: medium- to high-grade regional metamorphism (amphibolite facies).

ESSENTIAL MINERALS: hornblende, iron-rich garnet, plagioclase (oligoclase to bytownite).

ACCESSORY MINERALS: quartz, biotite, magnetite, hematite, sphene, apatite.

APPEARANCE: coarse-grained; massive structure with bands varying in grain size, color, and composition; microscopic examination shows an interlacing of elongate crystals of amphibole (diablastic texture).

COLOR: dark green with brown-red areas (garnet); reflections from the cleavage planes of hornblende appear whitish.

GEOLOGIC ENVIRONMENT: generally found in masses; small masses of eclogite are sometimes found associated with these rocks.

OCCURRENCE: in metamorphic mountain chains.

USES: used rarely in the building industry.

NOTES: Garnet amphibolite is a type of amphibolite which, as the name implies, consists of amphibole (hornblende) as well as plagioclase. When an accessory mineral is prominent, it is used to further specify the rock type: garnet amphibolite and diopside amphibolite are two examples.

▼ *Garnet amphibolite (sphene is visible) (ca. x 0.4), Val Malenco, Sondrio, Italy.*

ECLOGITE

PROTOLITH: rocks with a basic composition.

METAMORPHIC ENVIRONMENT: high-grade regional metamorphism with high pressure in an anhydrous environment (eclogite facies).

ESSENTIAL MINERALS: garnet (mixture of pyrope and almandine), omphacite (monoclinic pyroxene that is a solid solution of jadeite and diopside).

ACCESSORY MINERALS: kyanite, hypersthene or olivine, rutile, ilmenite, glaucophane, zoisite, albite, quartz.

APPEARANCE: very heavy, compact rock with a medium grain size; massive structure; granoblastic texture, sometimes with porphyroblasts (garnet).

COLOR: greenish and reddish variegated.

GEOLOGIC ENVIRONMENT: as boulders in glaucophane schist terrains, small lenses within metamorphosed peridotites, or with kimberlites.

OCCURRENCE: not very common.

USES: used rarely in the building industry; sometimes cut into slabs and polished.

NOTES: Eclogites formed in a high-temperature, high-pressure environment typically occur with migmatites, granulites, gneisses, and amphibolites. The garnet is intermediate in composition between pyrope and almandine, and the pyroxene is a jadeite-rich omphacite. Eclogites formed at lower temperatures occur with glaucophane schists and greenschists and contain almandine-rich garnet and omphacite composed of up to 40% jadeite.

▼ *Eclogite (ca. x 0.4), Italy.*

SERPENTINITE

PROTOLITH: ultramafic igneous rocks.

METAMORPHIC ENVIRONMENT: low-grade regional metamorphism with moderate pressure (greenschist facies), typical of mid-ocean ridges (spreading zones).

ESSENTIAL MINERALS: serpentine (antigorite, chrysotile).

ACCESSORY MINERALS: magnetite (even in large crystals), chlorite, talc, magnesite, calcite, dolomite, serpentine, brucite.

APPEARANCE: dense, fine-grained rock; massive structure; serpentinization begins along the edges and in internal fractures of crystals, producing a felted texture surrounding relict crystals of olivine, pyroxene, or garnet (pyrope).

COLOR: dark green, black, light green, or variegated.

GEOLOGIC ENVIRONMENT: masses of remarkable size.

OCCURRENCE: fairly common in mountain belts and in ocean crust.

USES: as an ornamental stone.

NOTES: Serpentinite is derived from the hydration of peridotite masses, as well as pyroxenite or amphibolite masses at temperatures below 930° F (500° C). Peridotites affected by the process of serpentinization constitute a major part of the oceanic crust.

▼ *Serpentinite, (ca. x 1), Italy.*

SERPENTINE SCHIST

PROTOLITH: ultramafic igneous rocks.

METAMORPHIC ENVIRONMENT: low-grade moderate pressure metamorphism (greenschist facies), typical of mid-ocean ridges (spreading zones).

ESSENTIAL MINERALS: serpentine, generally as the variety "antigorite."

ACCESSORY MINERALS: magnetite, chromium-rich spinels, chlorite, talc, magnesite.

APPEARANCE: soft, fine-grained rock; sometimes it can be broken into very thin sheets that are remarkably durable; schistose with light green, shiny planes in which relict lenses of massive serpentinite are visible; lamellar texture with bands and irregular aggregates formed by the antigorite.

COLOR: various shades of green.

GEOLOGIC ENVIRONMENT: masses of noteworthy expanse in metamorphosed ophiolitic complexes.

OCCURRENCE: in mountain chains.

USES: the compact varieties can be cut and polished and utilized for interior facing.

NOTES: Serpentine schist is derived from the hydration, at temperatures lower than 930° F (500° C), of peridotite masses (serpentinization) and also of pyroxenite or amphibolite masses. In border zones of serpentinized masses, tectonic deformation can be intense and results in oriented crystals that produce schistosity (serpentine schists). In the interiors of these bodies, the massive structure (serpentinite) is maintained.

▼ Serpentine schist (ca. x 0.5), Italy.

OPHICALCITE

PROTOLITH: ultramafic igneous rocks, siliceous dolomite.

METAMORPHIC ENVIRONMENT: low-grade, moderate pressure metamorphism (greenschist facies), typical of mid-ocean ridges (spreading zones).

ESSENTIAL MINERALS: commonly with angular fragments of serpentinite or gabbro.

ACCESSORY MINERALS: carbonate cement.

APPEARANCE: the fragments are easily visible, with sharp corners, welded by the cement.

COLOR: the coarse particles appear in various shades of green, in some cases they are red or violet; the cement is whitish, commonly with veining.

GEOLOGIC ENVIRONMENT: in masses of varying size, associated with metamorphosed ophiolites.

OCCURRENCE: commonly found in Alpine-type mountain chains.

USES: highly valued ornamental stone; can be cut into slabs and polished.

▲ Ophicalcite (ca. x 0.6), Liguria, Italy.

CHLORITE SCHIST

PROTOLITH: ultramafic igneous rocks, basalts, volcaniclastic rocks and marls associated with tuffaceous material.

METAMORPHIC ENVIRONMENT: low-grade regional metamorphism with low pressure (greenschist facies).

ESSENTIAL MINERALS: chlorite.

ACCESSORY MINERALS: actinolite, magnetite, glaucophane, rutile, epidote, pyrite, sphene, tourmaline, talc, serpentine and chloritoid (nesosilicate).

APPEARANCE: fairly friable, fine-grained, soft rocks; generally schistose, but not always well-developed; the microscope shows parallel, interlocking chlorite flakes (lepidoblastic texture), in some cases in the shape of a rosette, with highly modified crystals of accessory minerals, easily separable from the rock.

COLOR: dark green.

GEOLOGIC ENVIRONMENT: found in lenses within other rocks of greenscist facies including serpentinites; associated with phyllites and albite paragneiss.

OCCURRENCE: in metamorphic mountain chains, commonly with ophiolites.

USES: in the building industry; a variety of chlorite schist (potstone) with a fine, scaly texture and easily worked on a lathe, was formerly used in Alpine areas for making pots and vases.

NOTES: Chlorite is a mineral whose chemical composition varies: the alumina-rich variety is typical of chlorite schists derived from regional metamorphism of argillaceous rocks; in the chlorite schists derived from igneous rocks, the chlorite is rich in magnesium. In this case, magnetite is abundant which, together with the chlorite, is derived from the alteration of mafic minerals.

▼ Chlorite schist (ca. x 0.8), Roanenthal, Norway.

TALC SCHIST

PROTOLITH: serpentinites, siliceous dolomites.

METAMORPHIC ENVIRONMENT: low-grade regional metamorphism with moderate pressure (greenschist facies).

ESSENTIAL MINERALS: talc.

ACCESSORY MINERALS: chlorite, serpentine, magnetite, magnesite, calcite, dolomite, brucite, tremolite.

APPEARANCE: very soft rock, greasy to the touch, with a fine grain size; schistosity is obvious (resulting from the parallel arrangement of the talc flakes); microscopic examination shows parallel, interlocking flakes of talc (lepidoblastic texture) with granoblastic portions; commonly, relicts of olivine or pyroxene are present.

COLOR: gnerally light; whitish, light gray, sometimes speckled with green.

GEOLOGIC ENVIRONMENT: in beds or lenticular masses associated with serpentinites or other greenschist facies rocks (e.g., chlorite schist, tremolite schist)

OCCURRENCE: in metamorphosed ophiolite complexes which crop out in some mountain chains.

USES: the "steatite" variety is used in the manufacture of insulating materials, in paper mills; in the past, it was also worked to make containers of various shapes.

NOTES: The compact variety of talc schist is called steatite or soapstone; it has a fine grain size, is massive, greasy to the touch, and has a light color. It is differentiated from other varieties by its smaller grain size.

▼ Talc schist (ca. x 1), Val Malenco, Sondrio, Italy.

ACTINOLITE SCHIST

PROTOLITH: ultramafic rocks.

METAMORPHIC ENVIRONMENT: low-grade regional metamorphism with moderate pressure (greenschist facies).

ESSENTIAL MINERALS: amphibole (ferro-actinolite-tremolite), talc.

ACCESSORY MINERALS: magnetite, chlorite, serpentine.

APPEARANCE: medium- to fine-grained; schistose, more rarely fibrous, from the acicular habit of amphibole; microscopic examination shows parallel interlocking crystals of amphibole (nematoblastic texture).

COLOR: grass-green.

GEOLOGIC ENVIRONMENT: in small bodies associated with serpentinized peridotite masses.

OCCURRENCE: found in Alpine-type mountain chains.

USES: of exclusively scientific interest.

NOTES: Tremolite schists also exist. They are formed by metamorphism of siliceous dolomites or of some varieties of gneiss, rich in quartz, plagioclase, and diopside. The distinction between schists of different origins can be ascertained through chemical and mineralogical analyses, and in the field, by the examination of the metamorphic rocks with which they are associated.

▼ *Actinolite or tremolitic schist (ca. x 0.7), United States.*

GRANULITE

PROTOLITH: basic, acidic, or pelitic rocks.

METAMORPHIC ENVIRONMENT: high-grade regional metamorphism, typically at high pressure (granulite facies).

ESSENTIAL MINERALS: quartz, orthoclase (perthitic), plagioclase, garnet: in acidic granulites; plagioclase (antiperthitic), orthopyroxene (hypersthene, bronzite), amphibole: in basic granulites.

ACCESSORY MINERALS: orthopyroxenes, rutile, sillimanite, cordierite, biotite, spinel: in acidic granulites; clinopyroxenes, garnet, rutile, scapolite, spinel: in basic granulites.

APPEARANCE: compact rock with variable grain size; structure is generally massive; in some granulites the presence of flattened lenses of quartz, with dimensions in centimeters, gives them a slight schistosity; granoblastic texture, in some cases with porphyroblasts.

COLOR: light to dark, depending on the mineralogical composition.

GEOLOGIC ENVIRONMENT: stratiform with marbles, calcsilicates, and peridotites; on a regional scale it is associated with migmatites, paragneisses, and amphibolites.

OCCURRENCE: found in ancient Precambrian shields and in the mountains of Saxony.

USES: in the building industry.

NOTES: The name is derived from the Latin *granulum* (granule), and refers to the granular texture of the rocks. They are characterized by the absence of hydrated minerals, and acidic and basic varieties are differentiated on the basis of their mineralogical composition.

▼ *Granulite (ca. x 1), Valle Strona, Novara, Italy.*

AGMATITE (MIGMATITE)

PROTOLITH: regional metamorphic rocks.

METAMORPHIC ENVIRONMENT: medium- to high-grade regional metamorphism with variable pressure (amphibolite facies) under which partial melting occurs.

ESSENTIAL MINERALS: the rock is made up of two distinct parts: the neosome (recrystallized from a partial melt) and the paleosome (metamorphic part, not yet melted). In turn, the neosome is made up of two parts: the leucosome (light-colored) has quartz, microcline, and albite; the melanosome (dark-colored) has biotite, amphibole, and cordierite. The paleosome generally originates from a paragneiss, an orthogneiss or an amphibolite.

ACCESSORY MINERALS: none.

APPEARANCE: medium-coarse to very coarse-grained; it is a type of migmatite that has the appearance of a breccia: the paleosome represents the fragments which are "cemented" by the neosome.

COLOR: greenish, dark with light veining.

GEOLOGIC ENVIRONMENT: in masses associated with medium- to high-grade metamorphic rocks.

OCCURRENCE: fairly common rock in the deepest parts of metamorphosed mountain chains; found in Paleozoic basement rocks and Precambrian shields.

USES: used in the building industry for facings and floorings.

NOTES: Agmatite, from the Greek for "mixture," belongs to the migmatite group.

Outcrop of agmatite (ca. x 1), ▶ *Val Malenco, Sondrio, Italy.*

EMBRECHITE (MIGMATITE)

PROTOLITH: felsic regional metamorphic rocks (see agmatite).

METAMORPHIC ENVIRONMENT: medium- to high-grade regional metamorphism with variable pressure (initial stage of anatexis) (amphibolite facies).

ESSENTIAL MINERALS: quartz, potassium feldspar (microcline, perthite), plagioclase (oligoclase), biotite.

ACCESSORY MINERALS: zircon, apatite, magnetite, muscovite, garnet, cordierite.

APPEARANCE: medium- to coarse-grained; it is a migmatite with a striped, gneissic or augen texture in which the paleosome and the neosome are distinct; relicts (biotite amphibolites, calcsilicates) are common; granoblastic texture with poikiloblasts.

COLOR: light, whitish or grayish.

GEOLOGIC ENVIRONMENT: in masses associated with medium- to high-grade metamorphic rocks.

OCCURRENCE: in Precambrian shields and basements.

USES: used in the building industry for facings and floorings.

NOTES: Embrechite is a type of migmatite. It develops by the partial melting of metamorphic rocks; the melted portions have an aplitic-granitic composition and give rise to the leucosome.

▼ *Embrechite (ca. x 1), Sondrio, Italy.*

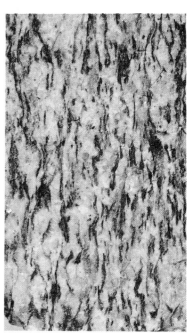

ANATEXITE (MIGMATITE)

PROTOLITH: felsic to intermediate regional metamorphic rocks (see agmatite).

METAMORPHIC ENVIRONMENT: high-grade regional metamorphism with variable pressure (advanced stage in the process of anatexis).

ESSENTIAL MINERALS: quartz, potassium feldspar (microcline, perthite), plagioclase (oligoclase to andesine), biotite.

ACCESSORY MINERALS: zircon, apatite, magnetite, garnet, cordierite.

APPEARANCE: medium-grained; it is a fairly homogeneous migmatite in which the paleosome (dark parts) and the neosome (light parts) are interpenetrated; granoblastic texture in some cases with macroscopic poikiloblasts and flow texture.

COLOR: light gray to dark gray.

GEOLOGIC ENVIRONMENT: in masses associated with medium- to high-grade metamorphic rocks.

OCCURRENCE: fairly common in Precambrian shields and basements.

USES: used in the building industry for facings and floorings.

NOTES: Anatexite belongs to the migmatite group and its name is derived from the Greek for "to melt." Anatexis is the process of partial melting of metamorphic rocks, which is the basis for the genesis of migmatites. In anatexite, the advanced degree of melting produces a greater degree of homogeneity in the rock.

▼ Anatexite (ca. x 0.5), Val Codera, Sondrio, Italy.

NEBULITE (MIGMATITE)

PROTOLITH: felsic or intermediate regional metamorphic rocks.

METAMORPHIC ENVIRONMENT: high-grade regional metamorphism with variable pressure (extremely advanced stage of the process of anatexis).

ESSENTIAL MINERALS: quartz, microcline, orthoclase, plagioclase (albite to andesine), biotite.

ACCESSORY MINERALS: apatite, zircon, magnetite, sillimanite, cordierite, garnet, amphibole.

APPEARANCE: medium-coarse-grained; the texture is fairly homogeneous, the neosome and the paleosome are intimately interpenetrated and no longer distinguishable; granoblastic texture.

COLOR: light gray to dark gray.

GEOLOGIC ENVIRONMENT: in masses associated with medium- to high-grade metamorphic rocks, especially anatexites.

OCCURRENCE: fairly common rock in the deepest parts of metamorphosed mountain chains; found in Precambrian shields and basements.

USES: used in the building industry for facings and floorings.

NOTES: Nebulite belongs to the migmatite group; in particular, the diatexites. In these rocks, the process of migmatization (partial melting or anatexis) is very advanced; there is a complete mobilization or homogenization of the neosome and the paleosome (see agmatite, p. 137), so that they are practically indistinguishable.

▼ Nebulite (ca. x 0.5), Novate Mezzola, Sondrio, Italy.

CALCIPHYRE

PROTOLITH: calcareous or calcareous-dolomitic rocks, with impurities of argillaceous minerals and flint.

METAMORPHIC ENVIRONMENT: medium- to high-grade contact metamorphism with low pressure.

ESSENTIAL MINERALS: calcite, wollastonite, garnet (grossular, andradite), plagioclase (labradorite, anorthite), scapolite, diopside.

ACCESSORY MINERALS: sphene, ilmenite, clintonite (calcium mica), graphite, quartz, phlogopite, brucite, spinel.

APPEARANCE: medium- to fine-grained; commonly massive; sometimes relict sedimentary structures are visible (stratification); small folds are not uncommon, due to the plastic behavior of the calcite even under low pressure; granoblastic texture.

COLOR: light; brown with pink shades, light green, bluish, whitish.

GEOLOGIC ENVIRONMENT: in metamorphic aureoles around plutonic bodies intruded at shallow depths (0.3–3 mi; 0.5–5 km); it can also be found as blocks torn out from a volcanic conduit as magma rises through it and erupts on the surface.

OCCURRENCE: found in Scotland, Ireland, Norway, Italy, etc.

USES: of exclusively scientific interest.

NOTES: Calciphyre is a metamorphic rock with a composition intermediate between hornfels and marbles. Metallic minerals and iron-rich silicates of metasomatic origin may be present. (see skarn, p. 139.)

▼ Calciphyre (ca. x 0.5), Albonico, Como, Italy.

ANDALUSITE BIOTITE HORNFELS

PROTOLITH: aluminum-rich pelitic sedimentary rock.

METAMORPHIC ENVIRONMENT: low- to medium-grade contact metamorphism with extremely low pressure (albite-epidote hornfels facies).

ESSENTIAL MINERALS: biotite, andalusite (commonly as chiastolite which has inclusions of carbonaceous substances).

ACCESSORY MINERALS: quartz, albite, epidote, muscovite.

APPEARANCE: fine-grained, in some cases with visible porphyroblasts; massive structure, lacks schistosity unless it was inherited from the original rock; under the microscope, grains of andalusite (chiastolite variety) are prominent.

COLOR: dark to light gray with dark, flattened or roundish spots.

GEOLOGIC ENVIRONMENT: in metamorphic aureoles of shallow plutonic bodies (0.3–3 mi; 0.5–5 km).

OCCURRENCE: fairly common rocks.

USES: used locally.

NOTES: It belongs to the aluminum-rich hornfels family. Hornfels farther away from the intruded mass generally retain their original textural features, and in some cases they appear spotted from the presence of graphite that was formed from the carbonaceous substances in the sediment. Hornfels closer to the intrusive body are commonly spotted with crystals of biotite and andalusite.

▼ Andalusite biotite hornfels, (ca. x 1), Germany.

CORDIERITE POTASSIUM FELDSPAR HORNFELS

PROTOLITH: pelitic to arenaceous sedimentary rock, in some cases with carbonates.
METAMORPHIC ENVIRONMENT: medium- to high-grade contact metamorphism with extremely low pressure (pyroxene hornfels facies).
ESSENTIAL MINERALS: potassium feldspar (orthoclase), plagioclase, cordierite, andalusite, biotite, sillimanite.
ACCESSORY MINERALS: garnet, epidote, hornblende, diopside, carbonates.
APPEARANCE: medium-grained compact rock; massive structure; microscopic examination shows equidimensional crystals (granoblastic texture), in some cases larger grains with inclusions (poikiloblasts).
COLOR: generally dark with pink, brown, green, and violet tones.
GEOLOGIC ENVIRONMENT: in metamorphic aureoles of shallow, usually basic, plutonic bodies (0.3-3 mi; 0.5-5 km).
OCCURRENCE: fairly common rocks; typical outcrops are found in France, Germany, Scotland, Norway, etc.
USES: used locally.
NOTES: It belongs to the aluminum-rich hornfels family; found in the interiors of contact metamorphic aureoles where temperatures reach 1112-1292° F (600-700° C). In contact with basic plutons, where temperatures exceed 1472° F (800° C), buchite (a glassy hornfels) is formed; the original argillaceous sediment melts to produce a glassy mass of a brownish color with small crystals of magnetite, cordierite, and pyroxene.

▼ *Hornfels (ca. x 0.5), Valle dell'Orba, Savona, Italy.*

SKARN

PROTOLITH: impure calcareous rocks or calcareous-dolomitic rocks.
METAMORPHIC ENVIRONMENT: medium- to high-grade, low pressure contact metamorphism with metasomatic reactions.
ESSENTIAL MINERALS: calcite, pyroxene (diopside, hedenbergite, hypersthene), wollastonite, metal-ore minerals (hematite, magnetite, pyrite and other sulfides) and iron silicates (andradite, Fe-amphibole).
ACCESSORY MINERALS: ilvaite, rhodonite, scheelite, fluorite, barite, scapolite, quartz.
APPEARANCE: grain size variable from coarse to fine; structure is massive or zoned and spotted, bands and lenses of differing mineral assemblages; granoblastic texture.
COLOR: spotted, reddish, violet, brownish, black, green.
GEOLOGIC ENVIRONMENT: near contact zones of calciphyres with an intrusive body; in some cases, extensive.
OCCURRENCE: associated with many intrusive bodies: in Sweden, Romania, the United States, Mexico, Bolivia, Peru, Japan, India, etc.
USES: has no direct use in the industrial field; however, they are important rocks since in many cases they form deposits of copper, iron, zinc, lead, or manganese.
NOTES: *Skarn* is a Scandinavian word that refers to metasomatic calciphyres. Metasomatism is a type of metamorphism that occurs by reaction of the original rock with hot circulating fluids that are usually water-rich and contain dissolved ions. The source of the fluid is generally a volatile-rich magmatic body, but circulating groundwater commonly contributes to these systems.

▼ *Skarn (ca. x 1), Island of Elba, Livorno, Italy.*

RODINGITE

PROTOLITH: gabbros, basalts.
METAMORPHIC ENVIRONMENT: low-grade metamorphism (greenschist facies) with metasomatic reactions.
ESSENTIAL MINERALS: calcium-rich garnet (grossular, andradite), pyroxene (diopside, fassaite), chlorite.
ACCESSORY MINERALS: sphene, amphibole, epidote, vesuvianite.
APPEARANCE: grain size is commonly coarse; zoned with parallel bands of different composition; cavities common; microscopic examination shows equidimensional grains (granoblastic texture), or elongated or interlocking crystals.
COLOR: light, brownish, reddish, greenish, rosy.
GEOLOGIC ENVIRONMENT: metasomatic alteration of gabbros and basalts enclosed by serpentinized peridotites.
OCCURRENCE: not very common rock.
USES: of exclusively scientific interest.
NOTES: Rodingite is formed in gabbroic dikes by reaction with fluids rich in calcium but poor in alkalis. These fluids are a by-product of the serpentinization of the associated peridotites.

Rodingite (ca. x 0.5), ▶
Valle dell'Orba,
Savona, Italy.

MYLONITE

PROTOLITH: any type of rock.
METAMORPHIC ENVIRONMENT: low-temperature high-strain metamorphism (rocks are strongly deformed but at low temperature).
ESSENTIAL MINERALS: variable, depending on the original rock; argillaceous minerals common; mylonites derived from igneous rocks generally contain chlorite, epidote, iron and titanium oxides.
ACCESSORY MINERALS: variable, depending on the original rock.
APPEARANCE: fine-grained rock; generally fairly compact, commonly schistose; generally not very durable and easily degradable unless new cements or recrystallization bond the crushed particles; microscopic examination shows some lenticular relicts of partially deformed crystals (porphyroclasts) immersed in a microcrystalline groundmass of finely pulverized, recrystallized, original minerals.
COLOR: black, dark gray, rarely light.
GEOLOGIC ENVIRONMENT: found in narrow, elongated bands parallel to fault zones; grade to undeformed rocks with increasing distance from the fault.
OCCURRENCE: rather uncommon rock; found in orogens.
USES: of exclusively scientific use.
NOTES: The name is derived from the Greek for "grindstone." Mylonitic texture occurs in cold brittle rocks subjected to shearing stresses. When the rock is lithified, the texture is clearly schistose; if the crushing is less intense, cataclasites occur.

▼ *Mylonite (ca. x 0.5), Val Masino, Sondrio, Italy.*

SEDIMENTARY ROCKS

SEDIMENTARY ROCKS

Top: detritus of a bed: an example of detrital material detached and accumulated at the foot of a crag.

Center: rocks fractured by tectonic movements.

Bottom: example of biochemical attack by plant life.

Sedimentary rocks are formed as a result of physical and chemical changes in rocks already existing on the earth's surface. As we know, this surface is characterized by low temperatures and pressures and by a hydrated, oxidizing environment. With few exceptions, sedimentary rocks are products of erosion, transport, and deposition, followed by lithification.

The sedimentary process begins with physical and chemical weathering and erosion. The following are important weathering and erosion processes.

a) Movements of the crust, called tectonic movements, cause fracturing in rocks and allow for infiltration of atmos-

ously, abrading the rocks they slide over and carrying fragments downslope. The fragments accumulate, forming a particular type of unconsolidated sediment called "moraine."

d) Running waters constitute a powerful mechanism for transport. They wash away slopes and transport enormous quantities of fragments towards watercourses. Once in the water, the distance the fragments travel depends on their size and the transport capacity of the watercourse.

e) The formation of new minerals can result from chemical alteration of rocks. Water may act as a slow but effective solvent, leaching more soluble compo-

nents from the rock and carrying them away. This is an important process in the formation of soils. Soil is a thin earthy layer produced by weathering of the rock on which it rests. In its uppermost part, it is generally lacking in the lithic fragments of the substratum. These fragments, however, increase towards the lower parts. In this soil, aqueous solutions and biological activity facilitate the destruction of some minerals and the formation of others.

f) Plants and, to some extent, animals have a twofold function: on the one hand, they contribute to weathering of rocks and soil; on the other hand, they can inhibit erosion. Plants, in particular, contribute in a direct way to the chemical reactions that lead to the formation of new clay minerals and to the accumulation of organic matter.

Dissolved and eroded components that have been carried away by water are eventually deposited in the ocean. As already noted, the particles are transported in ways that are strictly dependent on their size and the velocity (or energy) of the water. The smaller particles can flow along in suspension, while the larger fragments roll along the bottom. Rivers and streams have a remarkable selective capacity in relation to the particles. How many times have we happened upon a gravelly stream bed in the mountains or the sandy banks of a river on the plains? On its course, a river loses speed and energy, and the materials that become too heavy are deposited. However, the finer particles remain in suspension and are deposited in or close to the sea. Large fragments however, can reach the sea where outcrops are close to the

pheric water from the surface. Fracturing of surface rocks may also result in mass wasting phenomena such as landslides.

b) Insolation and freezing both contribute to the disintegration of rocks on the surface. Temperature fluctuations over the course of a day cause minerals to expand and contract, resulting in small cracks and subsequent erosion of individual crystals of rock fragments. Freezing acts in in a similar way. When water is present, it can easily seep into cracks and facilitate chemical weathering in hot climates. In cold climates, the increase in volume caused by the formation of ice facilitates physical disintegration.

c) Glaciers have enormous erosive powers: they move slowly but continu-

Top: development of life in the coral barrier reefs in tropical seas rich in dissolved minerals (in the photograph, Bahama Sea).

Center: satellite photo of the mouth of the Arno. The turbidity of the flow near the sea is shown by the particles in suspension.

Bottom: simplified diagram of the precipitation of salts in an evaporite basin.

sea and the flood plain is not particularly extensive, or where the energy of the river is high. The second case occurs during periods of heavy rain, when rivers are very full.

The sea also contributes to the erosion and deposition of particles. It does not have sufficient energy to keep the sandy particles carried by rivers in suspension, but the motion of the waves, which generally changes direction with the season, has the capacity to make them roll along the coast, especially during storms. On the contrary, the finer particles can reach the open sea while remaining in suspension due to the weak

this equilibrium, minerals precipitate out of sea water via organisms or simply as a result of oversaturation. Many marine organisms utilize dissolved compoinds to build their own shells or skeletons, a phenomenom which takes place in all the world's climates. However, this phenomenon is particularly striking in tropical climates where very specialized corals and madrepores (compound animals) live. They habitually build a common structure, sometimes of enormous size. These structures are the coral barrier reefs found along the coastline of continents such as Australia, islands like the Bahamas and Cuba, or

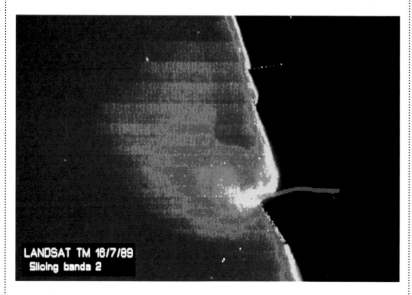

LANDSAT TM 16/7/89
Slicing bands 2

current generated by the flux of a river that empties into the sea. Once they are in the open sea, these particles can be deposited or be caught up by an ocean current, which disperses them over a broader area. The sea also has the capacity to sort rock particles: below the coastline, sands are predominant; towards the open sea, mud; and further out, clays predominate. In addition to its transport and sedimentation capabilities, the sea also has a strong erosional power, resulting from its abrasive action on the rocky shore where fragmented products are "worked on" by the aforementioned processes and by some marine organisms that slowly contribute to their disintegration.

Given that components in solution also reach the sea, it would almost seem that the salinity of the sea should increase over time. In fact, it remains relatively constant. In order to maintain

the atolls in the Pacific Ocean. In these environments the majority of the animals synthesize calcium carbonate, usually in the form of aragonite. These organisms need light, relatively warm, open waters which must be oxygenated, clean, and fairly shallow. Ancient reef rocks are generally massive and unstratified.

Direct precipitation of salts from sea water occurs in seas, or portions of seas, characterized by strong evaporation that is not balanced by input from from rivers or the open sea. Rocks formed in this environment are called evaporites, and their characteristic minerals include halite, gypsum, and calcite. The first salt to precipitate is the least soluble, since it reaches its saturation point first (in order: calcite, gypsum, halite). These sediments are stratified and are intercalated with argillaceous sediments typical of a closed, low-energy environment.

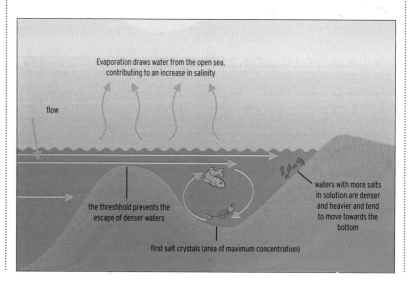

Evaporation draws water from the open sea, contributing to an increase in salinity

flow

the threshhold prevents the escape of denser waters

first salt crystals (area of maximum concentration)

waters with more salts in solution are denser and heavier and tend to move towards the bottom

ORTHOCONGLOMERATE

ESSENTIAL COMPONENTS: pebbles, generally rounded, of various lithologies.

ACCESSORY COMPONENTS: sandy matrix with calcareous-argillaceous cement.

APPEARANCE: the coarse grains are generally in direct contact within a medium-fine-grained matrix; fossils rare.

COLOR: variable, depending on the components.

MODE OF OCCURRENCE: in stratiform or lenticular masses.

GEOLOGIC ENVIRONMENT: in numerous sedimentary successions associated with sandstones or pelites or their unconsolidated equivalents.

USES: good material for cement, although the presence of sandstone fragments or a clayey mud matrix can reduce the quality.

NOTES: Conglomerates are terrigenous clastic rocks made up of at least 50% coarse grains (> 2mm) and are essentially consolidated gravels. The prefix "ortho" indicates that the pebbles are in close contact with one another, while the matrix is relegated to intergranular spaces; these rocks can also be defined as grain-sustained. Orthoconglomerates differ from paraconglomerates in that they are stratified and contain no mud.

▲ Polygenetic orthoconglomerate (ca. x 0.5), Trent, Italy.

PARACONGLOMERATE

ESSENTIAL COMPONENTS: pebbles, generally rounded, of various lithologies (metamorphic, igneous, sedimentary).

ACCESSORY COMPONENTS: muddy-sandy matrix with calcareous-argillaceous cement.

APPEARANCE: very obvious pebbles (size > 2mm) scattered and isolated in the matrix.

COLOR: varies according to the nature of the fragments; the matrix is generally grayish or brown from oxidation.

MODE OF OCCURRENCE: in masses of a wide range of dimensions.

GEOLOGIC ENVIRONMENT: in numerous sedimentary successions associated with sandstones or pelites or their unconsolidated equivalents.

USES: can be used as a low-cost concrete.

NOTES: A terrigenous clastic rock in which pebbles comprise more than 50% of the rock; the prefix "para" means that the pebbles that are not in direct contact; the pebbles are scattered within a fine-grained matrix. Paraconglomerates differ from orthoconglomerates in that they are not stratified, occurring as disorganized masses, and in that they contain mud. Conglomerates are typical of moraines: sediments produced by erosion and transported by glaciers. Glacio-marine (or glacio-lacustrine) conglomerates or pebble beds are formed when a glacier leaves behind fragments in an underwater environment. In these sediments, the finer-grained portions may be lacking because they have been carried away by the currents. If the conglomerate is made up of very rounded pebbles, the rock is called "puddingstone." Psephite (from the Greek for "little stone") or rudite (from the Latin *rudus*, "a fragment of rock") are synonyms for conglomerate.

Paraconglomerate ▶
(ca. x 0.7), Norcia, Perugia, Italy.

BRECCIA

ESSENTIAL COMPONENTS: angular fragments of various sizes, commonly of the same lithology.

ACCESSORY COMPONENTS: cement or matrix of argillaceous, siliceous, or carbonate composition.

APPEARANCE: sharp-cornered, coarse fragments are easily visible and welded by a fine-grained cement or matrix; fossils rare.

COLOR: varies according to the lithologies of the coarse elements.

MODE OF OCCURRENCE: in masses of various dimensions, located over small areas.

GEOLOGIC ENVIRONMENT: forms from fragments deposited close to the source area; for example, at the base of cliffs or within fractures as a result of tectonism (tectonic breccias).

USES: for the manufacture of concrete; if it is well-cemented, with the intergranular spaces occluded by cement, and has a pleasing esthetic appearance, it can be used as an ornamental stone (marble breccias).

NOTES: A terrigenous clastic rock; breccia is a variety of conglomerate made up of angular fragments (with sharp corners), indicative of limited transport prior to deposition.

▼ Breccia: polished slab (ca. x 0.5), Camaiore, Lucca, Italy.

QUARTZARENITE

ESSENTIAL COMPONENTS: quartz.

ACCESSORY COMPONENTS: feldspars, muscovite, glauconite, heavy minerals (zircon, garnet), and a generally siliceous cement.

APPEARANCE: compact, arenaceous rock with a medium-fine grain and a splintery or conchoidal fracture; in some places dense laminae (very thin, mm-thick layers) exhibit cross-bedding; quartz granules, commonly rounded and of equal size, are welded by quartzose, sericitic, more rarely carbonate matrix; fossils rare.

COLOR: commonly light; whitish, soft gray.

MODE OF OCCURRENCE: in stratiform masses, sometimes very extensive.

GEOLOGIC ENVIRONMENT: in Precambrian shields, commonly associated with metamorphic quartzites.

USES: used rarely in the building industry.

NOTES: A terrigenous clastic rock; quartzarenite is a variety of sandstone particularly rich in quartz (> 95%), which makes it very resistant to alteration. The term sandstone first appeared in the eighteenth century when it was used to describe rocks made up of such fine particles that they were not identifiable by the naked eye. Currently, scientists define "sandstones" or "arenites" as those rocks made up of sand-sized grains.

▼ Quartzarenite with cross bedding (ca. x 0.6), Libya.

ARKOSE

ESSENTIAL COMPONENTS: quartz, feldspars.
ACCESSORY COMPONENTS: lithic fragments (commonly granites and gneisses), micas, oxides, heavy minerals, cement (siliceous or carbonate).
APPEARANCE: sandstone with medium-coarse granules with a scanty, fine matrix; sedimentary structures are not very common; generally angular granules are poorly sorted and may be barely cemented making the rocky mass fairly friable; fossils are rare or absent.
COLOR: gray, pink or reddish (from hematite).
MODE OF OCCURRENCE: in stratiform masses.
GEOLOGIC ENVIRONMENT: formed near the major mountain chains.
USES: sometimes cut for building stone.
NOTES: A terrigenous clastic rock; the name is probably of Greek origin from the word for "primitive" or "old." Arkose is a sandstone with a chemical and mineralogical composition similar to that of granite and, when well compacted and cemented, it can be easily confused with granite. It is formed by the rapid erosion of granitic or gneissic masses. Eroded materials are generally transported short distances and deposited in a marine, fluvial, or lacustrine environment.

▼ *Arkose (ca. x 1), France.*

GRAYWACKE

ESSENTIAL COMPONENTS: quartz, feldspars, micas, various lithic fragments.
ACCESSORY COMPONENTS: heavy minerals (zircon, garnet, tourmaline, etc.), iron oxides and hydroxides, carbonate or argillaceous cement.
APPEARANCE: arenaceous rock of variable grain size, sometimes heterogeneous; the fine matrix can be abundant; sedimentary structures such as graded bedding and sole-marks; granules are generally angular, fossils are fairly rare and when present are, in general, broken.
COLOR: gray, greenish gray or brownish, rarely very dark.
MODE OF OCCURRENCE: in stratiform masses of remarkable expanse alternated with pelitic rocks.
GEOLOGIC ENVIRONMENT: very widespread rock; associated with orogens and forms sedimentary succession that can be as thick as a few thousand feet.
USES: used rarely in the building industry.
NOTES: A terrigenous clastic rock; the word is derived from the German *grauwacke*, originally probably a mining term from the Harz area of Germany, dating back to the eighteenth century. Graywackes are formed by the accumulation of detritus deposited by turbidity currents in a deep-sea environment. At one time, this stone was much used because it is easily worked. Currently, greywackes with a homogeneous grain and a uniform gray color are used in the production of ashlars and slabs that can be faced but not polished, unless they are treated with resins to fill the intergranular pores. Greywacke is generally not very durable.

GLAUCONITIC SANDSTONE

ESSENTIAL COMPONENTS: quartz, calcite, glauconite, feldspars, lithic fragments.
ACCESSORY COMPONENTS: micas, iron oxides and hydroxides, heavy minerals (zircon, garnet, tourmaline, etc.), carbonate or argillaceous cement.
APPEARANCE: fairly homogeneous rock with a medium to fine grain; the fine matrix can be abundant and sedimentary structures are rare; fossils, either whole or broken, are commonly present.
COLOR: greenish gray to dark green from the presence of glauconite.
MODE OF OCCURRENCE: stratiform; sometimes intercalated in other sandstones.
GEOLOGIC ENVIRONMENT: not very common rock; the greensands in England and on the Atlantic Coast of the United States are noteworthy.
USES: in general, it has no practical use.
NOTES: Glauconite sandstone is a sandstone with a silicate or carbonate composition, characterized by the presence of glauconite. Glauconite is a member of the phyllosilicate clay mineral group. It is rich in iron and is formed in the sea at depths of up to 656 ft. (200 meters).

Glauconitic sandstone ▶
(ca. x 0.7), the United States.

◀ *Graywacke (ca. x 1), Rhine Valley, Germany.*

GYPSUM ARENITE

ESSENTIAL COMPONENTS: quartz, feldspars, calcite, gypsum, lithic fragments.
ACCESSORY COMPONENTS: mica, clay minerals, heavy minerals, carbonate and/or gypsum cement.
APPEARANCE: similar to that of many sandstones; easily furrowed because of the presence of fine- to coarse-grained gypsum.
COLOR: light to dark gray with small whitish spots of gypsum.
MODE OF OCCURRENCE: stratiform.
GEOLOGIC ENVIRONMENT: not very common; can be found in proximity to evaporite rocks from which it is derived.
USES: rarely used.
NOTES: Gypsum arenite is similar to a sandstone, but contains a significant portion of gypsum rather than quartz. Gypsum is not commonly present in detrital sediments because it is easily soluble in water. Within the sediment, it may partially dissolve and reprecipitate in the form of cement. Psammite (from the Greek) or arenite (from the Latin *arena*), both meaning "sand," are synonyms for sandstone. Sand is used for sediments in a loose state.

▼ *Gypsum arenite with laminations (ca. x 0.6), Ascoli Piceno, Italy.*

SILTSTONE

ESSENTIAL COMPONENTS: quartz, calcite, feldspars, micas, clay minerals.

ACCESSORY COMPONENTS: carbonaceous material, pyrite, iron oxides, gypsum, argillaceous or carbonate cement.

APPEARANCE: fine-grained rock; sedimentary structures (stratification, lamination) are barely evident because of the fine grain-size; in constrast to the finer-grained argillites, not greasy to the touch and small grains are visible.

COLOR: dark gray, yellowish, brownish, greenish.

MODE OF OCCURRENCE: stratiform; in masses associated with other clastic rocks (sandstones, pelites).

GEOLOGIC ENVIRONMENT: very common rock in all sedimentary environments.

USES: can be used in the production of medium-quality baked bricks.

NOTES: A terrigenous clastic rock; siltstone is made up of granules between 0.063 and 0.002 mm; it corresponds to the grain size of mud (or silt), finer than sand and coarser than clay. Siltstone may form from a number of sediments in different sedimentary environments. For example: loess, friable with a yellowish color and massive structure, is a special mud without sand and with very little clay that is formed by aeolian accumulation in desert or glacial areas; varves are muddy sediments deposited by glaciers, identifiable by their typical alternating dark and light laminae; muddy sediments are also deposited on the banks of overflowing rivers.

▼ *Argillaceous siltstone (ca. x 0.8), Libya.*

DETRITAL ROCKS

Clastic rocks are formed by the transport and accumulation of small fragments in sedimentary environments (marine, lacustrine, fluvial, etc.). These rocks are classified according to the size and composition of the particles.

FROM "STONES" TO POWDERS

Grain-size analysis can be employed to study and compare loose rocks or "sediments." This is carried out by using a series of superimposed sieves each with an increasingly finer mesh than the one above it. Each sieve

Arkose: microphotograph of a thin section (ca. x 10; crossed nicols. 1 = quartz; 2 = orthoclase; 3 = plagioclase, twinned according to the albite law (alternate light and dark bands). The individual granules are cemented by secondary calcite (the bright colors) deposited after sedimentation, that changes the initial sand into lithified rock.

retains the grains that are larger than the openings in the mesh. For a heterogenous sediment, for example, one can use a sieve that separates the gravelly part from the sandy part and another sieve to separate the sandy part from the muddy and clayey parts.

The material retained in each sieve is weighed and recorded as a percentage of the total sample. To separate the muds from the clays sieves are not used, as the extremely fine mesh would tend to get stopped up. In order to

separate them, indirect methods are used that provide for the dispersion of the sample in water, taking advantage of their different settling rates. Many scales for granularity have been proposed; the one most often used in geology is the scale from MIT (Mass-

Macroscopic specimen of arkose (sandstone rich in quartz and feldspars), Alta Val Tiberina, near San Sepolcrio (Arezzo, Italy).

achusetts Institute of Technology) recorded in the table. In nature, it is quite improbable that sediments would be composed of a single grain-size class. Usually they are a mixture, in which one class predominates and the sediments are defined by the name of the most represented class, and if necessary, by the subsequent one.

For example, clayey "sand" indicates that sand is predominant and clay is present in significant quantity. The name corresponding lithified sediment is called "pelitic sandstone." The transformation of loose sediments to lithified rocks (diagenesis) happens through the cementing of granules by minerals (usually calcite or quartz) that precipitate in the pore space between the grains.

FROM SILICICLASTIC TO CARBONATE

Based on the mineralogical composition of the granules, detrital sedimentary rocks can be subdivided into two groups: siliciclastic (or terrigenous) and carbonate. The former are formed from the erosion of non-carbonate, silicate-bearing metamorphic, igneous, and sedimentary rocks. The latter have their origin in the erosion of other carbonate rocks. It is very common for there to be rocks of different types in an area subjected to erosion and it is, therefore, easy to find mixed sediments which may be referred to by the dominant component.

The term sandstone (or arenite) indicates a sediment rich in quartz clasts; while calcarenite indicates one that is rich in carbonate clasts.

In general, sandstones are a grayish color. They darken from alteration, and the individual clasts tend to be separate from one another. In contrast, calcarenites are more yellowish. They exhibit dissolution surfaces and the surface becomes more golden colored.

SIZE (MM)			GRAIN	LOOSE ROCK	LITHIFIED ROCK	GREEK DERIVATION	LATIN DERIVATION
	gravel	> 2 (0.1 in.)	coarse	gravel	conglomerate	psephite	rudite
2	> sand	> 0.063 (0.002 in.)	medium	sand	sandstone	psammite	arenite
0.063	> mud	> 0.002 (0.0001 in)	fine	mud (or silt)	siltstone	pelite	lutite
0.002	> clay			clay	argillite		

ARGILLITE

ESSENTIAL COMPONENTS: clay minerals.
ACCESSORY COMPONENTS: quartz, feldspars, carbonates, micas, carbonaceous materials, pyrite, nodules of gypsum, iron oxides.
APPEARANCE: very fine-grained rock, it crumbles easily in the hand and is greasy to the touch; very thin layers (laminae) are common; can be rich in both macro and micro (mollusk shells and foraminifera) fossils.
COLOR: extremely variable; various shades of gray, sometimes black from the presence of carbonaceous substances, reddish, brownish, yellow (oxidized ferriferous compounds), greenish (glauconite).
MODE OF OCCURRENCE: stratiform; in masses, sometimes of great thickness, associated with coarser sediments.
GEOLOGIC ENVIRONMENT: a very common sedimentary rock; formed in low-energy sedimentary environments
USES: primary material in the refractory industry, for ceramics and prized brickwork.
NOTES: Argillites are terrigenous clastic rocks in which clay-sized grains (< 0.002 mm) predominate. They are formed in a multitude of depositional environments: in the open sea, in lagoons, and in lakes. Residual argillites are formed by alteration of preexistent rocks from which the soluble compounds are separated. Pelite and lutite are synonyms for argillite.

▼ *Argillite (ca. x 1), Sassuolo, Modena, Italy.*

MARL

ESSENTIAL COMPONENTS: clay minerals, calcite.
ACCESSORY COMPONENTS: quartz, micas, carbonaceous materials, pyrite, gypsum.
APPEARANCE: fine to very fine grain, with conchoidal fracture; microfossils common; may flake off as a result of alteration or from an excess of clay minerals.
COLOR: gray, brown, beige and green.
MODE OF OCCURRENCE: stratiform; in masses of sometimes remarkable size, commonly alternating with argillites.
GEOLOGIC ENVIRONMENT: very common rock.
USES: principal material in the preparation of cement.
NOTES: Marl is a rock of mixed clastic and organic origin. It is formed by an accumulation of clastic material made up of carbonate-argillaceous mud in a marine or, more rarely, continental environment. The calcite is commonly made up of the shells of planktonic microorganisms. Marls are soft, easily altered rocks, and their surfaces become scaly from the dissolution of calcite. In the manufacture of cement, the rock is ground up and baked in rotating ovens; the high temperature drives off carbon dioxide, and small, roundish masses of calcium and aluminum silicates (clinkers) are formed. This process is used to produce various kinds of cement with different compositions and setting speeds.

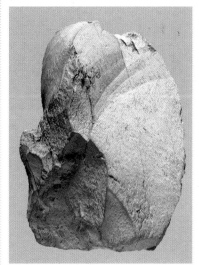

▲ *Pliocene marl (ca. x 0.5), Italy.*

CALCILUTITE

ESSENTIAL COMPONENTS: silt to clay-sized particles of calcite (micrite).
ACCESSORY COMPONENTS: shells (or fragments) of microfossils, carbonate cement, hematite, pyrite, quartz and orthoclase sometimes autogenous.
APPEARANCE: compact homogeneous rock with a fine grain (the particles are not visible, < 0.063mm); sedimentary structures are sometimes visible; the microfossils are sometimes visible with the aid of a lens and are enhanced by the recrystallization of calcite.
COLOR: various shades of gray, tawny, black.
MODE OF OCCURRENCE: stratiform; in masses, sometimes of great thickness, or intercalated in other carbonate or, more rarely, arenaceous rocks.
GEOLOGIC ENVIRONMENT: easily found in association with other carbonate rocks.
USES: in the preparation of various kinds of gravels; in cements; if it exhibits good aesthetic properties, for the manufacture of flooring and medium-quality facings.
NOTES: A clastic carbonate rock; when struck, it breaks along concave or convex surfaces (conchoidal fracture) producing chips from the sharp edge (the chips are generally not very durable). On breaking, some rocks of this type emanate a typical, fairly unpleasant odor from the presence of hydrocarbons.

▼ *Compact limestone (ca. x 1), Varese, Italy.*

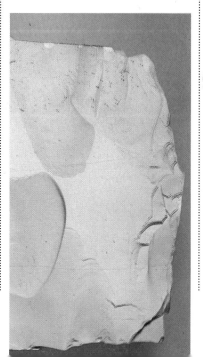

CALCARENITE

ESSENTIAL COMPONENTS: calcite particles, carbonate cement.
ACCESSORY COMPONENTS: fragments of fossils, clay minerals, sometimes autogenous quartz and orthoclase.
APPEARANCE: compact homogeneous rock with sand-sized grains (2–0.0063 mm) visible to the naked eye; internal deposition structures may be present.
COLOR: light to dark gray.
MODE OF OCCURRENCE: stratiform; also in large masses or intercalated with arenaceous or argillaceous-marly strata.
GEOLOGIC ENVIRONMENT: in carbonate successions.
USES: in the building industry for the production of gravels or cut into slabs for floorings and facings.
NOTES: When they are intercalated in an arenaceous formation, calcarenites are a good geochronological indicator because of the presence of fossils, generally absent in sandstones. Calcarenites are further classified according to the principal granular constituents; for example: oolitic calcarenite, composed of oolites (warm, rough sea environment, typical of barrier reefs), lithic calcarenite (rich in rock fragments indicative of the rocks from which the sediment was derived). Calcarenites generally exhibit textural characteristics similar to those of the non-carbonate sandstones.

▼ *Calcarenite (ca. x 1), Apuan Alps, Tuscany, Italy.*

CALCIRUDITE

ESSENTIAL COMPONENTS: carbonate lithic fragments, carbonate cement.

ACCESSORY COMPONENTS: fragments of fossil organisms, iron oxides and hydroxides, quartz.

APPEARANCE: compact rock with coarse elements; the granular components can be heterogeneous and constituted in large part of clastic granules, originating from the erosion of other carbonate rocks or coral reefs.

COLOR: more or less light hazel.

MODE OF OCCURRENCE: stratiform, in large masses or intercalated with carbonate and/or argillaceous-marly strata.

GEOLOGIC ENVIRONMENT: near carbonate deposits, either organic or lithic; commonly associated with turbidite carbonate sediments.

USES: in the manufacture of gravels; if the rock is compact and only slighted fractured, it can be cut into slabs and polished.

NOTES: A carbonate rock of detrital origin with a clastic element content greater than 50% of the total volume. The granules are of a size analogous to those of the conglomerates, bonded by a carbonate cement or immersed in a fine calcareous matrix. The constituent fragments are indicative of severe erosion in the areas of origin.

▼ *Calcirudite (ca. x 0.8), Tuscany, Italy.*

BIOCALCILUTITE

ESSENTIAL COMPONENTS: whole or fragmented microfossils, microcrystalline calcite.

ACCESSORY COMPONENTS: hematite, pyrite, sometimes autogenous quartz and orthoclase.

APPEARANCE: compact homogeneous rock with a fine grain (the particles are not visible, < 0.063 mm); sedimentary structures are sometimes visible; grains may be visible on a wet sample under a hand lens; the presence of microfossils is enhanced by recrystallized calcite.

COLOR: gray, tawny, black.

MODE OF OCCURRENCE: stratiform; in masses which may be very thick or intercalated in other carbonate or, more rarely, arenaceous rocks.

GEOLOGIC ENVIRONMENT: in association with other carbonate rocks.

USES: in the production of gravels and cement; if it is aesthetically pleasing, for the production of slabs for floors and medium-quality facings.

NOTES: A clastic carbonate rock; differentiated from calcilutite by the presence of fossils, but other properties are similar. When struck, it breaks along concave or convex surfaces (conchoidal fracture), producing chips from the sharp edge; the chips ae not very durable. On breaking, some rocks of this type emanate a typical, fairly unpleasant odor from the presence of hydrocarbons.

Biocalcarenite with a shark's ▶ tooth (ca. x 0.5), Modenese Apennines, Italy.

▼ *Biocalcilutite (ca. x 0.6), Apuan Alps, Tuscany, Italy.*

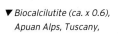

BIOCALCARENITE

ESSENTIAL COMPONENTS: fragments of fossil shells, carbonate lithic granules, carbonate cement.

ACCESSORY COMPONENTS: carbonate matrix, hematite, pyrite, autogenous quartz and orthoclase.

APPEARANCE: compact homogeneous rock with a medium grain (2–0.063 mm); grains visible to the naked eye; structures similar to those of the sandstones are present.

COLOR: light to dark gray, beige.

MODE OF OCCURRENCE: stratiform; in thick deposits or alternating with other limestones or with argillites.

GEOLOGIC ENVIRONMENT: common rock; for the most part associated with other carbonate sediments and, more rarely, with sandstones.

USES: in the production of gravels; rarely, cut into slabs for floors and facings.

NOTES: A carbonate clastic rock; biocalcarenite is essentially made up of well-sorted organic carbonate fragments (i.e., the fragments are all about the same size). The organic fragments are primarily from the shells of mollusks, but algal fragments can also be present.

BIOCALCIRUDITE

ESSENTIAL COMPONENTS: fragments of fossil remains of carbonate composition and lithic fragments.

ACCESSORY COMPONENTS: carbonate cement, hematite, pyrite, sometimes autogenous quartz or orthoclase.

APPEARANCE: compact rock with coarse components (> 2 mm), easily visible with the naked eye; the clastic granules are, for the most part, made up of fragments of fossil remains from the erosion and fragmentation of either a coral barrier reef or the shells of various species of mollusks.

COLOR: gray, tawny, black.

MODE OF OCCURRENCE: stratiform; in even, large masses or intercalated with carbonate and/or argillaceous-marly strata.

GEOLOGIC ENVIRONMENT: where large masses of carbonate rocks are, or have been, present, or near a coral barrier reef.

USES: in the production of gravels; if the rock is compact and only slightly fractured, it can be cut into slabs and polished.

NOTES: A detrital carbonate rock with over 50% of the fragments from organic remains. Biocalcirudite is indicative of severe erosion in the areas of origin. Some well-cemented kinds, without intergranular spaces, can be cut and polished for the production of slabs for floors and facings. The commercial value depends on the play of colors of the components. Biocalcirudites are commonly known by the name of the predominant fossil species; for example: coquina limestone (rich in mollusk shell fossils) and ammonite-bearing limestone (fossils from the Mesozoic era).

▼ *Biocalcirudite: coquina, small coastal breccia (ca. x 0.9), Italy.*

REEF LIMESTONE

ESSENTIAL COMPONENTS: fossil remains, calcite, dolomite.

ACCESSORY COMPONENTS: generally in very small quantities: clay minerals, pyrite, sometimes autogenous quartz.

APPEARANCE: variable; in some places the limestone is recrystallized with calcite crystals that cover the walls of the cavities which are commonly left empty, abandoned by dissolved fossil remains; whole fossils are commonly visible in a position of growth (still attached to the foundation); interstitial spaces are filled with secondary calcite, sometimes impure from the presence of clay, and by fragments of other organisms.

COLOR: whitish.

MODE OF OCCURRENCE: in masses of remarkable size; included among stratified carbonate rocks.

GEOLOGIC ENVIRONMENT: associated with other carbonate sediments.

USES: as a building stone and for gravels.

NOTES: A carbonate rock also known under the name biolithite; reef limestone is formed by the growth of colonial marine organisms fixed to a substratum (corals, calcareous sponges, algae, etc.) that assimilate the calcium carbonate in solution in the sea water and secrete it directly from their tissues. It has a current equivalent in the tropical coral barrier reefs. In a diagenetic environment, the whole mass may alter from calcium carbonate (calcite or aragonite) to dolomite.

▼ *Coral limestone (ca. x 1.5), Dolomites, Italy.*

TRAVERTINE

ESSENTIAL COMPONENTS: calcite or aragonite (hot springs).

ACCESSORY COMPONENTS: limonite, sandy granules.

APPEARANCE: compact, porous rock made up of dense, thin, wavy laminae formed by acicular, subparallel crystals of calcite; when formed by inititial precipitation, it is light, very porous (microporosity), with a brilliant white color; commonly contains impressions of plant remains whose organic material has been completely removed.

COLOR: banded, in ancient travertines: whitish, yellowish, rosy, brownish, commonly with slight differences in color depending on the limonite content; homogeneous brilliant white in those that are in the process of formation.

MODE OF OCCURRENCE: in not very extensive masses lacking stratification; in incrustations on other rocks.

GEOLOGIC ENVIRONMENT: not very widespread rock.

USES: in the building industry for floorings and facings.

NOTES: The name is derived from *lapis tiburtinus*, Tivoli stone, from the city where it was widely used by the ancient Romans. It is a carbonate rock of chemical origin that is formed by rapid precipitation of calcium carbonate from oversaturated waters (hot springs). Travertine also forms stalactites and stalagmites in limestone caves; in this case the speed of growth is much lower, probably because of the lower crystallization temperature.

▼ *Travertine with Ficus carica and Ocer Lobehi (ca. x 0.8), Ascoli Piceno, Italy.*

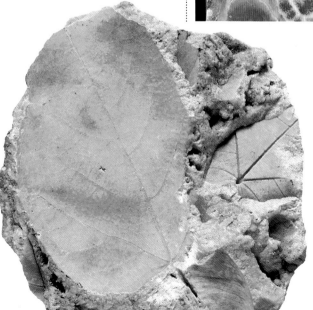

CALCAREOUS ALABASTER

ESSENTIAL COMPONENTS: concretionary calcite.

ACCESSORY COMPONENTS: iron oxides and hydroxides.

APPEARANCE: compact rock with a generally translucent, waxy look; structure is fibrous, radial fibrous, sometimes with concentric zoning.

COLOR: variegated with white, yellowish, brown, reddish brown (oriental alabaster), light blue, green, etc.

MODE OF OCCURRENCE: in concretionary masses.

GEOLOGIC ENVIRONMENT: in carbonate rocks.

USES: as an ornamental stone for slabs and objects; widely used by the ancient Romans.

NOTES: A carbonate rock of chemical origin; calcareous alabaster is formed by direct precipitation from waters very rich in calcium carbonate, commonly found within travertine deposits.

▼ *Calcareous alabaster (ca. x 0.3), Volterra, Pisa, Italy.*

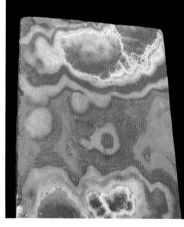

OOLITIC LIMESTONE

ESSENTIAL COMPONENTS: oolites, carbonate cement and/or matrix (micrite).

ACCESSORY COMPONENTS: siderite.

APPEARANCE: compact rock with easily visible grains; remains of fossil organisms are rare; the oolites are generally not in direct contact with one another and the intergranular spaces are filled with recrystallized mud or carbonate cement.

COLOR: whitish, yellowish, brownish.

MODE OF OCCURRENCE: in stratiform masses of variable thickness.

GEOLOGIC ENVIRONMENT: relatively common rocks; associated with ancient reef limestones; in current environments, they are in the loose state and look like a very white, medium-fine sand.

USES: of exclusively scientific interest.

NOTES: Oolitic limestone is a rock with a carbonate composition, made up of sub-spherical granules (oolites, also known as ooids or ooliths) with a shape analogous to that of sands, well sorted, of a size generally between 0.2 and 1 mm, up to a maximum of 2 mm. Oolites are formed in a shallow marine environment with high energy (waters continually agitated by the surf), through precipitation of calcium carbonate around an initial corpuscle. They have a concentric structure like the layers of an onion. Current oolites are made of aragonite. In ancient limestones, it is completely changed into calcite. Oolites can also be formed in lakes, caves, and near hot springs.

Oolitic ▶ limestone (ca. x 1), Bahama Islands.

PISOLITIC LIMESTONE

ESSENTIAL COMPONENTS: pisolites, sparry cement.

ACCESSORY COMPONENTS: dolomite, siderite, hematite, with a scanty calcareous matrix.

APPEARANCE: compact rock formed by the cementing of spherical to ellipsoid concretionary granules of large size (> 2 mm); sometimes smaller spheres are also present in the intergranular spaces; the pisolites may not be in direct contact and the interposed cement comes from chemical precipitation.

COLOR: whitish, yellowish, brownish.

MODE OF OCCURRENCE: rarely stratiform; in masses of reduced thickness.

GEOLOGIC ENVIRONMENT: fairly rare rocks, the ancient pisolitic limestones of England, Germany, and the United States are well-known.

USES: of exclusively scientific interest.

NOTES: A rock with a carbonate composition; the distinction between pisolites and ooliths is based on size (ooids are <2mm), but the origin of the two are different. Pisolites are formed by successive accretions of small needles of aragonite on granules of an inorganic nature in low-energy environments. They are found in caves, near hot springs, in bauxite or calcareous soils. These last are typical in areas with a semiarid climate. Pisolites also exhibit the typical concentric structure that indicates precipitation from waters rich in calcium carbonate.

DOLOMITE

ESSENTIAL COMPONENTS: dolomite, calcite.

ACCESSORY COMPONENTS: hematite, pyrite, autogenous quartz and orthoclase.

APPEARANCE: compact homogeneous rock with a generally coarse grain; the cavities are commonly impressions left by dissolved shells, and the walls are lined with small crystals of dolomite; fossil remains are rarely found.

COLOR: white, pink, tawny, and various shades thereof.

MODE OF OCCURRENCE: in accumulations that vary from stratified to massive.

GEOLOGIC ENVIRONMENT: common rock, easily found in association with other carbonate sediments. The Dolomite Alps are famous; their name is derived from the French naturalist Dolomien, who first described dolomite.

USES: In the building industry for the preparation of a variety of cements and gravels; for the extraction of magnesium; in the chemical industry and as a flux for working iron.

NOTES: in these rocks, a part of the calcium from the calcite or aragonite from preexistent calcareous sediments is replaced by magnesium, altering the calcium carbonate to calcium magnesium carbonate (dolomite). This process occurs in a diagenetic environment and, therefore, dolomites are only found in ancient geological formations.

SELENITE GYPSUM

 ...

ESSENTIAL COMPONENTS: gypsum (selenite variety).

ACCESSORY COMPONENTS: anhydrite, other halogen salts, calcite, dolomite, clay minerals, limonite.

APPEARANCE: coarse-grained rock with large crystals (selenite gypsum); massive structure; fossil remains are generally absent.

COLOR: white, if pure; gray from the presence of clay minerals; brown, black from the presence of hydrocarbons.

MODE OF OCCURRENCE: in stratiform masses; intercalated with argillaceous sediments.

GEOLOGIC ENVIRONMENT: common rock, found in evaporite successions.

USES: basic material in the paper industry; in the rubber industry; as a cementing material.

NOTES: Of chemical origin, selenite gypsum belongs to the evaporite rock group. The name is used to indicate either the mineral variety or the rock made out of it. The word is derived from the Greek for "moon," probably in reference to the pale color of the crystalline faces which recalls that of the moon. Selenite gypsum can precipitate directly from sea water as small crystals that can grow fairly quickly, or it is formed by replacement (pseudomorphism) of tabular crystals of anhydrite.

GYPSUM ALABASTER

ESSENTIAL COMPONENTS: microcrystalline gypsum.

ACCESSORY COMPONENTS: anhydrite, calcite, dolomite, clay minerals, limonite.

APPEARANCE: compact variety generally with translucent, waxy appearance.

COLOR: white, reddish, yellowish, brown.

MODE OF OCCURRENCE: in roundish masses in non-selenitic gypsum strata.

GEOLOGIC ENVIRONMENT: fairly rare; very beautiful material comes from Italy, Pakistan, Iran, etc.

USES: easily worked; cut into polished slabs and for decorative objects; particularly well-known is the gypsum alabaster of Volterra, Italy, where the work with this stone makes up a large part of the local economic activity.

NOTES: A rock of chemical origin, gypsum alabaster belongs to the evaporite rock group. Gypsum very commonly occurs in strata with a saccharoidal appearance, veined with gray from the presence of clay. Sometimes microcrystalline spheroids can be found inside it that are probably residue of the original gypsum (from the first precipitation) that has not recrystallized. Objects created with this material are very delicate, both because of gypsum's low durability (2 on the Mohs scale), and because it is easily soluble in water. Therefore, they are very easily furrowed and can lose their luster when wet.

▼ *Gypsum alabaster (ca. x 0.7), Volterra, Pisa, Italy.*

▲ *Pisolitic limestone (ca. x 1.5), Germany.*

▲ *Selenite gypsum (ca. x 0.5), Bolognese Apennines, Italy.*

◀ *Dolomite (ca. x 1), Arona, Novara, Italy.*

POLISHABLE LIMESTONES

In general terms, these rocks are defined as limestones or carbonates, but because they can be cut into slabs and polished, they are improperly called "marbles" in the world of ornamental stone materials. Because of their beauty of form and color these rocks can be quite valuable.

Violet breccia (Italy).

"Fior di bosco" (Italy).

Leopardo Salomé (Turkey).

"Breccia pernice" (partridge breccia), Italy.

Collemond red (the United States).

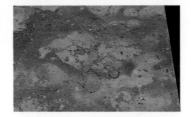

Verona red (Italy).

In the past, materials were not transported far from their sources. With the exception of some marbles such as statuary limestones, materials were usually employed for local use. Among the limestones that have always been used, portoro is one of the most famous. Like statuary marble it is only found in Italy, between Liguria and Tuscany and in the outskirts of La Spezia. It is a splendid example of a black microcrystalline carbonate rock, cut by veins of yellow calcite with some impurities from iron oxides. In some sections of the formation, the veins are white and the stone is not as valuable. Portoro is a limestone formed in a shallow sea environment by sedimentation of large quantities of carbonate mud, which, because of a momentary lack of oxygen that occurred during the Cretaceous Period, is rich in undecomposed organic matter. In various parts of the world during the Cretaceous Period, environments lacking in oxygen (anoxic) facilitated the formation of black limestones similar to portoro. Some of these are used, like portoro, to make polished slabs: in Spain, "Marquiña Black" and in France, "St. Laurent." Portoro has not yet been found in any

other part of the world. This rock can be considered to be among the most ancient Italian "marbles," since the Romans, and perhaps others before them, excavated it along with the Apuan whites. Today it has become remarkably rare. Probably the very last deposit, and not a very large one at that, was discovered a few years ago. It is a rock of very great value although, like all the carbonate rocks, it is not very durable.

THE ARABESQUES

Other rocks of remarkable beauty and with pleasing aesthetic and chromatic effects are the breccias. These may be of sedimentary or tectonic origin and may contain fragments of any kind of rock. Carbonates breccias are often cut and polished. An enormous variety of shapes and colors exist, both of fragments and cement, and they may

Portoro (Italy).

or may not be metamorphic. The breccias with white or gray elements immersed in an iron carbonate cement that ranges from white, to yellow, to red, to brown and black are very common. The names of breccias can be fanciful or tied to the color.

Violet breccia (Italy) has white elements and a reddish cement rich in iron oxides, that almost never achieves the color for which it is named. "Fior di bosco" (wood blossom) (Italy) and Leopardo Salomé (Turkey) are two typical breccias with gray elements, characterized by the variety of the shapes, and range of colors of the cement. "Breccia pernice" (partridge breccia) (Italy) has yellow fragments and white cement. Collemand red (the United States) and Verona red (Italy) have more or less intense brown-red elements and cement. Some commonly used non-carbonate breccias are the ophicalcites, made up of ophiolitic elements, bound with a white calcite cement like the "verde di St. Denis" (St. Denis green) (Italy). These rocks are commercially called "green marbles," because of their predominant color, without any distinction between breccias and other rocks. An intense red color characterizes the ele-

ments in "rosso antico" (Italy) as well as Rosso di Lepanto (Lepanto red) (Turkey) whose red fragments are jasper.

All ornamental polishable rocks must appear compact and nonporous. Travertine, having many empty spaces, is generally used for interior and exterior facings; rarely for floorings. In recent times, the empty spaces have been stopped up with preparations containing white cement, but the treatment has not improved its aesthetic properties. Only lately, a new methodology has been applied, using synthetic resins that can be polished like the rock. The transparent resins produce a pleasant sense of contrast and, overall, the stone is improved. As with all soft stones, the polish can be lost in areas of continuous traffic.

Verde di St. Denis (St. Denis green) (Italy).

Rosso antico (Italy).

Rosso di Lepanto (Lepanto red) (Turkey).

ANHYDRITE

ESSENTIAL COMPONENTS: anhydrite.

ACCESSORY COMPONENTS: gypsum, calcite, dolomite, clay minerals, bituminous substances, limonite.

APPEARANCE: medium-fine-grained; sedimentary structures are not very evident; fossil remains are absent.

COLOR: light gray, bluish; rarely yellowish, reddish, brown-black.

MODE OF OCCURRENCE: in stratiform masses; generally associated with gypsum and limestones.

GEOLOGIC ENVIRONMENT: fairly widespread rock; large deposits are found in Germany, Poland, the United States, etc.

USES: used in the paper industry; in the preparation of fertilizers.

NOTES: A rock of evaporative chemical origin; the term anhydrite is used to indicate both the mineral and the rock composed of it. It is formed in a lagoon or, more rarely in a lacustrine environment in areas with a hot, humid climate, through the precipitation of waters saturated with calcium sulfate. It is fairly rare on the surface as it hydrates into gypsum in the presence of water. This may even occur in the subsoil with the help of circulating waters. For this reason, excavations in tunnels can be very dangerous, since anhydrite, when changing into gypsum, swells up and the increase in volume causes overpressurization and, therefore, instability in the walls of the tunnel.

▼ *Anhydrite (ca. x 1), Maghreb.*

ROCK SALT

ESSENTIAL COMPONENTS: halite.

ACCESSORY COMPONENTS: gypsum, anhydrite, various halogen salts, clay minerals, bituminous substances, limonite.

APPEARANCE: medium-grained; sedimentary structures are rarely evident; fossil remains are very rare.

COLOR: generally light, gray, brown.

MODE OF OCCURRENCE: in masses that generally are not stratiform; commonly in the shape of a dome (diapirs).

GEOLOGIC ENVIRONMENT: not very widespread rock; because of its high solubility, it only outcrops in areas with an arid climate.

USES: utilized in the chemical industry; it has been an essential component of the human diet since antiquity; it may also be used to preserve foods.

NOTES: A rock of chemical origin; the term rock salt is sometimes used to indicate the mineral (halite) as well as the rock made out of it. It is an evaporite rock, formed in a lagoon environment in arid regions, through the precipitation of waters saturated with sodium chloride. Rock salt precipitates after calcite, anhydrite, and gypsum.

▼ *Rock salt (ca. x 0.6), Sicily, Italy.*

DIATOMITE

ESSENTIAL COMPONENTS: residue of diatoms (one-celled algae), sometimes with secondary siliceous cement.

ACCESSORY COMPONENTS: opal, sulfides.

APPEARANCE: friable rock with a very fine grain; light and porous; thickly laminated; sometimes compact from secondary cementation.

COLOR: whitish, yellowish, gray.

MODE OF OCCURRENCE: in banks and layers of moderate thickness intercalated with clays.

GEOLOGIC ENVIRONMENT: diatomaceous muds are found in the North Pacific, near Antarctica, and in small fresh water lakes in volcanic areas.

USES: because of its high absorbtion, it is used in the purification and bleaching of liquids; as a sound and thermal insulator.

NOTES: A siliceous rock of organic nature; diatomite is formed by the accumulation of diatoms in marine or lacustrine environments. Diatoms first appeared in the Triassic Period. They synthesize silica dissolved in the water in order to build their shells. They live on the surface, in clear waters rich in nutritive substances.

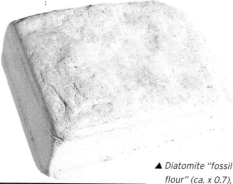

▲ *Diatomite "fossil flour" (ca. x 0.7), Grosseto, Italy.*

RADIOLARITE

ESSENTIAL COMPONENTS: shells and fragments of radiolarians (one-celled marine organisms); siliceous cement.

ACCESSORY COMPONENTS: carbonaceous substances, iron oxides, anatase.

APPEARANCE: very compact rock with a very fine grain, extremely hard, with conchoidal fracture; radiolarians can be observed only with the aid of a microscope.

COLOR: red, black, green.

MODE OF OCCURRENCE: in thin strata intercalated with argillites and associated with ophiolitic rocks.

GEOLOGIC ENVIRONMENT: fairly common rocks in mountain chains; radiolarian muds are present in the equatorial zones in the Pacific and Indian Oceans at depths greater than 131,233 ft. (4000 km).

USES: radiolarites with a medium-fine grain may be used as whetstones.

NOTES: A siliceous rock of organic nature; radiolarians are organisms that appeared in the Cambrian Period and are still present in the oceans today. They live in the superficial part of clear waters, rich in nutritive substances, in a marine (pelagic) environment. They synthesize the silica dissolved in the water to build their shells. When they die, their shells fall to the bottom of the sea where remarkable accumulations, called "radiolarian ooze," may be formed. Through successive diagenetic transformations, the sediment lithifies to form radiolarite.

Radiolarite ▶
(ca. x 1),
Tuscany, Italy.

FLINT

ESSENTIAL COMPONENTS: microcrystalline quartz.

ACCESSORY COMPONENTS: chalcedony, calcite, clay minerals, hematite.

APPEARANCE: forms small irregular masses, slightly flattened (on the order of a few inches); very hard, with a microcrystalline texture and with a conchoidal to splintered fracture.

COLOR: red, white, yellow, green, black.

MODE OF OCCURRENCE: in nodules or strips within compact rocks (for the most part calcareous ones).

GEOLOGIC ENVIRONMENT: found in carbonate rocks, commonly in a pelagic environment (deep sea).

USES: no practical use at present.

NOTES: The origin of flint nodules is similar to that of the stratiform flints (radiolarites). The difference consists in the quantity of silica of biological origin; when the remains of siliceous organisms are few and are scattered in the carbonate cement, during diagenesis the silica is chemically remobilized and concentrated in the form of nodules or strips of flint. This material has played an important role in the history of humankind: in the Stone Age, it was the most sought-after material for making arrow heads, scrapers, and other objects. It has been used for a long time as a lighter because of the ease with which it produces sparks when struck.

JASPER

ESSENTIAL COMPONENTS: quartz, chalcedony.

ACCESSORY COMPONENTS: hematite, pyrolusite, clay minerals, calcite rarely, anatase.

APPEARANCE: very compact rock with a conchoidal or splintery fracture; the radiolarians that are present are invisible without the aid of a lens.

COLOR: white, red, brown, gray.

MODE OF OCCURRENCE: very thin to massive strata.

GEOLOGIC ENVIRONMENT: in some mountain chains; associated with ophiolites, sometimes with phosphorites.

USES: in the past, it was utilized for ornamental objects and as a gravel for roads. However, because the fragments are excessively sharp, this practice was abandoned.

NOTES: A siliceous rock of organic nature, jasper is formed by an accumulation of organisms with a siliceous skeleton, essentially radiolarians, in a deep-sea environment rich in the dissolved silica necessary for these organisms. Probably, these environments are localized near submarine volcanoes whose eruptions enrich the circulating waters with silica.

BOG IRON ORE

ESSENTIAL COMPONENTS: goethite, siderite, hematite, limonite, chamosite (iron-rich chlorite), carbonate or limonitic cement.

ACCESSORY COMPONENTS: phosphates, flint.

APPEARANCE: fine-grained, fairly porous rock, sometimes with oolitic texture and fossil remains of plant matter and animals; ferriferous oolites (spheres with a concentric concretionary structure) are commonly squashed and made up of chamosite, sometimes altered to siderite or goethite and cemented.

COLOR: reddish (hematite), yellow, black.

MODE OF OCCURRENCE: in masses of limited expanse.

GEOLOGIC ENVIRONMENT: Current ferriferous sediments are formed as an accumulation of oolites and ferriferous pisolites in shallow lacustrine environments or in superficial strata rich in iron oxide. Earthy or pisolitic sediments in boggy areas result from the action of bacteria who use it to build their shells.

USES: formerly used for the extraction of iron.

NOTES: A rock of biochemical origin, the term bog iron ore applies to sedimentary rocks with iron oxides and iron minerals in a quantity greater than 15%. They are probably formed by the precipitation of ferric iron from acidic waters, rich in organic matter and with a low rate of clastic or carbonate sedimentation.

MANGANESE NODULE

ESSENTIAL COMPONENTS: manganese and iron oxides.

ACCESSORY COMPONENTS: clay minerals, quartz, calcite, copper oxides, zinc, nickel, cobalt, chromium, vanadium.

APPEARANCE: roundish shape; commonly incrusted with organisms such as foraminifera and bryozoa.

COLOR: dark gray, black.

MODE OF OCCURRENCE: in nodules and incrustations in the depths of modern oceans; in low concentrations corresponding to stratigraphic lacunae (episodes of interruption in sedimentation); in some ancient sediments cropping up on the continents.

GEOLOGIC ENVIRONMENT: concentrated deep in the Atlantic, Pacific, and Indian Oceans at depths of 6,562–16,404 ft. (2–5 km); in some areas of Mediterranean mountain chains.

USES: rocks of high economic value because of their high concentration of metals; at the moment, they are not used because current technology is not in a position to make their recovery feasible for industrial purposes.

NOTES: The process of formation of these rocks is not entirely clear. It is thought that it may be closely linked to submarine volcanic activity or to micrometeorites that regularly fall on the earth's surface. One of the few things known about their formation is their slow rate of growth (1 mm every million years). Therefore, they are found on sea bottoms where the sedimentary contribution is practically nil.

▼ *Bog iron ore (ca. x 1), Lorraine, France.*

◄ *Flint in limestone (ca. x 1), Gavarno, Bergamo, Italy.*

▼ *Manganese nodule (ca. x 1.5), Pacific Ocean.*

▼ *Jasper (ca. x 1), Sardinia, Italy.*

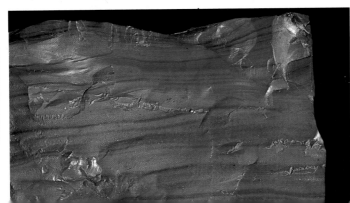

REEF LIMESTONES

We have all seen pictures, of the great coral barrier reefs. It is easy to be fascinated by this variegated animal and plant world with its extraordinary colors. But how many times have we asked ourselves about the size of this world or about the process of its formation?

It is not easy to comprehend the perfect equilibrium between the animate and inanimate worlds that nature has created in the coral barrier reefs.

Coral barrier reefs can cover enormous surfaces, on the order of thousands of square miles. The Great Barrier Reef in Australia extends along the northeastern edge of the continent for over 1200 miles. The barrier reefs that surround the islands of the Bahamas extend over a surface of about 60,000 miles. Other much smaller environments of this kind exist, like the atolls in the Pacific Ocean. However, all reefs have a formational environment in common, characterized by climatic conditions that permit the proliferation of such multicolored forms.

The main "builders" are small colonial polyps. In order to develop their framework, they need constant temperatures averaging around 71–75° F (22–24° C) and fairly shallow, well-oxygenated water which is continually in motion and rich in food and dissolved salts. Of these last requirements, dissolved calcium carbonate is indispensible since it is the principal component of the coral.

Drawing the Ca^{+2} and HCO^{3-} ions from the water, the polyps are able to synthesize the calcium carbonate ($CaCO_3$) necessary for the growth of their structures. The frameworks initially need a supporting foundation, which is commonly furnished by older corals.

How can these barrier reefs grow while always remaining at the same height with respect to sea level over the course of long geological periods? A concomitance of two growth systems probably occurs; one horizontal towards the open sea, the other vertical, both linked to subsidence. This phenomenon consists in the very slow, more or less continual subsidence of the ocean floor caused by the cooling of the lithosphere.

When there is a lull in subsidence, horizontal growth usually predominates, and when subsidence is marked, coral reefs grow upward. In the case of atolls, the coral barrier reef was probably based on preexisting islands, generally volcanic ones, surrounding

them like a ring. With the passage of time, the volcano sank below the surface of the sea, and the barrier reef grew vertically and expanded laterally until it finally covered up the volcano entirely. In this way, the initial ring-

shaped geometry is preserved and an interior lagoon is formed, characterized by a low energy marine environment. The speed with which the volcano subsides is sufficiently slow that it is balanced by the simultaneous growth of the coral structures.

A coral reef can die following climatic or geological changes. The former are easily deduced. The latter can be traced to the geodynamic evolution of the earth's crust. A rise in sea level (eustatic changes) or a lowering of the

foundation at a speed greater than the coral's potential for growth can be catastrophic for coral reefs.

Pollution caused by humans can also damage the extremely delicate environment of the coral barrier reef. For example, in the Bahamas, an area

Above: a group of common corals among which a semi-spherical mass of "cerebroidi" corals stands out. Below: oriented growth of elk's horn coral (the prevailing wave action is from the left).

with heavy tourist traffic, the coral habitat has been significantly jeopardized.

PHOSPHORITE

ESSENTIAL COMPONENTS: microcrystalline or amorphous apatite (collophane), carbonate or siliceous cement (microcrystalline quartz).

ACCESSORY COMPONENTS: limonite, bitumen, sometimes various kinds of fossil remains (mollusks, algae), phosphates, various arsenates.

APPEARANCE: porous, finely granular rock; commonly occurs with an oolitic texture, sometimes with cavities.

COLOR: yellowish, reddish, brownish; bluish white when altered.

MODE OF OCCURRENCE: rarely in stratified masses intercalated with sedimentary rocks; common in nodules or crusts.

GEOLOGIC ENVIRONMENT: large masses are found in North Africa, the United States, France, and China. Current phosphoritic deposits protrude off the coast of California and in Peru.

USES: utilized in the fertilizer industry.

NOTES: A sedimentary rock of organic nature; phosphorite is formed in a low-depth 197-984 ft. (60-300 m) marine environment particularly rich in nutritive substances and with a low rate of clastic sedimentation. Phosphorus is synthesized by plankton algae and is subsequently concentrated in the skeletons and excrement of the fish, birds, and crustacea that feed on it. When subjected to diagenetic processes, phosphatic sediments with this kind of origin are changed into phosphorites.

▼ *Phosphorite (ca. x 0.4), Morocco.*

PEAT

ESSENTIAL COMPONENTS: plant remains (mosses, sedges, heathers, willows, birches, pines) with a low carbon content (50-60%).

ACCESSORY COMPONENTS: quartz, calcite, animal remains, clay minerals.

APPEARANCE: slightly compact and porous; quite obvious felted texture in which plant remains can still be identified.

COLOR: chestnut-brown, brownish, black when soaked.

MODE OF OCCURRENCE: stratiform; alternating with argillaceous strata.

GEOLOGIC ENVIRONMENT: common in various areas of Western Europe (England, Germany, Poland, Denmark, etc.), the United States, and Canada.

USES: as a fuel with a low caloric capacity (4000 kcal), when placed in a furnace, it has a yield similar to that of lignite; mixed with soil, it is used as a fertilizer for plants.

NOTES: Peat belongs to the fossil coal group and, in particular, it is the initial product of the carbonization of plant remains. They are recent deposits, typical of the Quaternary Period; and are formed by an accumulation of plant matter in oxygen-poor environments (anoxic, euxinic), such as ponds, marshes, and narrow marine basins. In these environmental conditions, organic substances can be preserved because they are decomposed very slowly through the action of bacteria. Current deposits, peat bogs, are comparable to large sponges (they can contain up to 90% water). This amount can be reduced to 40% of its weight when the peat is dried. Peat bogs from the Quarternary Period are of interest for the environmental and climatic information furnished by the fossil species and pollens contained in them.

▼ *Peat (ca. x 1), Leffe, Bergamo, Italy.*

LIGNITE

ESSENTIAL COMPONENTS: plant remains enriched with carbon (60-70%).

ACCESSORY COMPONENTS: siderite, pyrite, calcite, clay minerals, phosphates.

APPEARANCE: different varieties exist that are recognizable at the macroscopic level: earthy lignite (homogeneous, friable, with fragments of plant matter); xyloid lignite (more compact, but similar); pitchy lignite (very compact with an opaque luster and conchoidal fracture).

COLOR: generally dark.

MODE OF OCCURRENCE: the compact varieties are generally stratiform.

GEOLOGIC ENVIRONMENT: very widespread; in particular in sediments from the Tertiary Period.

USES: as a fuel with a fairly low caloric capacity (5,000-6,000 kcal).

NOTES: Lignite belongs to the fossil coal group; it is formed by carbonization of plant remains of various kinds in oxygen-poor environments. At present, about 1,000 varieties of plants have been identified in coal deposits. It is distinguished from peat by its greater carbon content and a lower amount of water (it contains about 75% water).

▼ *Lignite (ca. x 0.9), Tuscany, Italy.*

ANTHRACITE

ESSENTIAL COMPONENTS: carbon (93-96%), with a low weight % of volatile elements.

ACCESSORY COMPONENTS: pyrite, quartz.

APPEARANCE: very compact and homogeneous rock with an almost metallic luster and conchoidal fracture; sometimes with opaque zones formed of scaly aggregates intersected by translucent veins (glassy).

COLOR: black, gray with yellow reflections.

MODE OF OCCURRENCE: in stratiform masses associated with sandstones and clays; commonly in blocks from the presence of many fractures, transverse to the stratification.

GEOLOGIC ENVIRONMENT: exclusively in Paleozoic rocks, especially from the Carboniferous and Permian Periods; large deposits are found in the United States, Europe, China, and Australia.

USES: high-quality fuel with a caloric capacity of 7,000-8,000 kcal; used in the iron and steel industry; in the dyestuff and synthetic rubber industries.

NOTES: Anthracite belongs to the fossil coal, group and it is the most advanced product of the carbonization of plant remains. It is a typical, very ancient coal in the rocks of the Carboniferous Period. Fossil coal with a carbon content intermediate between lignite and anthracite is called bituminous coal.

▼ *Anthracite (ca. x 1), Wales, Great Britain.*

CLASSIFICATION OF THE MINERALS, GEMS AND ROCKS

As an aid in consulting the classification schemes, classes are printed in bold-face upper-case letters; subclasses in plain upper-case; groups and series in italic lower-case; species in Roman lower-case.

MINERALS

Minerals described in the individual entries are classified below in subdivisions modified from those proposed by Hugo Strunz in 1970.

NATIVE ELEMENTS

Gold group:
- Gold
- Silver
- Copper

Platinum group
- Platinum

Mercury group:
- Mercury

Arsenic group:
- Arsenic
- Antimony

Sulfur group:
- Sulfur

Carbon group:
- Diamond
- Graphite

SULFIDES

Chalcocite
Bornite
Argentite (Acanthite)
Sphalerite
Chalcopyrite
Wurtzite
Enargite
Millerite
Galena
Cinnabar
Niccolite
Pyrrhotite
Covellite
Pyrite
Hauerite
Cobaltite
Marcasite
Arsenopyrite
Glaucodot
Molybdenite
Sylvanite
Skutterudite
Stibnite (Antimonite)
Bismuthinite
Orpiment
Realgar
Kermesite

Sulfosalts:
- Proustite
- Pyrargyrite
- Bournonite
- Tetrahedrite
- Geocronite
- Polybasite
- Jamesonite

HALIDES

Halite (Rock Salt)

Sylvite
Fluorite
Carnallite
Cryolite

Hydroxyhalides:
- Atacamite

OXIDES

Cuprite
Tenorite
Spinel
Chromite
Magnetite
Chrysoberyl
Corundum
Hematite
Ilmenite
Arsenolite
Rutile
Cassiterite
Stibiconite
Pyrolusite
Uraninite
Columbite

HYDROXIDES

Brucite
Goethite

CARBONATES

Calcite group:
- Calcite
- Magnesite
- Siderite
- Rhodochrosite
- Smithsonite
- Sphaerocobaltite

Dolomite group:
- Dolomite

Aragonite group:
- Aragonite
- Strontianite
- Witherite
- Cerussite

Azurite
Malachite
Aurichalcite
Phosgenite
Leadhillite

Hydrated carbonates:
- Artinite

NITRATES

Niter

BORATES

Nesoborates:
- Sassolite

Soroborates:
- Borax
- Ulexite

Inoborates:
- Colemanite

Tectoborates:
- Boracite

SULFATES

Anhydrous:
- Anhydrite
- Celestite
- Barite
- Anglesite

Hydrous:
- Gypsum
- Polyhalite

TUNGSTATES-MOLYBDATES

Crocoite
Huebnerite
Wolframite
Scheelite
Wulfenite

PHOSPHATES-ARSENATES-VANADATES

Monazite
Apatite

Pyromorphite series:
- Pyromorphite
- Mimetite
- Vanadinite

Adamite
Olivenite
Descloizite
Lazulite
Vivianite
Erythrite
Autunite
Turquoise
Wavellite

SILICATES

NESOSILICATES

Olivine
Zircon
Sphene
Almandine
Andradite
Uvarovite
Andalusite
Kyanite
Topaz
Staurolite
Phenakite
Datolite
Sillimanite

SOROSILICATES

Hemimorphite (Calamine)
Ilvaite
Epidote
Vesuvianite
Axinite

CYCLOSILICATES

Benitoite
Beryl
Cordierite (or Dichroite)
Tourmaline
Dioptase

INOSILICATES

Pyroxenes:
- Enstatite
- Diopside
- Hedenbergite
- Augite
- Aegirine
- Jadeite
- Spodumene

Amphiboles:
- Tremolite
- Actinolite
- Hornblende
- Glaucophane
- Riebeckite
- Arfvedsonite

Pyroxenoids:
- Wollastonite
- Rhodonite

PHYLLOSILICATES

Apophyllite
Prehnite
Serpentine
Chrysotile
Chlorite

Micas:
- Muscovite
- Paragonite
- Zinnwaldite
- Biotite

Clay minerals:
- Kaolinite
- Talc
- Pyrophyllite

TECTOSILICATES

Quartz
Tridymite
Opal

Alkali feldspars:
- Sanidine
- Orthoclase
- Microcline
- Albite

Plagioclases:
- Labradorite
- Anorthite

Barium feldspars:
- Hyalophane
- Danburite

Scapolite

Feldspathoids:
- Nepheline
- Leucite
- Analcime
- Sodalite
- Hauyne
- Lazurite

Zeolites:
- Natrolite
- Mesolite
- Scolecite
- Thomsonite

- Laumontite
- Heulandite
- Stilbite
- Phillipsite
- Chabazite

GEMS

Gems treated in the transparent minerals section are classified in order of decreasing hardness

DIAMOND
CORUNDUM:
- Ruby
- Sapphire
- Padparadscha
- Yellow sapphire
- Sunflower ruby
- Pink sapphire
- Sunflower sapphire
- Opalescent sapphire

CHRYSOBERYL:
- Alexandrite
- Cymophane

TOPAZ:
- Imperial topaz
- Golden topaz
- Sherry topaz
- Madera
- Blue topaz
- Pink topaz
- Colorless topaz

SPINEL:
- Blue spinel
- Nobel spinel
- Balas
- Picotite
- Pleonast
- Rubicelle
- Gahnite

PHENAKITE:
BERYL:
- Emerald
- Aquamarine
- Golden beryl
- Heliodor
- Morganite
- Bixbyite
- Goshenite

ZIRCON:
- Colorless zircon
- Starlite
- Hyacinth
- Jargon
- Malacon

UVAROVITE
ANDALUSITE
CORDIERITE
TOURMALINE:
- Rubellite
- Indicolite
- Achroite
- Dravite
- Schorl
- Elbaite
- Green tourmaline

DANBURITE
SILLIMANITE
QUARTZ:
- Amethyst
- Hyalite
- Citrine
- Pink quartz
- Smoky quartz
- Prase
- Cat's eye
- Hawk's eye
- Tiger's eye

ALMANDINE
OLIVINE:
- Chrysolite

SPODUMENE:
- Hiddenite
- Kunzite

VESUVIANITE:
- Californite

KYANITE:
- Sappare

JADEITE
BENITOITE
ANDRADITE:
- Topazolite
- Demantoid
- Melanite
- Schorlomite

ORTHOCLASE
MICROCLINE:
- Amazonite

LABRADORITE
ANORTHITE
OPAL:
- Precious opal
- Fire opal
- Black opal
- Harlequin opal
- Girasol opal
- Sard
- Cornelian
- Onyx
- Agate
- Chrysoprase

RHODONITE
EPIDOTE
GARNETS:
- Essonite
- Tsavorite
- Garnet jade (Transvaal jade)

LAZURITE:
- Lapis lazuli

SCAPOLITE
SODALITE
LAZULITE
TURQUOISE
DATOLITE
SPHENE
APATITE:
- Asparagus stone

MALACHITE
AZURITE
RHODOCHROSITE
ARAGONITE:
- Pearls

IGNEOUS ROCKS

Igneous rocks described in the individual entries are classified according to the scheme of Streckeisen; nomenclature according to Le Maître.

INTRUSIVE IGNEOUS ROCKS

Alkali feldspar granite
Granite
Granodiorite
Tonalite
Alkali feldspar syenite
Quartz syenite
Syenite
Quartz monzonite
Monzonite
Quartz monzodiorite
Monzodiorite
Quartz monzogabbro
Monzogabbro
Quartz anorthosite
Anorthosite
Quartz diorite
Diorite
Quartz gabbro
Gabbro
Feldspathoid Syenite
Shonkinite
Feldspathoid monzodiorite
Feldspathoid monzogabbro
Feldspathoid gabbro
Italite
Missourite
Dunite
Lherzolite
Pyroxenite

Hornblendite
Carbonatite

DIKE AND HYPABYSSAL ROCKS

Granophyre
Pegmatite
Aplite
Granite porphyry
Diorite porphyry

LAMPROPHYRITIC ROCKS

Minette (lampophyre)
Kersantite (lampophyre)
Spessartite (lampophyre)
Camptonite (lampophyre)
Kimberlite

EXTRUSIVE AND PYROCLASTIC IGNEOUS ROCKS

Alkali feldspar rhyolite
Rhyolite
Rhyodacite
Dacite
Alkali feldspar trachyte
Quartz trachyte
Trachyte
Quartz latite
Latite
Andesite
Basalt
Phonolite
Tephritic phonolite
Phonolitic tephrite
Tephrite
Basanite
Nephelinite
Leucitite
Melilitite
Hyaloclastite
Ignimbrite
Tuff
Peperino tuff
Obsidian
Pumice

METAMORPHIC ROCKS

Metamorphic rocks described in the individual entries are classified according to the type of metamorphism and the type of protolith.

REGIONAL METAMORPHISM

pelitic to arenaceous rocks:
- Slate
- Metagraywacke
- Phyllite
- Albite paragneiss
- Two mica schist
- Garnet mica schist
- Staurolite garnet mica schist

- Kyanite mica schist
- Two mica paragneiss
- Biotite paragneiss
- Garnet paragneiss
- Muscovite quartzite
- Quartzite
calcareous and calc-silicate rocks:
- Marble
- Cipolino marble
- Veined marble
- Bardiglio marble
- Statuary marble
- Pink marble
- Arabesque marble
- Marbled breccia
- "Calacatta"
- Calc-schist
acidic and intermediate volcanic or plutonic rocks:
- Sericite schist
- Metagranite
- Orthogneiss
- Augen gneiss
basic volcanic or volcaniclastic rocks:
- Glaucophane schist
- Prasinite (Chlorite-amphibole-epidote-albite schist)
- Epidote amphibolite
- Amphibolite
- Garnet amphibolite
- Eclogite
ultrabasic rocks:
- Serpentinite
- Serpentine schist
- Ophicalcite
- Chlorite schist
- Talc schist
- Actinolite schist
rocks of various composition:
- Granulite

ANATECTIC METAMORPHISM

rocks of various composition:
- Agmatite
- Embrechite
- Anatexite
- Nebulite

CONTACT METAMORPHISM

from carbonate rocks:
- Calciphyre
- Marble
- Cipolino marble
- Veined marble
- Bardiglio marble
- Statuary marble
- Pink marble
- Arabesque marble
- Marbled breccia
- "Calacatta"
siliceous rocks:
- Andalusite biotite hornfels
- Cordierite potassium feldspar hornfels

METASOMATIC ROCKS

Skarn (siliceous-carbonate rocks)
Rodingite (basic rocks)
DYNAMIC (HIGH STRAIN) METAMORPHISM
Mylonite (from any type of rock)

SEDIMENTARY ROCKS

Sedimentary rocks described in the individual entries are classified according to compositional and textural properties.

CLASTIC TERRIGENOUS ROCKS

Orthoconglomerate
Paraconglomerate
Breccia
Quartz arenite
Arkose
Graywacke
Glauconitic sandstone
Gypsum arenite
Siltstone
Argillite

CARBONATE ROCKS

Marl
Calcilutite
Calcarenite
Calcirudite
Biocalcilutite
Biocalcarenite
Biocalcirudite
Reef limestone
Travertine
Calcareous alabaster
Oolitic limestone
Pisolitic limestone
Dolomite

EVAPORITE ROCKS

Gypsum (selenite)
Gypsum (alabaster)
Anhydrite
Rock salt

ORGANIC AND CHEMICAL ROCKS

Diatomite
Radiolarite
Flint
Jasper
Bog iron ore
Manganese nodule
Phosphorite
Peat
Lignite
Anthracite

GLOSSARY

A

accessory minerals minerals that are present in small quantities in a rock and are not the major rock-defining phases (as opposed to essential minerals)

acicular crystals crystals with an elongated, slender, needle-like habit

acidic rock igneous rock containing more than 63% SiO_2

aggregate a mass of rock particles and/or mineral grains

alkaline or alkalic said of igneous rocks containing a high content of alkalis (sodium and potassium) relative to silica; alkali rocks may contain sodium and potassium feldspars or feldspathoids

allochromatic mineral a mineral that is found in different colors (i.e., there is no characteristic color to the mineral); the color differs as a result of elemental impurities in the crystal structure (e.g., amethyst) or of micro-inclusions of other minerals (e.g., blue quartz) (Ant., idiochromatic)

allotriomorphic anhedral

alluvial deposits or beds sediments deposited in a fluvial environment

alluvium unconsolidated clay, silt, sand, and gravel transported and deposited on land by running water

alteration a change in the mineral and chemical composition of a rock by chemical and/or physical processes (Cf., hydrothermal, pneumatolytic)

amygdule in an igneous rock, a formerly gas-filled hole or vesicle, now filled with secondary minerals such as quartz, calcite, or zeolites

anatexis the process of large-scale melting of preexistent rock; sometimes refers more specifically to melting of subducted continental (granitic) crust

angstrom unit of measurement equal to one ten-millionth of a millimeter

anhedral mineral a mineral with minimal development of its crystalline form (Syn., allotriomorphic, xenomorphic) (Ant., euhedral)

aphanitic texture texture of igneous rocks lacking crystals visible to the naked eye

aqua regia a mixture of nitric acid and hydrochloric acid in a 3 to 1 ratio

arborescent a three-dimensional branching, plant-like crystal habit (Cf., dendritic)

assortment the coexistence of a number of size classes within a sediment; poorly sorted or unsorted

asterism multi-pointed, star-shaped luminosity of a crystal which is a result of minute, regularly spaced, acicular inclusions

asthenosphere the ductile region of the mantle below the lithosphere where rocks behave plastically; lies at depths from 60 to 220 miles (100 to 350 km)

aureole the area which surrounds a magmatic intrusion and is affected by contact (thermal) metamorphism

autogenous mineral a mineral that forms within a rock after formation of the rock itself

autometasomatism replacement of minerals of magmatic origin by different minerals as a result of interactions with the residue of, or solutions derived from, the magma itself

B

bacillary crystals crystals with an elongated, prismatic habit

basement term referring to the undifferentiated mass of rock that lies beneath a series of sedimentary rocks and extends to the base of the crust

basic rock an igneous rock with low (45–52%) silica content; (e.g., basalts)

batholith large igneous intrusion the exact dimensions of which are unknown

birefringence the difference between the greatest and least indices of refraction of a non-isotropic (anisotropic) mineral

bituminous coal medium-grade coal with carbon content and heat content between lower-grade lignite and higher-grade anthracite

blast a suffix that refers to minerals formed in a metamorphic environment (e.g., granoblast, porphyroblast)

block a large solid fragment thrown from the conduit of a volcano

boiling structure see vesicular structure

boric fumarole a fumarole that emits boric acid

botryoidal refers to a globular growth of a mineral, similar to a bunch of grapes

burial metamorphism see regional metamorphism

C

cabochon a gem cut with a curved convex surface, used for opaque translucent or transparent stones of intense color

calcination the heating of a substance to the point of dissociation of its components (e.g., the calcination of $CaCO_3$ to CaO and CO_2)

calcsilicates medium- to high-grade metamorphic rocks, usually metamorphosed or metasomatized limestones, composed of calcium-bearing silicate minerals such as diopside and wollastonite

cap-rock the impermeable body of anhydrite, gypsum, and calcite which overlies the salt in a salt dome

carapace convex, upper part of a domical structure or form; the cooler, more solidified upper portion of an igneous intrusion

carat (c.) unit of weight used in gemology; equal to 0.2 grams and divided into 100 "points"

carbonization the transformation of an organic material into carbon, or the enrichment of carbon in the transformation of lignites into bituminous coal and anthracite

cataclasite a product of the brittle, mechanical fragmentation of rocks during the movement of large rock masses (tectonism)

cataclastic structure or texture texture typical of cataclasites: rocks with granular, fragmented, deformed, and strained mineral grains commonly flattened perpendicular to stress

cement a secondary mineral, usually a chemical precipitate that binds together the individual components of a sedimentary rock; commonly silica, carbonates, or iron oxides

chatoyancy an iridescent brightness similar to that of a cat's eye, caused by the presence of similarly-oriented, threadlike inclusions within a mineral

chemistry chemical composition of a rock or mineral

chimney a conduit through which magma reaches the surface

CIPW normative composition (or norm) hypothetical mineralogy of an igneous rock calculated from its bulk chemistry; used to classify and compare fine-grained igneous rocks (as opposed to mode)

class see symmetry class

clastic texture texture of sedimentary rocks composed of clasts

clasts fragments of rocks or minerals resulting from the mechanical breakdown of another rock

cleavage planes planar surfaces along which fracturing of a crystal tends to occur

colloid a suspension of extremely fine particles in solution

columnar crystals crystals with an elongated, prismatic, column-like habit

conchoidal fracture a break forming a curved surface similar in appearance to a shell (i.e., typical of glass)

concretion microcrystalline aggregates of minerals that have grown around a nucleus such as a piece of shell, bone, or mineral grain

conductor electrical: a body that allows the passage of electric current; or thermal: a body that allows the passage of heat; (Ant., insulator)

contact rocks rocks in contact with, or very near, an igneous intrusion that have been metamorphosed by the heat of the intrusion

continental shield a large region of stable continental basement (usually pre-Cambrian) of a tectonically stable "craton"

crust outermost, least dense layer of the earth; ranges in thickness from about 6 miles (10 km) in ocean basins to about 40 miles (60 km) at high mountain ranges

cryptocrystalline texture texture of igneous rocks whose crystals are too small to be distinguished except under very high magnification (such as through an electron microscope)

crystal lattice a three-dimensional, regularly arranged framework of points that describes the repetition of the crystal structure

crystal a solid, homogeneous body whose atoms are arranged in an orderly and regular fashion outwardly defined by planar faces

cumulate rocks igneous rocks characterized by granular mineral grains (cumulus minerals) that appear to have settled out or accumulated on the magma chamber floor, surrounded by anhedral interstitial (inter-cumulus) minerals

cut operation that enhances the aesthetic characteristics of a mineral or rock in order to transform it into an ornamental stone

D

decrepitation the breaking up of minerals during heating as a result of volatile release often accompanied by cracking sounds

dendritic a two dimensional branching, plant-like crystal habit (Cf., arborescent)

density mass of a body per unit of volume (usually expressed in g/cm^3 or kg/m^3)

deposition the natural process of depositing or placing down rocks and mineral grains by some natural force such as wind or water, usually in a sedimentary environment; sedimentation

diablastic texture metamorphic texture defined by intergrown, acicular or fibrous minerals oriented in all directions

diagenesis refers to the processes involved in the consolidation and lithification of a sediment

diapir a domical or mushroom-shaped structure assumed by upward flowing low density material (e.g., a salt dome or a magma body)

differentiation the process of chemical evolution in a magma during the course of its crystallization

dike a vertical to sub-vertical intrusive sheet

dimorphism see polymorphism

dome general term for a cupola-shaped structure

double refraction the splitting of a beam of light into two beams of unequal velocity (Cf., birefringence)

drag fold a fold that forms adjacent to a fault plane by the dragging of strata during movement along the fault

druse a cluster of small crystals

E

electrical conductivity the property of a material that allows electricity to pass through it (Cf., conductor, insulator)

endogenous or endogenic refers to processes that occur beneath the surface of the earth

endorheic basin a basin that lacks external drainages

environment complex of physical and chemical characteristics in which a mineral or rock is formed, or in which a geological process occurs

epithermal refers to the shallow sub-surface region above an igneous intrusion, characterized by abundant hydrothermal activity

eruptive see extrusive

essential minerals minerals that characterize the composition of a rock (as opposed to accessory minerals)

euhedral crystal a mineral in its characteristic crystal form (Syn., idiomorphic; Ant., anhedral)

eustatic changes world-wide fluctuations in sea level

evaporites sedimentary rocks or minerals that precipitate from solutions through evaporation (e.g., salt)

exogenous agent a superficial or external process that contributes to modification of the earth's crust (e.g., running water, glaciers, waves, wind, frost)

exogenous or exogenic that which occurs, or has its origins, on the earth's surface (Ant., endogenic)

exsolution or unmixing separation in the solid state of two or more mineral phases from an originally homogeneous solid solution

extrusive said of a magmatic process that occurs on the surface of the earth, or said of igneous rocks that solidify on the surface of the earth (Syn., volcanic, eruptive; Ant., intrusive, plutonic)

F

feldspars a family of tectosilicates (*see* systematic classification) including plagioclases, orthoclase, sanidine, microcline, and other less common minerals

felsic said of igneous rocks consisting primarily of light-colored minerals such as feldspars, quartz, feldspathoids, and muscovite (e.g., granite); (Ant., mafic)

felted texture a weaved, felt-like texture resulting from densely intertwined crystals

femic minerals normative iron and magnesium silicate, and iron and titanium oxide minerals (Ant., salic) (Cf., CIPW norm)

ferromagnetic mineral a mineral that is strongly attracted by a magnetic field (e.g., magnetite)

fissility rock property of splitting into thin sheets along almost parallel planes (e.g., shale along bedding planes or schist along cleavage planes)

flos ferri an arborescent type of aragonite which occurs in white masses

fluidal or flow texture texture of lavas characterized by elongated crystals oriented in the direction of flow

fluorescence temporary emission of radiation by a substance absorbing radiation; a type of luminescence

fold a bend or curve in a planar structure, such as sedimentary strata or foliation

foliation planar concentration of components or structures in a rock; refers to metamorphic or igneous, rather than sedimentary layering

formula the chemical composition of a substance as expressed in symbols

fracture a break in a rock

fragility the susceptibility to breaking

fumarole a late-stage volcanic vent that emits gas and vapor

G

gangue components of a mineral deposit that are not of economic interest

gel a semisolid colloidal solution

genesis the complex of processes involved in the formation of minerals and rocks

geode a cavity in a rock, lined with well-formed crystals

geothermal fields areas on the earth's surface characterized by high heat flow

geyser a hot spring that erupts intermittently

glass amorphous material produced when magma cools so rapidly that crystals cannot form

graded bedding sedimentary deposits exhibiting a stratigraphic gradation in grain size (e.g., from coarse gravel to fine sand)

grain a small particle of a rock or mineral; also refers to the structure of a rock according to grain size or arrangement of grains (e.g., foliation)

granular said of a rock composed of visible mineral grains of almost equal size

graphic structure an intergrowth of quartz within a single orthoclase, resulting in a pattern that resembles hieroglyphic writing

greisen a hydrothermally-altered granitic rock, composed of quartz and muscovite

groundmass fine-grained, micro-crystalline, or cryptocrystalline material which surrounds the larger phenocrysts in a porphyritic rock

H

habit characteristic shape of mineral crystal (e.g., prismatic, blocky)

hardness the resistance of a substance to scratching or penetration

heavy minerals minerals with a specific gravity greater than 2.9

high grade refers to metamorphic rocks or minerals formed at high temperature and/or high pressure; includes rocks of upper amphibolite, granulite, and eclogite facies (Cf., low grade, metamorphic grade)

holocrystalline texture texture of igneous rocks formed exclusively of crystals and no glassy parts; typical of plutonic rocks (Cf., hyaline)

hot spot a place on the earth's surface characterized by persistent volcanic activity and high heat flow, thought to be the surface manifestation of a geographically stable, anomalously hot region of the mantle

hyaline texture texture of igneous rocks composed of glass or amorphous minerals (Cf., holocrystalline)

hydration the introduction of water molecules into the structure of a mineral

hydrosphere the system of all the earth's waters including oceans, seas, lakes, rivers, ice, groundwater, and atmospheric water

hydrothermal alteration alteration of rocks and minerals by hydrothermal fluids; often occurs near intrusive bodies

hydrothermal veins fractures filled with minerals precipitated from hydrothermal solutions

hydrothermal refers to a process involving the action of hot water (usually heated by an intrusion), or to minerals formed from precipitation from hot-water solutions

hypabyssal in reference to a shallow intrusion

hypidiomorphic texture texture of igneous rocks composed of euhedral grains and surrounding interstitial anhedral grains

hypocrystalline an igneous rock that has crystalline components within a glassy groundmass

I

idiochromatic mineral a mineral that is only found in one color (Ant., allochromatic)

idiomorphic see euhedral

impurity an element foreign to the typical chemical composition of a mineral

inclusion solid, liquid, or gaseous material, organic or inorganic, encased in a mineral during its growth

index mineral a mineral that only forms under a particular set of temperature and pressure conditions and, therefore, can be used to determine metamorphic grade

index of refraction the ratio between the speed of light in a void to the speed of light in the substance of interest

insulator a material or body that does not allow electric current to pass through it or that does not conduct heat

interference colors the colors exhibited by birefringent minerals under cross-polarized light (Cf., birefringence)

intergrowth interlocking mineral grains such as those which crystallized simultaneously, or which have exsolved from one other

intermediate rock an igneous rock with a silica content between 52% and 63%

intersertal texture the texture of a porphyritic rock with a glassy groundmass, occupying the space between unoriented phenocrysts

intrusive magmatic process that occurs beneath the surface of the earth; igneous rocks that solidify beneath the surface (plutonic)

ion an atom in a charged state

ionic radius the radius of an ion, generally expressed in angstroms

iridescence the exhibition of the colors of the spectrum by a mineral; a result of interference of light by thin films or by layers of different refractive index

isomorphs minerals that have a similar composition and the same crystal form (isotypic minerals)

isostructural said of chemical compounds with identical crystal forms but completely different compositions (e.g., galena and halite)

L

laccolith a concordant, convex-up, lens-shaped intrusion

lamellar crystals bidimensional or plate-like crystals (e.g., micas) that form thin layers

leaching selective removal of soluble components in a rock by circulating water

lenticular shaped like a lens

lepidoblastic texture texture of metamorphic rocks rich in lamellar minerals

leucocratic a rock composed of light colored minerals such as quartz and feldspar (Ant., melanocratic)

leucosome the light-colored, quartz and feldspar-rich part of a migmatite

lithic inclusion a rocky piece encased in a magma

lithosphere the rigid, outermost 62 miles (100 km) of the earth; includes the crust and uppermost mantle; makes up the "tectonic plates"

lopolith a concave-up, lens-shaped intrusion

low grade refers to metamorphic rocks or minerals formed at low temperature and low pressure; includes facies such as zeolite, prehnite-pumpellite (Cf., high grade, metamorphic grade)

luminescence the emission of light from a material as a result of some stress such as pressure, or as a result of its being subjected to radiation

luster the appearance of a mineral surface under reflected light (e.g., glassy, metallic, dull)

M

mafic magnesium and iron-rich minerals such as pyroxene and olivine, or rocks primarily composed of these minerals

magmatic said of a rock or mineral crystallized from a magma; a process involving magma (Syn., igneous)

magnetic susceptibility when a magnetizing field is applied, the ratio between the induced magnetization (Amps/m) of a material and the applied magnetic field (Amps/m)

mammillary aggregates with a rounded structure similar to breasts

mantle the intermediate layer of the earth which lies between the crust and the core

massive structure large-scale homogeneous texture

matrix the detrital, finer-grained intergranular portion of a rock (e.g., mud and clay in sand, or sand within a gravel)

melt or magma liquid rock, either intrusive or extrusive, from which igneous rocks originate

mesocratic rock an igneous rock containing roughly equal portions of light and dark minerals (Cf., leucocratic)

metabasite metamorphosed basic igneous rock

metalliferous seam a planar deposit of metallic minerals

metamict a mineral whose crystal structure has been destroyed by radiation emitted by radioactive atoms within the mineral

metamorphic grade refers to the pressure and temperature conditions of formation of metamorphic minerals and rocks (Cf., high grade, low grade)

metamorphic rock a rock formed by metamorphism

metamorphism the process of mineralogical and/or textural changes in any rock (sedimentary, igneous, or metamorphic) when it is subjected to changes in temperature and/or pressure; may involve changes in bulk chemistry of the rock

metasomatism a chemical change in a rock that entails the addition or loss of chemical components (e.g., hydrothermal alteration)

metastable form the form of a mineral outside the pressure and temperature conditions of its normal stability

meteorite a rock that has fallen from space onto a planetary surface

miarolitic cavities in plutonic rocks, small irregularly shaped cavities lined with minerals characteristic of the rock

microcrystalline texture texture of rocks composed of crystals visible only under a microscope

microlitic structure or texture structure or texture of a porphyritic rock whose groundmass is composed of minute crystals or microlites

micron one-thousandth of a millimeter

migmatite a rock containing a mixture or mingling of granitic and mafic metamorphic material

migmatization the process of formation of a migmatite; partial melting of a metamorphic rock

mineral deposit a naturally occurring mass of one or more usable minerals

mineral a naturally occuring inorganic substance with a definite chemical composition that occurs in a crystalline structure

mineralogical species see mineral

mineralogical variety refers to minerals of the same species which are differentiated by color or by minimal compositional variations that do not justify separation of the mineral into several species

miscibility the capacity of two minerals to mix to form one phase of intermediate composition

mode the actual percentage of a particular mineral in a rock (Cf., CIPW norm)

Mohs scale scale of relative hardness of minerals

mylonite a fine-grained rock characterized by extensive grain-size reduction via ductile deformation during tectonism (Cf., 'cataclasite' = same, but by brittle deformation)

N

nanometer (nm) unit of measurement equal to one-millionth of a millimeter or 10 angstroms

native element an element, such as copper or gold, that is found in nature in an uncombined state

nematoblastic texture texture of metamorphic rocks with homogeneously sized, prismatic or fibrous, recrystallized minerals oriented parallel to one another

neoform crystal a new crystal formed during diagenesis or metamorphism (as opposed to a mineral that was recrystallized)

neosome the newly formed granitic material in a migmatite

nodule a small, roundish lump of a mineral or aggregate of minerals (e.g., chert nodule, manganese nodule)

nugget a small lump of precious metal in a secondary deposit

O

ocellar texture texture of igneous rocks with small spheroidal "phenocrysts," which are actually aggregates of smaller crystals arranged radially or tangentially around larger euhedral crystals

oolith or ooid a spherical granule (< 2 mm in diameter) consisting of a mineral or organic nucleus surrounded by concentric layers of a mineral such as calcite or hematite

ophiolite a metamorphosed, tectonized suite of rocks consisting of cherts, gabbros, dunites, sheeted dikes; thought to be sections of oceanic lithosphere that have been obducted or thrust onto continental crust

ophitic texture texture of igneous rocks that have euhedral plagioclase crystals included in, and surrounded by, larger pyroxene crystals

orbicular structure structure characterized by concentric layers of different compositions or color

orogenesis the formation of mountains

orogenic pertaining to orogenesis

orthomagmatic stage the main stage of crystallization of a magma

oxidation the process in which oxygen combines with, or hydrogen is removed from, a substance; any reaction involving the loss of an electron from an atom (Ant., reduction)

P

paleosome the preexisting, unmelted portion of a migmatite (as opposed to neosome)

paragenesis the characteristic association of minerals in a rock

pegmatites very coarse-grained rocks usually found as dikes within, or just beyond, igneous intrusions; the large crystals are a result of crystallization of a magma rich in water which inhibited nucleation sites for crystal growth

pelagic sediment fine-grained, deep marine deposit of organic origin

pelitic rock a sedimentary rock; also refers to a metamorphic rock whose protolith was a shale, mudstone, or clay-rich sediment (e.g., pelitic schist)

penetration twins twinned crystals that have grown through each other

perlitic texture texture consisting of small, almost spherical, glassy lumps resulting from concentric cracking during rapid cooling of a volcanic rock

phenocrysts in a porphyritic igneous rock, crystals that are larger and more conspicuous than the rest of the rock (or groundmass)

phosphorescence luminescence that lasts even after cessation of exposure to radiation

piezoelectricity the ability of certain crystals (e.g., quartz) to develop an electrical potential (a positive electrical charge at one end of the crystal and negative charge at the other) when compressed, or to vibrate when an electrical potential is applied

pillow structure a globose, pillow-shaped structure of volcanic rocks, generally formed by rapid cooling of lava under water or in water-saturated sediments (also referred to as pillow basalt or pillow lava)

pilotaxitic texture texture of igneous rocks with a groundmass of lath-shaped microlites (minute crystals) intergrown in a random fashion

pisolite a nearly spherical granule similar to an ooid, but larger (>2 mm in diameter), composed of an organic or inorganic nucleus covered with concentric layers of calcium carbonate or ferrous oxides

placer a sedimentary concentration of minerals

plankton animals and plants that float (rather than propel themselves) on or near the surface of the ocean

plateau basalt a basalt flow or series of flows that extends horizontally over an extremely large area (several hundred square miles) (e.g., the Columbia River plateau basalts)

pleochroic a characteristic of a mineral that absorbs different wavelengths of transmitted light depending on its crystallographic orientation and thus shows different colors when light is transmitted along different crystallographic directions

plutonic rocks see intrusive rocks

pneumatolytic vein a vein of minerals formed by pneumatolysis (Cf., pneumatolytic)

pneumatolytic said of the process of crystallization of species transported as a gas and fluid phase, or of alteration of minerals by a gaseous solution derived from a magma during the last stages of its crystallization

poikilitic texture said of igneous rocks where large crystals contain inclusions of small crystals of a different mineral; the large crystals are called "oikocrysts"

poikiloblastic texture said of metamorphic rocks whose porphyroblasts contain inclusions of smaller crystals of a different mineral

polarized light light that has passed through a material (a polarizer) in such a way that its vibrations are restricted to a single plane

polarizer a lens or prism that restricts transmission of light waves to those in a single plane

polymorphic said of a chemical substance that is found in more than one crystalline form or symmetry class; i.e., it forms more than one mineral (e.g., $CaCO_3$ is hexagonal as calcite and orthorhombic as aragonite)

polysynthetic twinning multiple twinning (of three or more individuals) on parallel planes

polyptism a type of polymorphism common in micas in which the two polymorphs differ only in the stacking of identical two-dimensional sheets

porphyritic texture said of an igneous rock that has both large crystals (phenocrysts) and a finer-grained groundmass

porphyroblast a relatively large crystal among smaller crystals in a metamorphic rock

porphyry copper a large, low-grade deposit of copper disseminated in a hypabyssal porphyry

precipitation formation of an insoluble compound out of solution

primary said of a deposit containing minerals crystallized *in situ*

pseudomorph a replacement mineral that retains the shape of the mineral it has replaced (e.g., limonite pseudomorphs of pyrite)

purity absence of fluid or gaseous inclusions or of other minerals in the interior of a crystalline structure

pyroclastic rocks aggregates formed and deposited as a result of explosive volcanic eruptions

pyroelectricity the development of an electrical potential in a crystal as a result of changes in temperature

R

radial fibrous texture texture consisting of oblong fibrous crystals converging on a central point

radioactivity the emission of energetic alpha and beta particles and gamma rays during the spontaneous disintegration of an atom

recrystallization the growth of new mineral grains in the solid state from old mineral grains, during diagenesis or metamorphism

reducing environment an oxygen-poor environment

reducing flame a flame that is capable of reducing, or extracting oxygen from, metals

reflectance the ability of an opaque mineral to reflect an incidental beam of light

refraction the deviation of a beam of light from a straight path as it changes velocity upon passing from one medium into another

regional metamorphism metamorphism that affects vast areas of the earth's crust; associated with orogenesis

regression retreat of the sea from the land (Ant., transgression)

relict crystal, mineral, or structure premetamorphic mineral, fabric, or structure present within a metamorphic rock

reorientation a new arrangement of the structures and textures in a rock as a result of deformation and metamorphism

residual liquid last fluid part of a magma before complete solidification

residual mineral a mineral that has resisted alteration

retrograde metamorpism metamorphism that results when rocks are subjected to lower pressures and temperatures (as opposed to prograde metamorphism)

rift a depressed crustal lineament bound by inward-facing faults, commonly a focus of volcanic and seismic activity; a manifestation of large-scale extension in the crust (e.g., the Rio Grande Rift)

rock a natural aggregate of minerals or clasts in either a lithoid (e.g., granite) or unconsolidated (e.g., sand) state

rotary polarization the ability of some crystals that lack a center of symmetry to cause the plane of vibration of polarized light to rotate

S

saccharoidal appearance the look of marble that is composed of minute, equidimensional crystals which resemble sugar

saturated rock an igneous rock with quartz crystals (saturated with respect to silica) (Cf., undersaturated, supersaturated)

saturated solution a solution in which the concentration of a dissolved component has reached the limit of solubility, so that further additions of the component will not dissolve

schistosity the characteristic foliation in a schist, resulting from the parallel arrangement and planar orientation of platy minerals such as mica

secondary deposit an accumulation of selectively transported, strong, hard minerals

secondary minerals minerals formed by alteration of preexisting, primary minerals

sectility the ability of a rock or mineral to be cut with a knife

sedimentary rocks rocks that form from the deposition and consolidation of fragments of other rocks or organic material, or that form through precipitation of minerals from solution *sedimentation* refers to the general processes of initial erosion, transport, deposition (or precipitation), and lithification of mineral grains and rock fragments in the formation of sedimentary rock

segregation separation of crystals from a magma in the early stages of crystallization

sericitization the replacement of a mineral with sericite mica during deuteric (by groundwater or surface water) or hydrothermal alteration or metamorphism

serpentinization hydrothermal alteration of olivines and pyroxenes into serpentine minerals, transforming basalt into serpentinite

sialic minerals minerals rich in silica and alumina; minerals common in the continental crust

silicification the replacement of rocks, wood, and other organic tissue by silica

silky silk-like luster characteristic of interwoven fibrous minerals

sill an intrusive sheet emplaced concordant to the intruded layers

single refraction refraction of light in an isotropic crystal; in isotropic minerals, the speed of light is the same in all directions (Cf., double refraction, birefringence)

sole mark a small irregularity found on the underside of a coarser-grained sediment in contact with a finer-grained sediment, a result of the overlying material filling in cracks or depressions on the surface of the underlying material; used as a directional indicator

solfatara a type of fumarole that emits sulfurous gas

solid solution a single crystalline phase that varies in composition as a result of ionic substitution

sparry said of a mineral that resembles spar; any transparent or translucent, relatively lustrous, easily cleavable mineral

specific gravity the dimensionless ratio of the mass of a substance to the mass of an equal volume of water; equivalent to density in g/cm^3

spherulitic texture texture of rocks composed of spherulites; spherical granules with a radialfibrous structure

stalactite a conical or cylindrical carbonate deposit hanging from the ceiling of a cave

stalagmite conical or cylindrical carbonate deposit that develops upwards from the floor of a cave

star *see* asterism

stock an igneous intrusion smaller than a batholith

stratification layering according to texture, grain size, composition; usually refers to sedimentary deposits

striped texture a texture resulting from thin strata of different compositions (compositional layering) or appearance (i.e., color, grain size, resistance to erosion)

structure sum of the observable characteristics of a rock

subduction plane the plane along which a tectonic plate subducts, or descends into the mantle beneath another plate

sublime to pass directly from the solid to gaseous state, and vice versa

subsidence the gradual sinking of a continental or submarine region

supersaturated rock a rock that has an excess of silica and is rich in quartz

symmetry class the association of elements of symmetry by which a group of crystals can be described

symmetry the correspondence in size, arrangement, and form of parts of a crystal structure on opposite sides of a point, line, or plane; the property exhibited by crystals whose structure is identical when rotated, reflected, translated, or inverted

synthetic stone precious stone with the characteristic hardness, optical properties, and composition of gems, but which is produced in the laboratory and has not been found in nature

system group of symmetry classes

T

tabular crystals crystals with a flattened form

tectonic plates distinct slabs of lithosphere which move as single rigid bodies over, under, past, and away from each other

tectonics the branch of geology dealing with the large-scale movements of rocks and the resulting structures

texture sum of the shape, size, and arrangement of minerals in a rock

thermoluminescence property of a substance of emitting visible light when heated to relatively low temperatures

toughness a mineral's resistance to breaking by cutting

transgression sea level rise; encroachment of the sea onto the land (Ant., regression)

transgressive conglomerates conglomerates, coarse-grained sediments deposited during transgression

transparent mineral a mineral that allows light to pass through it

trap a dark-colored hypabyssal or extrusive rock

triboluminescence luminescence produced when a mineral is rubbed, scratched, or crushed

tubes, open or closed instruments used to perform chemical analyses

turbidites sediments or rocks deposited by a turbidity current

turbidity current an underwater landslide of large masses of unconsolidated sediment along a continental slope

twin two or more crystals of the same mineral that have intergrown in the most stable configuration for that mineral

U

ultrabasic rock rocks containing less than 45% silica

ultramafic rock igneous rocks composed primarily of mafic minerals such as olivine and pyroxene

ultraviolet rays electromagnetic waves with wavelengths slightly shorter (1–400 nm) than those of visible light (400–700 nm)

undersaturated rock a rock that has a low silica content and contains no quartz (e.g., syenites)

uniaxial said of a birefringent crystal (a crystal with two refractive indices) that has only one optical axis (e.g., minerals in the tetragonal, hexagonal, and trigonal systems)

V

vacuolar texture *see* vesicular structure

vein a thin fracture filled with minerals deposited from solution; commonly, quartz or calcite

vesicular the structure or texture of a volcanic rock characterized by abundant vesicles, bubbles formed by the expansion of gases in the magma

viscosity a measure of a material's resistance to flow

volatile a gaseous component of a magma

volcanic eruptive; extrusive

X

x-rays electromagnetic waves with wavelengths slightly shorter (.0001–10 nm) than ultraviolet rays

xenolith a foreign piece of rock in an igneous rock; a piece of the intruded rock which was not part of the original magma

xenomorph anhedral; allotriomorphic

Z

zonal structure or zoning more or less concentric variations in composition of minerals (*see* zoned crystals) or large-scale (outcrop to regional) variations in mineralogy and texture of metamorphic rocks

zoned crystal a mineral crystal that varies in composition from the outside (rim) to the inside (core)

BIBLIOGRAPHY

Bates, R.L. and J.A. Jackson. *Glossary of Geology*. Falls Church, Virginia: American Geological Institute, 1980.

Bernardini, G.P., G. Carobbi, C. Cipriani, and F. Mazzi. *Trattato di mineralogia*. Vols. 1–2. Florence: USES, 1971.

Borelli, A. and N. Cipriani. *Guida al riconoscimento dei minerali*. Milan: Arnoldo Mondadori Editore, 1987.

Bosellini, A., E. Mutti, and F. Ricci Lucchi. *Rocce e successioni sedimentarie*. Turin: UTET, 1989.

Cipriani, C. and A. Borelli. *Pietre preziose*. Milan: Arnoldo Mondadori Editore, 1984.

D'Amico, C. *Le rocce metamorfiche*. Bologna: Pàtron, 1973.

D'Amico, F. Innocenti, and F.P. Sassi. *Magmatismo e metamorfismo*. Turin: UTET, 1989.

Deer, W.A., R.A. Howie, and J. Zussman. *Introduzione ai minerali che costituiscono le rocce*. Bologna: Zanichelli, 1992.

Fry, N. *The Field Description of Metamorphic Rocks*. New York: Geological Society of London Handbook Series, 1984.

Hyndman, D.W. *Petrology of Igneous and Metamorphic Rocks*. New York: McGraw-Hill Book Co., 1985.

I marmi apuani: schede merceologiche. Florence: Regione Toscana, 1980.

Kearey, P. and F.J. Vine. *Tettonica globale*. Bologna: Zanichelli, 1994.

Le Maître, R.W. *A Classification of Igneous Rocks and Glossary Terms*. Oxford: Blackwell Scientific Publications, 1989.

Minerali e rocce. *Enciclopedia Italiana delle Scienze*. Vol. 1. Novara: Istituto Geografico De Agostini, 1972.

Mottana, A., R. Crespi, and G. Liborio. *Minerali e rocce*. Milan: Arnoldo Mondadori Editore, 1987.

Perini, G. *Gemme, pietre dure e preziose*. Milan: Arnoldo Mondadori Editore, 1994.

Pettijohn, E.J. *Sedimentary Rocks*. New York: Harper & Row, 1975.

Press, F. and R. Siever. *Introduzione alle Scienze della Terra*. Bologna: Zanichelli, 1985.

Speranza, C. and M. Bignami. *Gemmologia*. Vols. 1–2. Milan: Hoepli, 1972.

Strunz, H. *Mineralogische Tabellen*. Leipzig: Akademische Verlagsgesellschaft, 1970.

Tucker, M. *The Field Description of Sedimentary Rocks*. New York: Geological Society of London Handbook Series, 1982.

Valloni, R. et al. *Proposta di classificazione macroscopica delle areniti*. Parma: L'Ateneo Parmese, 1991.

Volcano. Amsterdam: Time-Life Books, 1982.

Zampieri, D. *Grande atlante delle rocce e dei minerali*. Milan: Arnoldo Mondadori Editore, 1984.

INDEX OF NAMES

For the minerals and rocks, the page number refers to the individual entry in which the species is described; for the gems, the page refers to the individual mineral entry or to the box in which the species is discussed.

A

Acanthite, 89
Achroite, 67
Actinolite, 69
Actinolite schist, 137
Adamite, 55
Aegirine (Acmite), 68
Agate, 76
Agmatite (Migmatite), 137
Albite, 76
Albite paragneiss, 126
Alexandrite, 40
Alkali feldspar granite, 102
Alkali feldspar rhyolite, 114
Alkali feldspar syenite, 103
Alkali feldspar trachyte, 115
Almandine, 58
Amazonite, 76
Amethyst, 75
Amphibolite, 135
Analcime, 78
Anatexite (Migmatite), 138
Andalusite, 60
Andalusite biotite hornfels, 138
Andesite, 117
Andradite, 59
Anglesite, 51
Anhydrite (mineral), 50
Anhydrite (rock),152
Anorthite, 77
Anorthosite, 106
Anthracite, 155
Antimony, 88
Apatite, 54
Aplite, 112
Apophyllite, 71
Aquamarine, 64, 65
Arabesque marble, 131
Aragonite, 45
Argentite (Acanthite), 89
Argillite, 147
Arkose, 145
Arsenic, 88
Arsenolite, 41
Arsenopyrite, 93
Artinite, 48
Asparagus stone, 53
Atacamite, 39
Augen gneiss, 133
Augite, 67
Aurichalcite, 47
Autunite, 57
Axinite, 64
Azurite, 46

B

Balas (ruby), 40
Bardiglio marble, 131
Barite, 50
Basalt, 117
Basanite, 119
Benitoite, 64
Beryl, 64
Biocalcarenite, 148

Biocalcilutite, 148
Biocalcirudite, 148
Biotite, 72
Biotite paragneiss, 128
Bismuthinite, 95
Bixbyite, 65
Black opal, 76
Blue spinel, 40
Blue topaz, 61
Bog iron ore, 153
Boracite, 50
Borax, 49
Bornite, 89
Bournonite, 94
Breccia, 144
Brecciated marble, 131
Brucite, 43

C

Calacatta (Clastic marble), 132
Calc-schist, 132
Calcarenite, 147
Calcareous alabaster, 149
Calcilutite, 147
Calciphyre, 138
Calcirudite, 148
Calcite, 43
Californite, 63
Camptonite, 113
Carbonatite, 110
Carnallite, 38
Carnelian, 76
Cassiterite, 41
Cat's-eye, 75
Celestite, 50
Cerussite, 46
Chabazite, 81
Chalcedony, 76
Chalcopyrite, 89
Chlorite, 73
Chlorite-amphibole-epidote-
    albite schist, 133
Chlorite schist, 136
Chromite, 96
Chrysoberyl, 40
Chrysolite, 58
Chrysoprase, 76
Chrysotile, 73
Cinnabar, 35
Cipolino marble, 130
Citrine, 75
Clastic marble, 132
Cobaltite, 91
Colemanite, 49
Colorless topaz, 61
Colorless zircon, 59
Columbite, 97
Copper, 86
Cordierite, 64
Cordierite potassium feldspar
    hornfels, 139
Corundum, 40, 42
Covellite, 91
Crocoite, 51

Cryolite, 39
Cuprite, 39
Cymophane, 40

D

Dacite, 114
Danburite, 77
Datolite, 62
Demantoid, 59, 63
Descloizite, 55
Diamond, 34
Diatomite, 152
Diopside, 66
Dioptase, 66
Diorite, 106
Diorite porphyry, 112
Dolomite (mineral), 44
Dolomite (rock), 150
Dravite, 65, 67
Dunite, 109

E

Eclogite, 135
Elbaite, 65
Embrechite (Migmatite), 137
Emerald, 64, 65
Enargite, 89
Enstatite, 66
Epidote, 63
Epidote amphibolite, 134
Erythrite, 57
Essonite, 63

F

Feldspathoid gabbro, 108
Feldspathoid monzodiorite, 108
Feldspathoid monzogabbro, 108
Feldspathoid syenite, 107
Fire opal, 74, 76
Flint, 153
Fluorite, 38

G

Gabbro, 107
Gahnite, 40
Galena, 90
Garnet, 63
Garnet amphibolite, 135
Garnet jade, 63
Garnet mica schist, 127
Garnet paragneiss, 128
Geocronite, 94
Girasol opal, 76
Glaucodot, 93
Glauconitic sandstone, 145
Glaucophane, 69
Glaucophane schist, 133
Goethite, 97
Gold, 86
Golden beryl, 65
Golden topaz, 61

Goshenite, 65
Granite, 102
Granite porphyry, 112
Granodiorite, 102
Granophyre, 110
Granulite, 137
Graphite, 88
Graywacke, 145
Green tourmaline, 67
Gypsum, 50
Gypsum alabaster, 150
Gypsum arenite, 145

H

Halite (Rock salt), 37
Harlequin opal, 74, 76
Hauerite, 91
Hauyne, 91
Hawk's-eye, 75
Hedenbergite, 67
Heliodor, 64, 65
Hematite, 96
Hemimorphite, 62
Heulandite, 81
Hiddenite, 68
Hornblende, 69
Hornblendite, 110
Huebnerite, 52
Hyacinth, 59
Hyaloclastite, 120
Hyalophane, 77

I

Ignimbrite, 120
Ilmenite, 96
Ilvaite, 62
Imperial topaz, 61
Indicolite, 65, 67
Italite, 108

J

Jadeite, 68
Jargon, 59
Jamesonite, 95
Jasper, 153

K

Kaolinite, 73
Kermesite, 37
Kersantite, 113
Kimberlite, 113
Kunzite, 68
Kyanite, 60
Kyanite mica schist, 127

L

Labradorite, 77
Lapis lazuli, 79
Latite, 116
Laumontite, 80

Lazulite, 55
Lazurite, 79
Leadhillite, 47
Leucite, 78
Leucitite, 120
Leucosapphire, 42
Lherzolite, 109
Lignite, 155

M

Madera, 61
Magnesite, 43
Magnetite, 96
Malachite, 47
Malacon, 59
Manganese nodule, 153
Marble, 130
Marcasite, 92
Marl, 147
Melanite, 59, 63
Melilitite, 120
Mercury, 87
Mesolite, 80
Metagranite, 132
Metagraywacke, 126
Microcline, 76
Migmatite, 137, 138
Millerite, 90
Mimetite, 54
Minette, 112
Missourite, 109
Molybdenite, 93
Monazite, 53
Monzodiorite, 104
Monzogabbro, 105
Monzonite, 104
Morganite, 64, 65
Muscovite, 72
Muscovite quartzite, 128
Muscovite schist, 132
Mylonite, 139

N

Natrolite, 79
Nebulite (Migmatite), 138
Nepheline, 78
Nephelinite, 119
Niccolite, 90
Niter, 48
Noble spinel, 40

O

Obsidian, 121
Olivenite, 55
Olivine, 58
Oolitic limestone, 149
Onyx, 76
Opal, 74, 76
Opalescent sapphire, 42
Ophicalcite, 136
Orpiment, 36
Orthoclase, 75

Orthoconglomerate, 144
Orthogneiss, 133

P

Padparadscha, 42
Paraconglomerate, 144
Paragneiss, 126, 128
Paragonite, 72
Pearls, 45
Peat, 155
Pegmatite, 110
Peperino tuff, 121
Phenakite, 62
Phillipsite, 81
Phonolite, 117
Phonolitic tephrite, 119
Phosgenite, 47
Phosphorite, 155
Phyllite, 126
Picotite, 39, 40
Pink marble, 131
Pink quartz, 75
Pink sapphire, 42
Pink topaz, 61
Pisolitic limestone, 150
Platinum, 86
Pleonaste, 39, 40
Polybasite, 36
Polyhalite, 51
Prase, 75
Prasinite (Chlorite-amphibole-
   epidote-alb ite schist), 133
Precious opal, 74, 76
Prehnite, 71
Proustite, 36
Pumice, 121
Pyrargyrite, 36
Pyrite, 91
Pyrolusite, 97
Pyromorphite, 54
Pyrophyllite, 71
Pyroxenite, 109
Pyrrhotite, 90

Q

Quartz, 74, 75
Quartz anorthosite, 106
Quartz arenite, 144
Quartz diorite, 106
Quartz gabbro, 107
Quartz latite, 116
Quartz monzodiorite, 104
Quartz monzogabbro, 104
Quartz monzonite, 103
Quartz syenite, 103
Quartz trachyte, 116
Quartzite, 129

R

Radiolarite, 152
Realgar, 37
Reef limestone, 149

Rhyodacite, 114
Rhyolite, 114
Rhodochrosite, 44
Rhodonite, 70
Riebeckite, 69
Rock crystal, 75
Rock salt (mineral), 37
Rock salt (rock), 152
Rodingite, 139
Rubellite, 65, 67
Rubicelle, 40
Ruby, 42
Rutile, 41

S

Sanidine, 74
Sappare, 60
Sapphire, 42
Sard, 76
Sassolite, 49
Scapolite, 78
Scheelite, 52
Schorl, 65, 67
Schorlomite, 63
Scolecite, 80
Selenite gypsum, 150
Serpentine, 73
Serpentine schist, 136
Serpentinite, 135
Sherry topaz, 61
Shonkinite, 107
Siderite, 43
Sillimanite, 66
Siltstone, 146
Silver, 86
Skarn, 139
Skutterudite, 94
Slate, 126
Smithsonite, 44
Smoky quartz, 75
Sodalite, 79
Spessartine, 113
Sphaerocobaltite, 44
Sphalerite, 34
Sphene, 58
Spinel, 39, 40
Spodumene, 68
Starlite, 59
Statuary marble, 131
Staurolite, 61
Staurolite garnet mica
   schist, 147
Stibiconite, 41
Stibnite, 95
Stilbite, 81
Strontianite, 46
Sulfur, 34
Sunflower ruby, 42
Sunflower sapphire, 42
Syenite, 103
Sylvanite, 93
Sylvite, 37

T
Talc, 71
Talc schist, 136
Tenorite, 95
Tephrite, 119
Tephritic phonolite, 117
Tetrahedrite, 94
Thomsonite, 80
Tiger's-eye, 75
Tonalite, 102
Topaz, 60, 61
Topazolite, 59, 63
Tourmaline, 65, 67
Trachyte, 116

Travertine, 149
Tremolite, 68
Tridymite, 74
Tsavorite, 63
Tuff, 121
Turquoise, 57
Two mica paragneiss, 128
Two mica schist, 127

U
Ulexite, 49
Uraninite, 97
Uvarovite, 60

V
Vanadinite, 54
Veined marble, 130
Vesuvianite, 63
Vivianite, 56

W
Wavellite, 57
Witherite, 46
Wolframite, 52
Wollastonite, 70
Wulfenite, 52
Wurtzite, 34

Y
Yellow sapphire, 42

Z
Zinnwaldite, 72
Zircon, 58, 59

PICTURE SOURCES

The photographs in this volume were taken by Roberto Germogli or were supplied by the Mondadori archives with the exception of the following:

Agenzia Contrasto, Milan: p. 16, p. 85 upper right, pp. 98–99 macrophotograph.

Agenzia K & B, Florence: p. 22, p. 24, p. 25, p. 26 top, p. 101, p. 125, p. 134.

Fabbri Archives: p. 10 bottom.

Nicola Cipriani, Florence: p. 29, p. 124, pp. 140–141 macrophotograph, p. 142, p. 143, p. 146, top, p. 154.

Alessandro Stefani, Maestri orafo-argentieri, Bologna: p. 87 bottom left.